Fostering
Brighter Futures

How small groups of international community psychologists are healing our
polarized societies

First edition, January 2025

Donata Francescato is the first
professor of community psychology
in Italy and cofounder of ASPIC and
ECPA.

Manuela Tomai is an associate
professor of clinical psychology at
Sapienza University,Rome, specializing
in community education and health

For more informaton
https://www.donatafrancescato.it/

Donata Francescato   Manuela Tomai

Fostering
Brighter Futures

How small groups of international community
psychologists are healing our polarized societies

First edition, January 2025

1st edition, January 2025 WM Books
© copyright 2025 by Donata Francescat and Manuela Tomai

Translation, editing, proofreading and formatting by: Mavenhill Books
www.mavenhill.com

ISBN 978-1-940387-22-2

Table of contents

Introduction

This handbook aims to disseminate the values, theoretical constructs, and intervention methods developed by community psychologists to address individual and collective, local and global challenges and to promote better futures. Community psychology examines the interactions among individuals, small groups, organizations, networks, and both physical and virtual communities to enhance psychosocial congruence – that is, the alignment between people's expectations and capabilities and the demands and resources of their social contexts. We hope this book will be valuable not only to students and community psychologists but also to mayors, managers, and social and health service workers interested in implementing social projects that foster individual and collective well-being.

Part 1 begins with the origins and development of community psychology in the United States of America, Italy, Europe, and globally (chapter 1). In chapter 2, we briefly explore the scientific paradigms – positivist and constructionist – and the core values that guide both moderate and radical community psychologists. These values include social justice and equity, respect for diversity, inclusion of minorities and marginalized groups, and the promotion of individual and collective empowerment, social support, a sense of community, and active participation. We then present key theoretical constructs developed in the United States, Europe, and various developing countries in Africa, Oceania, and Latin America, examining how the internationalization of community psychology has facilitated the exchange, confrontation, and sometimes integration of different theoretical and value-based perspectives. Finally, we lay the groundwork for a theory of technique in community psychology that incorporates theoretical constructs from both moderate and radical approaches.

Community psychology formally emerged in the 1960s in the United States within a cultural and historical context marked by minority rights movements. It forcefully highlighted the need to explore the social origins of human distress and mental disorders, challenging the prevailing individualistic approach that dominated psychiatry, clinical psychology, and psychology in general at the time. Traditionally, universities treated the individual and social dimensions separately, with the psychological sciences focusing on the former and the social sciences (such as economics, political science, sociology, etc.) on the latter. Community psychologists proposed a new paradigm that values and examines the relationship between individuals and their contexts through an interdisciplinary approach.

As an empirical and applied discipline, community psychology has evolved over recent years in response to concrete situations of social distress and the growing demand for a better quality of life. This evolution has unfolded in both similar and divergent ways across different nations, leading to intriguing perspectives that emphasize various intervention strategies. These developments have enriched the theoretical heritage, professional skills, and values of the discipline. For instance, in the United States, women, ethnic minority groups (e.g., African Americans, Hispanic Americans, Asian Americans), and other marginalized communities (such as the LGBTQI+ community) have gained greater visibility and influence among community psychologists, addressing the persistent issues of racism and sexism in society. In Europe, British community psychologists have developed both a radical wing of the discipline and the theoretical constructs of the moderate wing. For example, Jim Orford (1992, p. 6) argues:

Community psychology seeks to understand people in their social worlds and to use this understanding to improve their quality of life. It aims to understand and help. Thus, it is both a research area and a branch of the academic study of psychology and a branch of the helping professions. It stands in a bridging position between the psyche and the social, the private and the public.

Orford has focused on developing a general theory for community psychology that emphasizes the circular relationship between *the individual and the environment*. He formulates "key concepts" that bridge the fields of the person and the environment, facilitating connections between the two. In the domain of the person, Orford (1998) identifies identity, status, and feelings of self-worth as central constructs. These concepts, widely studied in clinical psychology, are complex enough to link personal and social dimensions. For example, a person's sense of identity and self-esteem are influenced and shaped by the social contexts they inhabit. In the social field, Orford highlights four areas where the environment can either support or hinder individuals: in the development of their sense of identity, in achieving and maintaining social status, and in fostering self-esteem.

In Italy, Piero Amerio, a renowned social psychologist, has stressed the importance of integrating traditional clinical psychology's focus on change-oriented helping attitudes and attention to individual cases with a broader focus on the political and historical dimensions of the social context, considering both its material and symbolic aspects. According to Amerio (2000, p. 208), community psychology centers on the individual's sense of self within its social dimension. It also emphasizes the sense of participation in constructing the common good, which underpins the community as a repository of human values. Furthermore, Amerio highlights the constructive sense of action as a process that integrates mental and practical activities, connecting the individual with the social sphere. This approach not only enables individuals to adapt to their context but also empowers them to work toward changing it.

European community psychologists have delved deeper into the negative aspects of past community movements than their American counterparts. They have examined the historical evolution of the concept of community, uncovering how communitarian ideals have been transformed across different historical periods – sometimes becoming sources of oppression, as well as vehicles for emancipation and personal growth. Community utopias have been used both to foster solidarity and the emancipation of individuals and to create totalitarian environments where individuals were stripped of many freedoms (Amerio, 2000; Arcidiacono *et al.*, 2021).

The 20th century was a period of opposing extremes in its various economic and political forms: *anarcho-liberalism*, which emphasizes individual rights at the expense of communal duties, and *totalitarianism*, as seen under fascist and communist regimes, which sacrifices individual needs. This juxtaposition has contributed to the rise of dualistic thinking, particularly in capitalist nations, where excessive faith in individual initiative has led to a diminishing value placed on solidarity. Modern communities are increasingly marked by exclusionary attitudes rather than welcoming ones, with a widely recognized lack of social bonding. Some authors have termed this phenomenon an "epidemic of narcissism" (Twenge, Campbell, 2009). While people may *cognitively understand* their interdependence with others, there is a growing inability to *emotionally perceive* the interdependence that underpins the social fabric. The emotional dimension, not just in conscious terms, has been increasingly recognized by European community psychologists over the last decade. Many authors, in the process of understanding phenomena, have progressively given more space and credibility to the shared, often unconscious, emotional meanings with which individuals represent themselves and make sense of their experiences and relationships. In chapter 4, we discuss the gradual "rapprochement" between community psychology and dynamically oriented clinical psychology, demonstrating how integrating the two disciplines enriches the complexity of event interpretation and enhances community psychology's intervention models (e.g., Di Maria, 2005; Liang, Tummala-Narra, West, 2011; Francescato, Aber, 2015; Francescato, Zani,

2017; Mannarini, Salvatore, 2019; Caputo, Tomai, 2020; Caputo *et al.,* 2020; Di Maria, Falgares, 2021).

In Part Two, we delve into various community psychology intervention methodologies, highlighting their multidimensional and complex nature, including community analysis, multidimensional participatory organizational analysis, working groups, self-help groups, network work, empowering training, and social-affective education (Chapters 5, 7, 8, 9, 10, 11, 12). In response to the expansion of narcissistic cultural dimensions, we emphasize group intervention modes throughout this book. Small groups, precisely because they consist of a limited number of individuals who know each other, represent intermediate structures between the individual and the broader community. Within these groups, people actively experience conditions of operational interdependence, fulfilling needs for belonging, individuation, and emotional solidarity. Working to foster the growth of these groups contributes directly to community development. We also draw on the important contributions of Italian occupational psychologist Enzo Spaltro (1977, 1985, 1999), who argued that developing a sense of community requires a shift from a "couple culture" to a "group culture." Couple culture, based on an oedipal relationship, is characterized by a singular command structure and values such as competitiveness and dependence. In contrast, group culture values leadership differentiation, pluralism, cohesion, and change.

Today, we seek a new model that balances individual needs with societal goals, values each person's diversity, protects against the tyranny of the majority, and encourages cooperation for common purposes. Throughout this handbook, we describe the political, economic, and cultural changes that have reshaped the contexts in which we live and in which we must act to improve psychosocial congruence by listening to diverse voices. For instance, in chapter 3, we explore the criticisms of radical community psychologists who argue that moderate community psychologists have not sufficiently denounced the European countries that colonized parts of Africa and South America or the American government, which has oppressed and continues to exploit Indigenous Peoples for the benefit of big business. Ecofeminists and radical community psychologists (Raworth, 2017; Stevens, 2018; Ndlovu-Gatsheni, 2020; d'Eaubonne, 2022) also critique autocratic regimes, where political leaders seek historical recognition as heroes avenging past national defeats by engaging in arms races and undeclared wars for economic and power gains.

The 21st century has brought eco-political and cultural shifts that have increased anxiety and fear of the future, from the collapse of the Twin Towers in New York in September 2001 – ushering in the era of global terrorism – to the massive influx of asylum-seeking migrants from nations engulfed in armed conflicts, such as Somalia, Libya, Syria, Ukraine, Israel, Palestine, Lebanon and-or countries like Bangladesh, devastated by climate change. This century has also seen the decline of mass political parties and the utopian ideals of the 20th century, alongside a rise in political apathy and voter abstention, as many citizens lose faith in government and politicians or retreat into themselves, confined to the present. However, there is also an increase in those who express their anger and hostility toward politicians, briefly demonstrating in public squares over factory closures, the killing of African Americans by police (as seen in the Black Lives Matter movement in the United States), or in solidarity with Afghan women barred from universities or Iranian girls imprisoned for pursuing freedom. Every day on social media, many voices express deep dissatisfaction with this unjust society.

It is the task of the political class to establish new rules for the monetary and economic system, but as political scientist Sabino Cassese (2020) points out, politics is in decline even in the world's largest democracies. This degradation of politics is dangerous for the democratic growth of society, as political scientist Zagrebelsky noted back in 1995:

Suppose you 'depoliticize' people through fictitious and elusive messages about the complexity of problems, endearing them to laziness and passivity, which as a temptation inhabits all of us. In that case, you are not eliminating politics, but you are dispossessing them in favor of small circles of power where real politics is made (Zagrebelsky, 1995, p. 4).

Indeed, we live in a historical period where it is increasingly important to understand economic, political, and cultural changes to identify emerging opportunities and obstacles. At the same time, we must develop a greater awareness of the link between individual and family well-being and technological, political, and socioeconomic changes in our local, national, and international contexts.

Over the past decades, leaders of the neoliberal economy have heavily influenced those elected to political office, both locally and nationally, to secure opportunities for limitless profit. Markets dictate government economic policies, and financial groups wield excessive power. Big finance has absolute control over interest rates, setting the cost of money and exchange rates to its advantage. Neoliberal capitalism has concentrated more power and wealth in the hands of the top 1 percent of the population, particularly the plutocrats at the top, while further impoverishing the most disadvantaged. Various economists (Raworth, 2017; Butera, 2021) argue that large firms source raw materials and labor at low prices, aim to generate high profits for shareholders and lavish salaries for executives, and often disregard ethical or environmental considerations. For years, many economists claimed that ever-increasing economic growth would eliminate scarcity for all until environmental scientists documented that this mode of production was endangering the planet's future, destroying ecosystems, increasing environmental degradation, and harming the health of individuals and entire communities. Much of what is produced is designed to become quickly obsolete: when products break down, they are not repaired but discarded, creating mountains of waste that pollute the seas and damage farmland. Human behavior is melting the polar ice caps, threatening the future of entire generations.

It is no coincidence that teenagers have spearheaded the Fridays for Future movement, advocating for a circular economy based on the "four R's" principle: reduce, reuse, repair, and recycle. Recognizing the gravity of environmental problems that threaten to destroy our planet, we have included – for the first time in a community psychology handbook – an entire chapter dedicated not only to climate change but also to how community psychologists have developed strategies to raise awareness of environmental issues and encourage individuals to act, fostering a planetary sense of community (chapter 15). As we have mentioned, making these changes requires simultaneous exploration and action on conscious, unconscious, individual, and collective dimensions that influence both our choices and behaviors.

This book is intended as a valuable resource for persons with diverse professional orientations and interests who nevertheless share a commitment to creating a society where people enjoy individual freedoms and rights while also striving to increase social solidarity, reduce socioeconomic inequalities, assume mutual community obligations, and develop a strong sense of community. Achieving this goal in the 21st century, where narcissism is rampant and exacerbated by new media, social media, and the pervasive culture of selfies, is no small task. We need competent and courageous political leaders who can stand up to big business to address systemic problems such as climate change, racism, sexism, migration, inequality, and more. These issues are making the Earth and its inhabitants sick, fomenting divisions and fragmenting identities, which in turn hinders efforts to implement projects for the common good.

However, Robert Putnam (2023) has just published the results of an intriguing survey that underscores the importance of developing a heightened sense of community in our contexts – an insight that enhances the content of our book. Putnam argues that the major economic, political, cultural, and familial changes in the Western world have led to an alternation between a **society of the self** – highly individualistic, socioeconomically unequal, politically polarized, and fostering public and private narcissism – and a **society of the we** – more egalitarian, collaborative, and willing to prioritize social responsibility over self-interest. Examining American history from the late 19th century to 2020, Putnam documents alternating periods dominated by individualism (such as the last decades of the 19th century) and periods focused on developing the "we," such as the years from 1900 to 1960. However, "we" societies also tend to be conformist and, in the Western world, have historically been dominated by white males who excluded women and people of color.

According to Putnam, a political-cultural shift occurred in 1968 that mobilized women and excluded minorities to fight together for their rights, creating a new union and a great utopian hope for change, but still based on individual's freedom to gain more rights for themselves. This focus on individual rights has contributed to the growth of a "**society of the self**" since that period. Putnam notes that many of the young activists of the 1960s who were elected to political office at the local, state, and national levels have become more extreme over time. A similar process has occurred within the evangelical and economically conservative right-wing factions of the Republican Party, leading to increased political polarization in many centers of government.

This polarization makes it difficult for representatives of opposing political views to respect each other and collaborate on projects for the collective good. According to Putnam, the growth of today's "I" society has been fueled by technological development, particularly social media, which has exacerbated the fragmentation of identity groups and increased opportunities for casual sexual encounters, but reduced face-to-face contact and diminished relational capacities for resolving conflicts and building long-term relationships. As a result, the number of people living alone has increased in the United States and several other countries (ISTAT, 2023). Putnam believes that to shift toward a more cohesive, community-centered society, leadership alone is not enough; throughout history when America has leaned toward a more community-oriented society, it has been due to thousands of ordinary citizens creating grassroots projects and building networks to bring about the desired changes. The creation of a bottom-up, community-based welfare system is already in progress, as documented by Arcidiacono and collaborators (2021). These volumes illustrate how community psychologists are addressing the needs of the homeless, women victims of violence, immigrants, and the lonely elderly while promoting relational well-being and mutual help, even within apartment buildings. Other are fostering solidarity and a planetary sense of community, which is essential for combating climate change and environmental degradation (Francescato, 2020; 2022). Additionally, some are developing outreach programs to tackle social issues such as school dropout, cyberbullying, and drug use while also supporting ground-level networking, educating about democracy, environmental respect, and conscious, respectful sexuality. Others are creating innovative online experiences, (Francescato, Putton,

2022), to assist mayors, managers, and service operators in implementing programs that enhance well-being and foster a strong sense of community, respect, and mutual aid.

Despite the impactful work being done by a minority of community psychologists, their efforts remain relatively unknown in Italy, where most psychologists are focused on psychotherapy. This is why we dedicate Part Three of our book to community psychologists's experiences, narrated directly by the practitioners themselves (Chapters 16-17) . The authors are 29 community psychologists, ranging in age from 26 to 69, and mostly women with a notable male presence, contribute to these chapters. They narrate their stories individually, in pairs, and in two cases, as small groups working together in enterprises and services they have established in Florence and Turin. Their accounts are compelling, as they explain why they chose to become community psychologists, challenging themselves in various contexts while sharing a deep passion and, we might say, a sense of pride and fulfillment in their complex career choices – whether as clinical community psychologists (Chapter 16) or as researchers and social community psychologists with little or no clinical training (Chapter 17).

In chapter 16, young colleagues express the belief that all psychologists should master new online technologies, which could be instrumental in facilitating participatory action research, primary prevention, and the development of community psychology that not only addresses problems but also promotes well-being and happiness. Among the experiences shared, young psychologists recount how online platforms aided them during the COVID-19 pandemic, providing training in different ways (e.g., in Padua) or increasing online involvement of husbands in a research study on violence against women in Naples, effectively opening their homes and lives to the study. Clinical community psychologists, freelancers, and seasoned professionals share decades of experience: some have utilized all the tools discussed in this volume; others specialize in affective emotional education, working with students, teachers, and parents in schools; and some have taken leadership roles in family counseling centers to implement larger projects. Other psychologists, embedded in public agencies, such as the National Institute for Public Policy Analysis and the National Agency for Active Employment Policies, have been involved in research and active employment policies. They have applied research-intervention methods, empowering training, and socio-affective education to issues such as gender differences and school-to-work transition. They have also contributed to the development of orientation models and tools for various target groups, including youth and employed and unemployed adults. These models and tools have been tested in schools and employment services across the country by trained operators. Still, others have served as consultants to mayors and policymakers, explored how new technologies can foster empowerment, and organized "future workshops."

In chapter 17, we meet the psychology of social community entrepreneurs. A passionate professor, Patrizia Meringolo, and a small group of community psychologists created the LabCom enterprise in Florence through an academic spin-off. In Turin, four community psychologists have established new services for migrants, women victims of violence and the homeless. Others have revitalized social connections among the elderly in nursing homes and helped inmates in prisons envision possible futures.

Chapter 17 also introduces community psychologists from social psychology backgrounds who have participated in urban redevelopment projects, founded new associations, and engaged in social and cultural planning, project monitoring and evaluation, and social reporting. Other female psychologists work in various fields simultaneously, but their primary goal is to become research fellows and university lecturers, training the next generation of community psychologists – a critical need. They work on projects related to renewable energy communities and land regeneration through reforestation, promoting sustainability and energy transition. They participate in research projects funded by the European Union and are also activists involved in global movements advocating for Indigenous Peoples, opposing early marriages, and combating human trafficking Some are involved in the European Society for Community Psychology (ECPA), others work with public health departments, and still others become activists in regional organizations, addressing issues such as migration and violence against women. Many of them have gained experience studying and working abroad (in Portugal, Brazil, the United States, etc.) and have maintained international contacts for years. Some began their careers by volunteering and working in disadvantaged suburbs of large cities before encountering a lecturer or book on community psychology that deepened their understanding of the work they had already been doing. Reading their stories is as captivating as enjoying a novel.

The volume is the result of the common work of the two authors; however, some chapters were developed more by one of us. Donata Francescato wrote chapters 1, 2, 3, 7, 11, 12, 15, 16, 17, and Manuela Tomai wrote chapters 4, 5, 6, 8, 9, 10, 13, 14.

Rome, January 2025

Part 1
Values and Theories

1
The origins and development of community psychology

1. Introduction

Understanding how and why a new scientific discipline arises and evolves is a process that requires a careful assessment of the historical and cultural context in which this happens, and of the interplay that always exists between a country's sociopolitical environment and the dominance of specific aspects of a social discipline, such as community psychology. For a documented analysis, it is necessary to examine:

(*a*) how, for example, psychologists have conceived of their profession, and with what social mandate they have operated in various historical periods; (*b*) how their theoretical postulates have gradually changed as society has changed; and (*c*) how the development and dissemination of certain psychological theories and different modes of intervention have in turn contributed to this change in the psychosocial climate (Francescato, 1977a, p. 15).

In the case of community psychology, this analysis is indispensable because it is a discipline that places a strong emphasis on the social context in understanding individual psychological functioning. In fact, according to Orford (1992, p. 6), "community psychology seeks to understand people in their social worlds and to use this understanding to improve the quality of life. It aims both to understand and to help. Thus, it is both a research area and a branch of the academic study of psychology and a branch of the helping professions. It stands in a bridging position between the psyche and the social, the private and the public."

As an empirical and applied discipline, community psychology develops because of the stimulus of concrete situations of social distress and expectations for a better quality of life, similar but also different across countries, giving rise to intriguing perspectives that privilege divergent intervention strategies, enriching, as we shall see in this chapter, the theoretical heritage, professional skills, and values of the discipline.

In the United States of America, community psychology emerged under the impetus of clinical psychologists and other practitioners who progressively distanced themselves from an individual, biological, and intrapsychic view of distress and sought etiological explanations and forms of intervention within the individual-environment relationship. At first, innovation remained confined to the field of psychiatric services and the treatment of mental illness, then the perspective broadened to other areas of social and community life, setting the goal of preventing distress, promoting social resources, and changing social and institutional realities. A thorough examination of the historical roots of community psychology in the United States of America and the political and social influences on its development has already been made by various authors (Katz, Bender, 1976; Francescato, 1977a, 1977b; Rappaport, 1977; Francescato, Contesini, Dini, 1983; Heller *et al.*, 1984; Levine, Perkins, 1987; Francescato, Ghirelli, 1988). In this context, we will simply outline the main lines.

2. Birth and evolution of community psychology in the United States

The birth and evolution of community psychology are preceded by the development of psychiatry and mental hygiene. The first significant stage in this process can be traced back to the late 18th century, during which the consolidation of the Industrial Revolution allowed a slow but progressive conquest of elementary social rights by the disadvantaged masses. Thus, the denunciation by former patients or intellectuals of the most inhumane abuses to which inmates in mental hospitals were subjected began to find space. A new climate of social reform enabled the development of welfare programs of various kinds. It will suffice here to recall the establishment of juvenile courts (1899), the founding of organizations still active today such as the Young Men Christian Association (YMCA; 1906) and the Wood Badge training program for Scout leaders (1919), or the establishment of the National Committee for Mental Hygiene (1909) by Clifford Beers, a former inpatient who was highly critical of mental hospitals. Following the economic crisis after World War I, which resulted in the depression of 1929, American social initiatives and services experienced a significant downturn. Under Franklin Delano Roosevelt, the federal government prioritized public works over the expansion of psychology and mental health services. After World War II, there was a renewed optimism on the part of the nation about its ability to cope with major social problems, chief among them the reintegration of war veterans. However, there is a perceived lack of theories and methods appropriate to the management of more serious mental illnesses and cases of more pronounced social distress. The federal government appeared to have little involvement in the planning of social services, however, the passage of the National Mental *Health Act* in 1946 and the subsequent founding of the National Institute of Mental Health as the federal structure charged with combating mental illness and promoting mental health was significant. The 1950s were characterized by a climate of a race for affluence and the attainment of a certain social status. The response to distress is increasingly the psychoanalytic practice, the private clinic, and the psychiatric hospital. At the same time, however, the spread of behaviorism, the so-called "second force," and of the humanistic-existential orientation, known as the "third force," challenges the dominance of the psychoanalytic orientation, places emphasis on the development of the individual's personal resources, and facilitates the spread of the relational model, which shifts attention from the study of individual variables to the analysis of the communication network and context.

In the early 1960s, several politicians and professionals saw their faith in a technical and neutral science within an open society soured. Above all, it is the increasingly widespread social movements and struggles – such as the Black, student, women's, and anti-militarist movements against the Vietnam War – that undermine the myth of an "equal opportunity" society. The critique of traditional bourgeois and individualistic values is accompanied by the recovery of some "human" values and a new confidence in community and collective resources. Among psychologists and mental health workers, there is growing attention to socioeconomic, ecological, and cultural factors, as well as to methodologies focused on relational, behavioral, humanistic, and socio-environmental orientations. The social and cultural climate, along with the idealistic drive of some Kennedy-era reformers, contributed to the passage of the *Community Mental Health Act* in 1963, which established the principle of territorial organization of psychiatric services. The territorialization of psychiatric services gives a strong impetus to already socially oriented practitioners, stimulating research and reflection on new methodologies of analysis and intervention.

The "official" designation, community psychology, was coined in 1965 in the United States of America at a conference in Swampscott, Massachusetts, on community mental health training sponsored by the American Psychological Association (APA) and the National Institute of Mental Health (NIMH). According to Bennett (1965, p. 833) what united the participants was an awareness of a gap between the professional developments that had seen an increasing number of psychologists working in the community and the training that was continuing in university programs in clinical psychology. The frontier of mental hygiene was shifting from the amelioration and treatment of illness to preventive intervention at the community level.

The conference emerges as a foundational moment for community psychology as an autonomous field, emphasizing the importance of community-based preventive interventions, the de-medicalization of psychiatric services, and a broad interdisciplinary approach. In this context, the community psychologist is viewed as a participant-conceptualizer (or "participant-theorist"): "an agent of social change who actively engages with the community and its subsystems, bringing a scientific attitude, a commitment to research, evaluation, and theoretical development" (Spielberger, Iscoe, 1972, p. 8). The formal establishment of community psychology as a distinct discipline was further solidified in 1966 with the creation of the Division of Community Psychology within the American Psychological Association (APA). During this period, significant advances were made in the field of mental hygiene, along with the first efforts to develop alternative intervention tools to traditional clinical practices, such as:

- Crisis intervention, a strategy designed to prevent a temporary or transitory hardship from turning into a permanent difficulty;
- Mental health counseling, which is the provision of technical assistance by an expert to an individual or organization concerning psychological problems that may arise in the workplace;
- Planned change, which is the implementation of systems and community intervention programs through training courses and surveys in which innovations are planned, implemented, and evaluated (Mannino, Maclennan, Shore, 1975).

The late 1960s marked a positive period for community psychology; the spread and prestige of this discipline were bolstered by several social laws passed under the administration of President Lyndon B. Johnson. To increase domestic support, which had been shaken by the events of the Vietnam War, the government launched numerous compensatory programs. These included initiatives such as Head Start, aimed at providing early childhood education to disadvantaged children; various programs targeting unemployed youth; and efforts focused on the recovery of drug addicts and alcoholics. Scientifically, too, community psychology begins a period of critical revision by both conservative psychiatrists, who advocate the recovery of traditional methods of treatment, and community psychologists belonging to the radical wing of the discipline.

The radical movement emerged in the 1970s under the impetus of the journal "Radical Therapist" and denounced a lack of citizen participation in the management of the new services and an insufficient incisiveness of the mental hygiene movement in effecting changes in the American sociopolitical system. Authors such as Kunnes (1972) and Ryan (1971) blame the interventions carried out by psychologists belonging to the moderate area of community psychology because they judge them to be aimed more at compensating for distress than at promoting structural changes in the environment. According to members of the radical wing, the proposed interventions were not going to make a dent in political-economic and social problems, which were mainly responsible, instead, for determining distress. The radical wing pointed out that many forms of mental malaise were found among the destitute, as the malaise was influenced by lack of power and poor accessibility to economic and social resources on the part of different strata of the population, and therefore reproached moderate psychologists for absolving the social system producing the malaise and activating policies that discriminate against the victim.

On a practical level, the radicals' contestation resulted in numerous initiatives to combat institutionalization, the creation of alternative settings, self-help experiences, knowledge socialization, and *community organizing*. A plethora of new programs were born, each attempting to respond to an unmet need, each as an indication of past neglect and a challenge to prevailing conditions. Before long, the Peace Corps, Vista, Head Start, and the War on Poverty made their appearance. In each project, despite differences in implementation, there was a commonality of purpose: each program was an attempt to change the status quo and to provide a vehicle through which individual action could become an instrument of social change. The point of reference was the condition of man, the focus shifted to the community of men and by implication to the community in which man lived (Goldenberg, 1971, pp. 474-5).

Therefore, since then, disagreements between "politically radical" and "politically moderate" community psychologists about the roles of activism and professionalism have deepened.

Those who advocated a technical professional role conceived of a community psychologist who helps to better understand a problem by collecting data, co-designing and evaluating programs with community members, and offering advice, but leaving decisions to institutional authorities. The "politicians" wanted the psychologist to take an activist role, seeking to influence decisions by participating alongside marginalized groups in their struggles (Spielberger, Iscoe, 1972). In the 1970s, a strong minority of the radical wing developed, aiming for a social revolution through movements to liberate the oppressed, following Ryan's (1971) theories on blaming the victim and Steiner's (1971, p. 5) view that psychiatrists and psychologists helped to deceive human beings by "interpreting as deficiencies, defects, individual problems the symptoms of a condition of oppression".

In the first half of the 1970s, power struggles between the Republican and Democratic parties in the United States of America resulted in political and institutional crises (culminating in Watergate), while international events such as the energy crisis led to alternating inflation and recession and periods of high unemployment. As a result, Republican administrations severely restricted funding for social services, and the expansion of community programs was scaled back. Community psychologists during that decade progressively abandoned mental hygiene terminology to delve into concepts proposed by Lewin in his psychological field theory, which postulates a continuous and reciprocal transaction between the individual-subjective and social-objective spheres, and an active subject who can perform transformative actions. Such actions are considered psychosocial processes because they combine the subjective world of intentions and motivations with the objective world of environmental resources. Barker (1978) proposes the construct of *behavior setting,* which examines how human behaviors are influenced by temporal contexts-spaces (see also chapter 15).

Above all, several authors construct systemic-ecological theoretical models, which see the community as a network of social systems (including various levels, from the individual to micro, meso and macro-levels) capable of creating well-being or distress depending on whether risk or protective factors prevail in them (Mills, Kelly, 1972; Murrell, 1973; Bronfenbrenner, 1986). Others reworked the theoretical principles of primary, secondary, and tertiary prevention formulated in the medical-psychiatric and health fields (Caplan, 1964; Cowen, 1973; Korchin, 1977).

On the legislative side, after Nixon's resignation, the Democratic majority in 1975 passed two laws with innovative content: the first sought to establish real alternatives to the psychiatric hospital, and the second attempted to promote coordination among the various public welfare facilities and to increase opportunities for citizen participation.

The early 1980s constituted a new period of crisis for community psychology, both because of the drastic cuts in welfare programs made by the Reagan administration and because of the changed social climate favoring a revaluation of individual success and narcissistic gratifications. To break the impasse, some community psychologists proposed mediating strategies, hypothesizing proposals for reading and intervention that were less socially critical (Repucci, 1981; Serrano-García, 1982a, 1982b). Other prominent community psychologists, including Rappaport (1980), argued instead that community psychology should respond to the crisis by returning even more rigorously to its original commitment to social system transformation. Still, others sought to predict which areas of intervention might become more prominent by the 1980s. Glenwick (1982) pointed to program evaluation, counseling of spontaneous groups and self-help groups, promotion of coordination among different public agencies, and counseling of private business groups to promote the health of their employees as areas of the future. In response to these various needs, the Division of Community Psychology promoted a whole series of symposia, discussions, and special issues of the division's bulletin in those years.

In the 1980s, community psychologists in the United States of America sought to respond to emerging social needs: they collaborated with architects and urban planners on the design of new buildings and new neighborhoods; they interacted with Department of the Environment officials to study how to deal with the consequences of environmental disasters on populations. For ideological reasons, the corporate world in the 1960s and 1970s had not been a favored place for community psychologists to intervene. In the more versatile 1980s, a minority of professionals ventured into this area by carrying out health education campaigns for employees, working on the psychological implications of unemployment, and designing innovative programs for at-risk workers. Finally, many efforts have gone into more general community development programs. Community psychologists sought to work with members of a system to establish needs, identify resources, and provide the training and coordination necessary to facilitate the implementation of desired change. In this decade there was a clear effort to create new intervention settings beyond the traditional venues designated for mental hygiene and/or educational function. Underlying this new outreach was an important and further theoretical development of the discipline that began in these years and would continue and have its greatest depth in the following decades.

There are essentially four concepts that appear to have dominated community psychology in the 1990s: empowerment, self- and mutual help, social support, and a sense of community. These are highly innovative concepts as they integrate the professional contributions of the radical and moderate wings. We can call them four bridging concepts that mediate between the personal and the political, and that can only be understood in the interaction between a person and their context. Working on distress from a community psychology perspective means moving beyond ideological alignments centered on individual faults or social causes and adopting a model, of reading and intervention, that is complex and interactive. Sharing this model, community psychologists in the 1990s opted for a variety of strategies to strengthen positive interactions between individuals and the environment through the promotion of individual and community empowerment. The concept of empowerment is understood as a goal to be achieved through forms of self-help, which empower and value the individual's contribution, and various forms of social support, which recognize the value of solidarity and the importance of environmental interactions

Programs centered on "empowerment" adopt the moderate stance of aiming to empower individuals but with a shift from the therapeutic-rehabilitative approach of the 1970s to a political-emancipatory perspective. The unequal distribution of resources and differences in access to sources of power (force, law, money, and knowledge) in different social and ethnic groups are recognized. At the same time, it is accepted that the person who perceives himself as powerless is often unable to identify and use both personal and social resources accessible to him. Therefore, empowerment-centered programs aim to increase a sense of personal power and the ability to read different social systems to understand the constraints they place on our daily lives, but also the opportunities (in terms of services, environmental resources, etc.) they offer.

In the 1990s, multidimensional intervention projects based on complex causality and oriented on the concepts of empowerment, social support, and self- and mutual help became progressively more widespread because they were shown to be able to integrate the two ideologies prevalent in the United States of America. With respect to the fields of application, we can say that community psychologists, through value choices have always preferred to intervene on behalf of the most marginalized groups in society, supporting the various political and sociocultural movements that have progressively emerged in the Western world. Interventions on minorities panned out to be one of the most studied emerging areas in the 1990s, during which community psychologists sought to foster the empowerment of ethnic and social minorities to enhance cultural pluralism.

This tendency to deal more with the most marginalized groups has strengthened since the beginning of the new century, not only in the United States of America but also in Italy and across Europe, partly due to a series of momentous sociopolitical changes. Economic, political and environmental changes have also fostered greater internationalization of community psychology. In countries of the Global South and on various continents, community psychologists have developed theoretical constructs and sociopolitical analyses of liberation psychology and critical community psychology that share analyses from the radical area of American community psychology.

3. First steps of community psychology in Italy

Donata Francescato studied at four American universities – University of Texas, Rice University, University of Houston, and Brandeis University – from 1964 to 1972. She earned a doctorate in clinical psychology and completed her internship at the Southshore Mental Health Center near Boston, where the first community psychologists were working. Back in Italy, she became an adjunct professor in the new psychology graduate program at La Sapienza in Rome in 1973. Many of her students, already working in social and health services, were drawn to her lectures on community psychology, particularly to the workshops and guided practice experiences (GPE). In her role teaching personality inquiry techniques, Francescato introduced community intervention methods she had learned in the United States. Meanwhile, as the co-founder of "Effe," a feminist magazine on the newsstands from 1973 to 1982,,[1] Francescato was often invited on television to talk about, abortion, affective and sex education, all politically contentious topics on a faculty where conservative and progressive professors were often rivals. Francescato was repeatedly reprimanded by more conservative colleagues, with one even telling her to choose between being a university lecturer or a TV showgirl. That said, she also found some supporters in Mario Bertini, Luisa Camaioni, Marisa D'Alessio, Eraldo De Grada, Caterina Laicardi, and Riccardo Venturini, who began to fight with her to introduce the discipline of community psychology. It took twelve years as the subject was considered too "political" by the most conservative.

Among the factors hindering the broader spread of community psychology was the strong opposition from fascism and the persistent distrust of psychology within large sectors of Italian culture, even in the postwar period. Additionally, the absence of degree programs until 1971 and the lack of a professional register and order for many years further limited the status of psychologists, especially those working "in the field."

The development of community psychology in Italy has occurred within a context lacking a strong tradition of the psychological profession. This, in our view, explains why practitioners have largely adhered to the well-established medical model, primarily identifying as psychotherapists focused on individual clinical cases or, at most, small groups of patients.

Not surprisingly, great support was provided instead by Augusto Palmonari and Bruna Zani, who were teaching social psychology in a progressive city like Bologna. They invited Francescato to give a seminar in 1976 and encouraged her to publish. The first article on community psychology was published in the "Italian Journal of Psychology" in April 1977 (Francescato, 1977b) and the first volume for the Feltrinelli publishing house (Francescato, 1977a). Both describe the development of community psychology in the 1960s in the United States of America, the theoretical approaches and intervention strategies, from both the moderate and radical wings, that Francescato believed could also be useful in Italy. In the 1970s, under the impetus of social, ideological, and cultural changes, numerous provisions were passed: law No. 118 of March 30, 1971 (establishment of territorial rehabilitation units); Presidential Decree No. 416-420 of May 31, 1974 (delegated decrees in schools); law No. 382 of July 22, 1975,(transfer of state competencies

---

[1]. See https://efferivistafemminista.it/.

to regions and local authorities); law No. 354 (prison reform); Law No. 405 of July 29, 1975 (establishment of counseling centers); Law No. 685 of December 22, 1975 (new services for drug addiction); Law No. 517 of August 4, 1977 (inclusion of handicapped persons in regular classes); Law No. 180 of May 13, 1978 (psychiatric assistance); Law No. 194 of May 22, 1978 (voluntary interruption of pregnancy); Law No. 833 of December 23, 1978 (the establishment of the national health service).

Several psychologists and social workers began to advocate community psychology. In 1979, Guido Contessa in Brescia invited Francescato as a speaker at the first community psychology conference, and in 1980 the Division of Community Psychology was formed in the Italian Society of Scientific Psychology, which elected Francescato as coordinator of a small group of pioneers. With a passionate professional coordinator, Marco Traversi (who sadly passed away prematurely), in a few years the Division of Community Psychology reached about 400 members, almost all professionals, working in territorial services (SERT, DSM, consultancies), social cooperatives, voluntary associations, school organizations, and trade unions. Several articles and books (Palmonari, Zani, 1980; Contessa, Sberna, 1981) began to compare intervention models and experiences. In the 1980s and 1990s, master's programs offered outside the traditional university system emerged in cities like Milan, Rome, Padua, Lecce, Catania, and Pisa. These two-year, fee-paying training courses were designed for psychologists working in territorial health services (such as SERT, DSM, and counseling centers), social cooperatives, voluntary associations, and primary and secondary school teachers. About half of the approximately 2,000 trainees received subsidies from their local authorities. Unfortunately, cuts in health, public education, and welfare funding have since reduced public subsidies, leading to the closure or decline of these non-university master's programs. However, training opportunities within universities have increased.

In 1985, a five-year bachelor's degree program with a clinical and community focus was established in several universities, which formally introduced and made compulsory the subject of community psychology, which was in high demand among students. Other reforms in the university subsequently introduced community psychology into both bachelor and master's degrees. The first doctorate in community psychology and educational processes was organized by Bianca Gelli with the Universities of Turin, Lecce, and Rome. At the same time, Francescato became the first full professor of community psychology in Italy, followed by Miretta Prezza. Several highly qualified and passionate Ph.D. graduates aspire to become researchers (see chapter 17), but competition for scarce university positions is intense, and the most deserving candidates do not always succeed in advancing their careers. Moreover, the placement of the discipline within the fields of Dynamic Psychology and Social Psychology has led to the development of diverse theoretical approaches and intervention methods. While these have enriched the discipline, they have also created challenges in academic competitions. In most universities worldwide, community psychology is typically a component of clinical psychology, which also encompasses dynamic, cognitive, humanistic, and systemic-relational orientations (historically referred to as the "first," "second," and "third forces," with family therapy being the "fourth force"). In Italy, however, dynamic psychology, thanks to the tireless efforts of Nino Dazzi, became an autonomous field, leading to more professorships but also intensifying competition and power struggles.

In 1995, with Marisa D'Alessio, Edoardo Giusti, and Claudia Montanari, Francescato founded a four-year private graduate school in clinical-community psychology and integrated psychotherapy at ASPIC (Association for the Psychological Development of the Individual and the Community), which has trained about 650 psychologists and a few physicians to date. Italy has now 400 private graduate four year schools, approved by the State, that teach all forms of mostly individual psychotherapies.

Following the establishment of the Order of Psychologists and the dissolution of the Italian Society of Scientific Psychology, the Italian Society of Community Psychologists (SIPCO) was founded in 1994. During these years, SIPCO was given a statute, bylaws, and a code of ethics. It has a newsletter, a journal, "Community Psychology," and a dedicated website.[2] The first presidents, Caterina Arcidiacono, Patrizia Meringolo, and Elena Marta, together with members of their boards, organized national conferences, seminar initiatives, training meetings, and summer schools for young members and fostered generational change. Fortuna Procentese, has promoted during the period of the COVID-19 pandemic several online meetings, thematic appointments, and webinars on how community psychology can promote environmental sustainability, advocated the creation of a research and intervention agenda for community psychology of climate change, and launched a special issue of the journal "Community Psychology" on these issues.

The 2000s saw a significant increase in the number of universities offering courses in community psychology. In 2002, there were only 11 universities with such courses (Francescato, Tomai, Ghirelli, 2002), but by 2012, this number had grown to over 30 (Zani, 2012), and by 2023, around 40 universities offered community psychology courses. Additionally, five master's programs now specifically include community psychology in their titles: Milan's "Clinical Psychology, Health, Family Relations, and Community Interventions" (since 2003), with about 100 students enrolled per year, often exceeding application limits; Bologna's "School and

---

2. https://www.SIPCO.it/

Community Psychology" (since 2008), with around 50 students enrolled annually; Turin's "Clinical and Community Psychology" (since 2010), with approximately 270 students, reaching its capacity limit; Brescia (since 2010), with enrollment ranging from 45 to 55 students; and Cesena, a campus of the University of Bologna, offering community psychology instruction and first-year training workshops. These developments reflect the growing recognition and integration of community psychology within academic institutions across Italy.

Since 2010 in Padua, Massimo Santinello, Alessio Vieno, and other colleagues have created a master's degree program in community psychology, according to what appears on the website,[3] "of the promotion of well-being and social change"; this educational path is characterized by a strong multidisciplinary approach, aimed at the design of professional interventions [...] able to act on the interaction between the individual and the environment with specific skills. The object of study and intervention are therefore people in contexts, communities in the broadest sense: cities, neighborhoods, local and virtual communities, work and school contexts. The main objective is to foster civil coexistence, improvement of the quality of life, and individual and collective well-being [...] by using and activating the resources present in the community itself."

Since 2015, Terri Mannarini and Caterina Arcidiacono have established a new English-language journal, "Community Psychology in Global Perspective," which documents the continued marginalization of community psychology within mainstream psychology. Despite this, the discipline has gained a presence not only in the United States and Europe but also in nearly every democratic country worldwide, addressing common global challenges

In recent years, "Community Psychology in Global Perspective" has devoted several special issues to environmental issues, showing the accelerating interaction between global macro-changes and the development of these issues in some disciplines such as community psychology. Indeed, the interplay between economic and political macro-changes fostered by globalization and major technological innovations such as social networks and the development of artificial intelligence has created new challenges for psychologists. The globalization processes of the late 20th century have energized hope for an extension of welfare, skills, resources, and rights worldwide. To some extent, this has happened in some developing countries, which have seen several million people move out of absolute poverty and millions more citizens enter the middle classes (Raworth, 2017), but in the West, many workers lost their jobs when factories were relocated to countries where labor was underpaid.

---

3. https://testweb.psicologia.unipd.it/offerta-didattica/corsi-di-laurea-magis-trale/?tipo=LM& sort=2023&key=PS2381

In 2001, another significant event occurred: the demolition of the Twin Towers in New York ushered in the era of global terrorism and multiplied the number of armed conflicts, in predominantly Muslim countries, accused of favoring terrorist groups. These wars have strengthened the arms industry and fostered the rise of right-wing politicians. In addition, in the economic sphere, neoliberal globalization adopted in both capitalist and socialist countries has fostered not only a freer exchange of goods and services but a greater financialization of the economy, which has multiplied income and wealth inequalities between nations and within each nation (Lilla, 2017).

4. Development of a European community psychology

4.1. First steps toward ENCP, the first European network of community psychologists Several European community psychologists, such as Italy's Donata Francescato, Portugal's José Ornelas, and Germany's Wolfgang Stark, who met in the 1980s at various conferences of the American organization Society for Community Research and Action (SCRA), agreed on the need to promote the development of a community psychology better suited to the diverse European context. In 1992, Ornelas organized the first international conference in Lisbon, where various shortcomings of American community psychology were critically examined. The discussions highlighted that the field in America, dominated by its moderate wing, was often too abstract, politically naive, and more focused on action than reflection. Additionally, the intervention tools were still largely aimed at individuals or small groups rather than broader community contexts. The conference called for increased exchanges among European community psychologists, who had developed diverse theoretical approaches and intervention strategies across different countries.

In 1995 Arcidiacono and Francescato organized the first European conference on community psychology in Rome, where professionals from various European countries compared notes (Arcidiacono, Gelli, Putton, 1996). Around the same time, the European Network of Community Psychologists (ENCP) was founded, bringing together three Italian, two Portuguese, two Norwegian, two English, one Scottish, one Spanish, one Dutch, one German, one Austrian, and one Belgian psychologist. Francescato was elected as the coordinator, acknowledging Italy's leading role in the theoretical and methodological development of the discipline. The network proposed organizing a biennial conference in different European cities to raise awareness of community psychology alongside an annual working meeting limited to professional and academic members. These gatherings aimed to foster the exchange of experiences, and the dissemination of tools developed by European psychologists, such as community profiles, participatory multidimensional organizational analysis, empowering training, future workshops, centers for forming mutual aid groups, and early workshops on environmental sustainability. Since group work is a crucial tool for community psychologists (see chapter 8), we share how this group of pioneers evolved in promoting community psychology across Europe.

In 1996 Stark organized the first seminar in Munich, where Ornelas proposed that the network should become a legally recognized association to obtain funds for research and interventions from national bodies and the European Union. David Fryer, a Scottish lecturer, strongly opposed this by proposing that the group become a radical movement to fight against the inequalities promoted by European governments that supported the ruling classes. Furthermore, Fryer wanted no hierarchies to be created and no member to have more power than the others, but instead to include a few citizens from the most marginalized groups in the community participate in each seminar, and ENCP to support the critical community psychology that was developing in Britain. As coordinator, Francescato pointed out that both scholars had expressed important points of view, and that reflection was needed. Indeed, the issue was discussed, briefly, in our small group on the sidelines of each conference organized in the following years – in Lisbon (1998), Bergen (2000), Barcelona (2002), and Berlin (2004) on issues important to ENCP – and at greater length during working meetings held in Vienna (1997), Stirling, Scotland (1999), Lecce (2001), Paris (2003), and Leuven, Belgium (2005). At these meetings, we were able to learn about and disseminate the most significant experiences developed in each nation, to learn about and appreciate militant viewpoints such as Fryer's, which seemed to us to be indispensable for countries in the Global South plagued by problems caused by colonization, extreme inequalities and discrimination of Indigenous Peoples, but also to consider the intervention strategies proposed by Ornelas and almost all other ENCP members to be more appropriate for our European countries. **Ornelas** also promoted the **first European master's degree** in community psychology in Lisbon, where for years Stark, Orford, and Francescato taught. After long and animated discussions, all ENCP members decided to launch the European Community Psychology Association (ECPA).

4.2. Formal birth and evolution of ECPA In 2005, Orford, Francescato, and Stephan Van den Broucke signed on as "legal founders" of ECPA. Ornelas was elected its first president. The ECPA aimed to spread community psychology "as an academic discipline and as a profession, promoting health, human development and social justice" (Ornelas, 2000, p. 384).

4.3.   Ornelas quickly took action to support networks of researchers who secured European funding to empower and integrate people with mental disorders, foster the inclusion of immigrants and minorities, and provide housing for the homeless. In the years that followed, Stark, Fryer, Caterina Arcidiacono, Serdar Degirmencioglu, Liz Cunningham, and Maria Vargas-Moniz each served as chairpersons, continuing the tradition of promoting conferences and working groups in various European cities. They critically debated theories, concepts, methodologies, and approaches used in different contexts. Professional community psychologists provided local communities with support, advice, and tools to improve well-being and promote sustainable development, health, and education (Francescato, Putton, 2022). ECPA has promoted European research on countering gender-based violence, refugees, inequality, sociocultural factors that cause people with right-wing political orientations to hold hostile attitudes toward immigrants, the impact of cultural and symbolic variables on individual and collective psychological processes, and the effects of *service learning* (see chapter 14). With the election of Cinzia Albanese and a board composed primarily of young women, the anticipated generational shift began in 2019. Leveraging their digital skills and increased use of social media – accelerated by the COVID-19 pandemic – Albanese and the board expanded online events, hosting dozens of webinars. The first, in 2020, focused on how community psychologists can combat climate change, particularly by fostering a sense of global community (Francescato, 2020). Since then, there has been a surge in the production of articles and special issues in various journals (see chapter 16), exploring how different community perspectives (clinical, political, social, ecological, multidisciplinary, action-oriented, etc.), key concepts (resilience, coping, prevention, sense of community, well-being, networks, participation, empowerment), and the social justice and climate-oriented values of community psychology can contribute to addressing the climate crisis through diverse interventions. In addition, there have been several interesting webinars in recent years on burning issues such as detention centers for migrants arriving illegally, the dramatic events surrounding refugees in Europe, the need to promote the deinstitutionalization of the elderly and disabled, and the impact of climate change on children. Reports were also presented on projects that experiment with strategies to address homelessness, such as a community advocacy program to combat gender-based violence or explore interactions between social media and local communities. ECPA's website[4] provides information on the statutes, various initiatives (future webinars, summer schools, conferences, etc.), awards to promising young people, or the careers of European Community psychologists.

ECPA also wants to disseminate research findings and community interventions that can be relevant in developing new directives, policies, and reforms, and have a political and social impact in addressing both general

4. https:// www.ECPA-online.com

problems and promoting greater well-being and better quality of life in our local communities.

5. Global expansion of the discipline

Irma Serrano-García organized the first international conference on community psychology in Puerto Rico in 2006. At the conference, Donata Francescato, invited as a keynote speaker, emphasized that European community psychologists could not uncritically adopt U.S.-oriented concepts and intervention strategies, which often exalt the freedom and individual success of the self-made man. She argued that Europe, with its long tradition of promoting social capital and collective welfare, required a different approach. Francescato also highlighted how a vision of the community psychologist was emerging in Europe, one that integrates the role of a "radical activist" with that of a politically moderate professional (Francescato, Zani, 2013; 2017).

The "radical activist," as we will explore further in the next chapter, is a concept promoted by liberation and decolonization psychology, as well as critical community psychology. This role involves denouncing the material and psychological harm caused by European colonization, which led to significant economic imbalances and social disparities between white oppressors and Indigenous Peoples and fostered the **internalization of negative views by the oppressed**. The activist fights against poverty, racism, and sexism alongside the most marginalized groups. On the other hand, the "moderate practitioner" uses a diverse skill set within local communities to foster individual and collective well-being, support mutual-help groups, and empower organizations and communities. In her keynote address, Francescato highlighted three innovative methodologies developed by European community psychologists that merge traditional scientific positivist and narrative-constructivist approaches: community profile analysis, empowering training, and participatory multidimensional organizational analysis. These methodologies garnered significant interest from South American community psychologists, who hoped to establish a global association to share experiences across different world contexts. While no formal association has been founded, an international network was informally established, coordinated by Serrano-García and the organizers of the ten international conferences held every two years.

Ornelas, a passionate advocate for a global association, organized the second conference in 2008 in Lisbon. Following that, international conferences were held every two years in various locations: Puebla, Mexico (2010), Barcelona (2012), Fortaleza, Brazil (2014), Durban, South Africa (2016), Santiago, Chile (2018), Melbourne (2020), and Naples (2022), and Montevideo (2024). These conferences provided opportunities to compare the strengths and challenges of community psychology in different contexts. During the COVID-19 pandemic, the 2020 Melbourne conference was held entirely online, and the 2022 Naples conference featured global roundtables. Each conference attracted between 400 and 800 participants from 50 countries across all continents, increasing the patrimony of shared experiences worldwide.

Unfortunately, in recent years armed conflicts and environmental disasters have led to a massive influx of refugees and economic migrants into Europe, primarily from predominantly Muslim countries, which are also the origin of many terrorists who have carried out attacks in various European cities. Ambitious politicians have exploited understandable fears, portraying migrants as the cause of societal hardships. This has fueled the rise of populist and nationalist parties, which have exacerbated conflicts among different groups of citizens and deepened divisions between the elites (the top 1 percent) and other social classes, between immigrants and locals, men and women, young and old, and between urban and rural populations. Additionally, the COVID-19 pandemic, which began in China in 2019, has intensified public resentment against ruling elites, as seen in movements like the "yellow vest" protests in France against President Macron. The pandemic also sparked hostility toward governments that mandated vaccines, leading to clashes between vaccine supporters and the "anti-vax" faction, often supported by right-wing groups. These tensions likely contributed to the rise of center-right parties in Italy and across Europe in 2023. The return of war to Europe, with the conflicts in Ukraine in 2022, has prompted some political scholars (e.g., Moavero Milanesi, 2022) to call for a revival of the European spirit that, in May 1950, forged alliances between long-standing enemies like France and Germany and laid the foundation for a European Union that has ensured seven decades of peace and progress in rights. However in the last two years Putin invaded Ukraine on 2022 and after the October 7 attack of Hamas in Israel in 2023; wars and conflicts have increased the wealth of weapons producers and diminished the funds for health, education and welfare; killed thousands of people, mostly civilians women and children, increasing the need for peace and the diffusion of the principles, and values of community psychology.

In the following chapters, we will explore various theoretical constructs developed in different environmental contexts, delving into the richness of community psychology's foundational ideas and the diverse intervention methodologies that have been developed, which are especially needed today in our troubled world.

2
Scientific paradigms and aims of community psychology

1. Peculiarities of the approach and prevailing goals

In this chapter, we briefly present the two scientific positivist and constructionist paradisms and the values that moderate and radical community psychologists are inspired by. Then we illustrate some of the theoretical constructs developed in the United States of America, Europe, and several developing countries in Africa, Oceania, and Latin America, and how the internationalization of community psychology has facilitated the meeting, confrontation, and sometimes integration of theoretical and value viewpoints. Finally, we describe the premises for a theory of technique in community psychology that includes some of the theoretical constructs developed by both moderates and radicals.

In Bergen in 2000, during the opening paper at the Third European Congress of Community Psychologists, Donata Francescato argued that to create robust theoretical foundations for community psychology focused on both individual and social change, it is essential to creatively and pragmatically integrate the positivist scientific paradigms of traditional psychology with the insights of social constructionist theorists (Bruner, 1988; Sarbin, Kitsuse, 1994). The positivist approach can be valuable for understanding how adaptation processes occur through the repetition of certain interactions between individuals and their environments. It is particularly useful when identifying recurring elements and regularities in transactions between individuals, groups, and the material and structural aspects of social reality, as well as when examining how people generally behave in specific environmental settings or how certain environmental factors and social contexts influence individual, group, and organizational behaviors. Positivist-oriented psychologists maintain that human behavior can be predictable if one understands both the person and the situation.

We believe this is especially true for habitual behaviors in familiar settings, where the positivist approach proves most useful.

When we want to understand the meanings that individuals attach to their interactions with the material environment or explore how change can be affected, it is beneficial to also use postmodern or social constructionist paradigms. These approaches are valuable for community psychologists in generating alternative interpretations of events, exploring the world of meanings rather than just the world of facts, and understanding how intentional changes can be initiated by individuals or groups.

We argue that a theory of technique in community psychology should unite what Jerome Bruner (1988) called the "two ways of doing science": the paradigmatic and the narrative. The paradigmatic model is based on positivist theories and is characterized by the formulation of hypotheses that need to be tested, the search for invariances, and the decontextualization of the phenomenon studied; it aims at the construction of laws that make it possible to predict the evolution of a phenomenon. Moreover, the underlying epistemology is that there is a correspondence between studied phenomena and an objective reality. In contrast, the narrative model is predominantly employed by historians and biographers, clinical psychologists, psychotherapists, writers, and filmmakers. The goal is to tell a story about a phenomenon in its context. Criteria for acceptance or non-acceptance are based on the coherence of the story. Users of the narrative model are prepared to incorporate the influence of multiple contexts into their stories. Indeed, narratives are seen as influenced by the perspectives of the narrators who produce them: "The logic of constructionism promotes the introduction of multiple perspectives to oppose positivist assumptions of a uniform, objective social reality" (Sarbin, Kitsuse, 1994, p. 8). Proponents of social constructionism reject the idea underlying traditional science that there are social objects, out there in the real world, waiting to be discovered, measured, and understood; instead, they believe that social objects are created. Humans, in this view, are not seen as adapting or processing information derived from a given environment, but as social agents who construct meanings in their interaction with others.

The roots of social constructionism can be found in the symbolic interactionism of the Chicago School, according to which the basis of action is the meaning that human beings attach to interaction, and this meaning is largely determined by a continuous, mutual interpretation of actions, performed by actors in each situation. According to Amerio (2000), the social constructionism of the interactionist perspective has been important in showing that many psychic phenomena, especially those of the "pathological" type, are produced by interaction rather than intrapsychic processes. For example, in studying deviance, these scholars have shifted the analysis from traditional individual etiological factors to socio-relational factors, showing how various deviant behaviors do not derive from the character or psychological variables of some individuals but are socially constructed by the actions that others manifest toward them. Amerio writes:

"the social world exists because we don't question its existence; we continuously describe and explain it through our routine daily practices. Ultimately, it is these everyday practices, carried out in the small acts of existence, that constitute social reality. This reality is not governed by systems, rules, structural architectures, or any other explanatory principles. Instead, it is shaped by the tacit consensus through which social actors accept the conventions they are immersed in, using them to navigate the challenges of their lives "(*ibid.*, p. 169).

It becomes crucial for community psychologists to find ways to challenge the tacit consensus that leads individuals and groups to accept prevailing systems of conventions. Generally, constructionists reject the notion of certain scientific knowledge or categories as objective truths, instead choosing to construct and evaluate diverse stories about ourselves and our world. We are not passive "discoverers" of a single reality but are instead architects, builders, and custodians of multiple realities. The work of constructivists on the significance of language and narratives can be harnessed to facilitate personal and social change.

In many of the community psychology tools described in this book – such as community profiles, participatory multidimensional organizational analysis (PMOA), research–intervention, empowering training, and group and network work – desired changes are achieved by promoting the creation of new metaphors and narratives by individuals or small groups. These new perspectives offer alternative interpretations of social situations (Berger, Luckmann, 1967). By breaking the tacit consensus that made conventional interpretations the only conceivable social reality, new scripts and roles for individuals and social groups become possible, creating new foundations for legitimizing the changes to be promoted.

Julian Rappaport (2000) is particularly attentive to this socio-constructivist aspect, suggesting that community psychologists must critically analyze dominant cultural narratives as well as local or personal histories. He emphasizes the importance of transforming oppressive narratives – stories of terror that have marginalized certain groups or minorities – into stories of joy that foster a renewed sense of community and hope for the future. Rappaport notes that our personal stories are shaped by and intertwined with the narratives of the communities in which we live, providing both the text and context of our culture.

Understanding community narratives is one way to understand culture and context and its profound effects on individual lives. Shared stories are the substance of our social world. People who have stories in common know where they come from, who they are, and where they are going; they are a community. A community cannot be such without a shared narrative. Shared narratives are the current in which our individual lives flow in the river of time. They are resources that oppress or facilitate. They give direction and meaning to our lives" (*ibid.*, p. 6), as narratives are resources that can be conceived of as potential tools for empowerment. "Some personal stories are more empowering than others. But we can, if we are part of a group that does, mutate and recreate our community narratives even as they affect us" (*ibid.*).

Rappaport particularly values the role of life stories and community narratives as empowerment tools to counter dominant class narratives. The author also emphasizes how the narrative resource is unequally distributed across social classes, as the dominant social classes have more tools to spread and create their narratives. In this position, in our opinion, the author exaggerates the importance given to only changing community narratives as a tool for social change. While Rappaport gives proper weight to socio-constructivist variables, he seems to give too little to socioeconomic, political-legal, and environmental variables, which constitute the material context of the world imagined, constructed, and explained in the narratives. Changing only the narratives, important as it is, perhaps leads to hopes destined to be dashed if one does not also try to understand what changes need to be implemented in the other domains.

Before presenting our approach to the theory of technique in community psychology, which is particularly influenced by European theoretical contributions due to the context in which we authors live, we would like to briefly discuss some theoretical contributions from both the moderate and radical wings, developed in various parts of the world. We believe these contributions significantly enrich the theoretical, methodological, and value-based foundations of community psychology.

2. Theoretical contributions, values, and goals of moderate and radical community psychologists

2.1. Energizing values and goals

Community psychologists operate by adhering to certain values with which everyone identifies social justice and equity, respect for diversity, inclusion of minorities and the marginalized, and promotion of individual and collective empowerment, social support, a sense of community, and active participation. These values are the motivational nourishment and support of community psychologists in their complex, exciting, and difficult work of pursuing challenging goals of preventing distress and promoting individual and collective well-being. In addition, the presence of moderate and radical psychologists provides ongoing opportunities to sharpen conflict management skills and listen to divergent viewpoints, enriches social capital, and sometimes creates lasting friendships among very different people. In short, we, the authors, hope that readers will be inspired to become community psychologists, as our world greatly needs more of them. Community psychology is a truly fascinating field, offering numerous opportunities to contribute to building a better world through hope and trust.

2.2. Prevention of distress and promotion of well-being: key goals of community psychologists The etymology of the word "prevention" refers to a temporal concept: coming before or anticipating the onset and consequences of a physical illness, psychological distress, social problem, or ecological crisis. Community psychology interprets the preventive strategy in an even more radical sense, attempting, when possible, not only to avoid or reduce the manifestation of distress but to improve the living conditions of the community and foster positive and dynamic integration between the individual and the environment. Caplan (1964) was the first to explicitly apply these concepts in the field he called "preventive psychiatry," identifying some possible interventions to work in a preventive sense.

Two prevailing views can be recognized in the understanding of hardship: the *exceptionalist* theory and the *universalist* theory. According to the exceptionalist or social selection theory, distress and/or pathology are determined by the presence of occasional and random individual factors (genetic, character, personality) that place people at a disadvantage, excluding them from the mechanisms of evolution and success used by others. Those who adhere to this thesis tend to see discomfort as a problem or disease, an accident of the individual's path, and point to the treatment (therapeutic, pharmacological, or rehabilitative) of the discomfort itself and/or to the early identification of people at risk, predisposed to the development of conditions of discomfort or pathology.

Proponents of the universalist theory, or social causes, consider social distress not as an exception, an anomaly of the normal state of health, but as a function of a community's social relations, linked to an unequal distribution of resources, to conditions that are not unusual and occasional, but predictable and therefore preventable. It thus becomes essential not only to "repair" the consequences of critical situations but, where possible, to anticipate them, to address them by developing systemic resources and creating the conditions that promote quality of life and well-being. There is a need to shift from focusing on risk factors (measurable individual or environmental characteristics whose presence is associated with an increased likelihood of developing distress) to protective factors (modifiable individual or contextual characteristics that increase the likelihood of promoting a state of well-being). Some moderate-wing community clinical psychologists adopted a primary, secondary, and tertiary prevention strategy as early as the 1960s and 1970s, drawing inspiration from the patient-centered medical model that dominated the healthcare settings where they worked (Korchin, 1977; Albee, 1996). However, in recent years, many community psychologists working in locally-focused services in various countries, including Italy, have adopted a preventive and health and well-being promotion approach that aims to increase the active participation of the community fabric and the development of its skills as important resources for achieving health and well-being outcomes for citizens (Laverack, Wallerstein, 2001; Laverack, 2006; Cicognani *et al.*, 2020; Walker, Zlotowitz, Zoli, 2022).

Orford (1992, p. 48) believes that there is a reciprocal relationship between the two sets of individual and environmental variables to which the exceptionalist and universalist theories refer:

The relationship between stressful life events and psychological distress is reciprocal: not only can the presence of stress increase the likelihood of psychological distress, but the presence of distress can elevate the likelihood of a stressful life event happening. [...] Belonging to an affluent class can mitigate the influence of stressful events on distress. Stressful life events can in turn influence social class: losing a job is an obvious way in which this can happen. The transactional model is further complicated by introducing the idea of resources, personal social, and economic, whose lack generates further vulnerability.

The need to adopt a transactive model also emerges from the findings of various transformative research that has promoted the prevention of problem behaviors or promoted positive development (Santinello, Vieno, Martini, 2006; Schwartz, Brownell, 2007; Vieno *et al.*, 2008). Santinello and collaborators (2018) note that the primary goals of prevention interventions correspond to the secondary goals of positive developmental promotion interventions and vice versa. Francescato and Putton (2022) also found that several interventions that aimed to increase positive relationships and problem-solving and teamwork skills among pupils in schools from kindergarten to secondary had also decreased incidents of bullying and other problem behaviors. In schools with high dropout rates, we found that the use of social-affective education and PMOA techniques promoted better relationships with parents, teachers, and peers but also decreased dropout rates and created mutual-help groups among parents, and networks among associations in the local area. Once empowered, participants innovate by using what they have learned in different contexts (see chapter 17), increasing community competence.

2.3.   Evolution of the systemic ecological approach Community psychology uses the concept of systems when referring to "social systems," which are conceived as a set of relationships between elements of increasing complexity: individuals, small groups, organizations, and communities. The community is configured as a network of both formal (structured organizations) and informal (spontaneous groups) systems. Several authors have drawn inspiration from biological ecology, attempting to identify principles for understanding the mechanisms of mutual influence between environments, groups, and individuals. Mills and Kelly (1972) identify four: interdependence, resource cycling, adaptation, and succession. For Bronfenbrenner (1986), human behavior is determined by the influence of a series of concentric structures that are interdependent and how these environmental dimensions are perceived and experienced by the individual. The structures are as follows:

- The *micro-level*, composed of the systems the individual has direct experience of, which includes physical spaces, people, and the interactions between them and the individual;

- The *meso-level*, which includes two or more microlevel systems and the links between them;

- The *exo-level*, which includes the systems with which the individual does not interact directly, but which affect the lives of people who interact with him; and

- The macro level, which includes the extended social context and the superstructures that have the power to influence all the levels below.

The person is constantly changing roles and must always renegotiate their position in different environmental situations. Bronfenbrenner calls this process "ecological transition" and stresses the importance of grasping the interconnections between different levels and the circular relationship that exists between individual and environmental variables. The author proposes the concept of ecological niche, which he defines as the environmental, physical, and relational context capable of facilitating or hindering the development of the people who frequent it based on the combination of personal factors and environmental characteristics (Zani, 2012; Santinello *et al.*, 2018; Arcidiacono *et al.*, 2021). Levine and Perkins (1987) and Murrell (1973) offer valuable insights into how to intervene to promote changes at different levels.

Levine proposes five practical principles:

1. A problem arises in a specific setting or situation: situational factors cause, trigger, exacerbate, and/or maintain the problem;
2. A problem arises because the adaptive capacity of the setting to solve problems is blocked;

3. To be effective, an aid must be strategically placed with respect to the emergence of the problem. One must consider the temporal and spatial aspects of the problem and the other types of interventions already in place;
4. The goals and values of those offering help must align with the goals and values of the setting. Sometimes, an intervention may provoke conflicts between the values of the community psychologist or the service attempting to provide help to the setting, and this can become a learning opportunity for both the community psychologist and the members of the setting. However, it is crucial to identify the right moment and provide an effective way to manage conflicts (see chapters 8 and 11);
5. The form of help must be established systematically by utilizing the resources already present in the setting or by introducing resources that can become institutionalized as part of the setting.

The proposed changes must be sustainable over time and continue to serve as a resource for problem-solving. For instance, it is essential to enhance the skills of teachers and parents in using social-affective education strategies, enabling them to independently adopt new communication methods. Ideally, these acquired skills could then be shared with other teachers and parents through mutual-help groups. Community psychologists must promote empowerment by recognizing that good practitioners do not foster dependency but instead plan for their exit (Francescato, Putton, 2022; see chapter 12). Murrell, although not widely cited, is an author who we believe deserves greater recognition even in the 21st century. The concepts he proposed remain valuable for understanding what it means to work from an ecological perspective and to promote social change at various levels. He defines community psychology as:

"the area within psychology that *studies* transactions between networks of social systems, populations, and individuals; that *develops* and *evaluates* methods of intervention that improve person-environment *"fits"*; that *plans* and *evaluates* new social systems; and that from this knowledge and change seeks to *increase* the psychosocial opportunities of the individual (Murrell, 1973, p. 39)."

Murrell believes that personality itself is a function of active transactions between the individual and the environment, in which the individual is not only a reactant but also an agent. He starts from a complex concept of being human: individuals are motivated according to the specific experiences, expectations, and values they have developed during a lifetime in which they have encountered problems and attempted to solve them. There are no universal hierarchies of needs but rather problem areas, i.e., several areas of psychosocial interest to the individual, who seeks to achieve their goals in them. To achieve these goals, they must come into contact, in constant negotiation, with their environment, which can help or hinder them.

Everyone develops an order of priorities, varying over time, for these key areas and creates their preferred solutions (*individual problem management*), their own "program" related both to personal experiences and to the obstacles and opportunities offered by the system. Psychological well-being depends on psychosocial agreement, which refers to the degree of alignment between the demands of the social system and the individual's approach to managing their key areas of life. The system's demands consist of the limits and expectations imposed by the social system, determining whether a behavior is acceptable based on the individual's social roles and their interpretation of the system's demands and resources.

Because everyone belongs to multiple social systems, Murrell uses the concept of intersystem accommodation to describe the degree of compatibility between these different systems as they interact with the individual. This concept captures how the network of systems affects the person: for instance, when different systems impose conflicting demands, the individual may find themselves in paradoxical situations that can diminish their well-being. For example, when children begin attending school, they are exposed to new demands and expectations, some of which may differ from or even conflict with the expectations and norms of their family context.

Building on this psychological framework, Murrell seeks to provide appropriate tools and variables to describe different social levels, including the individual, small group, system, and network of systems. He suggests incorporating concepts from general systems theory, organizational psychology, Barker's (1968) ecological psychology, and the literature on small groups. This approach allows for the analysis of variables such as the distribution of decision-making power, organizational climate, communication types within a system, or the "unit of environmental strength" characteristic of a particular group setting. The goal of this analysis is to identify aspects of complementarity and congruence, or discrepancy and conflict, in the *transactions* between different social levels.

By examining these transactions, it becomes possible to plan and implement the most effective interventions to enhance psychosocial agreement. This involves selecting the level where action can be most effective while also avoiding the mistake of "punctuating" (Watzlawick, Beavin, Jackson, 1971) any one level in isolation. Instead, it is crucial to understand and address the interaction and interdependence among all the levels involved.

Murrell describes six levels of intervention, varying in breadth and complexity from one focused on the individual to comprehensive programming for an entire community.

1. Individual outplacement. This refers to the idea that not every individual can harmoniously integrate into all social systems, and not every social system can adequately address the needs of every individual. When the interaction between an individual and a system is so incompatible that improvement on either side seems unlikely, it may be advisable to relocate the individual to another system. For example, placing a child with adoptive parents if the biological parents are unable to care for them. However, if this relocation is temporary, there is a risk that the individual may adapt well to the new system but then face even greater difficulties when returning to the original system. A common example of this is the role of a therapeutic community: while it may be highly effective in removing an addict from a harmful environment and instilling new values and rules, the process of reintegrating into broader society remains the most challenging aspect.

2. Interventions on the individual. Here the goal is to change or develop the person's resources or strategies so that they can better fit into the system. Examples of this strategy are technical training, behavior modification programs, psychotherapy, crisis intervention in place, etc. These are potentially effective interventions, provided the person requests the intervention and is willing to remain in the system. Both individual outplacement and interventions on the individual, however, have a considerable risk from the perspective of community psychology: that of "blaming the victim" (in Ryan's terms, 1971) i.e., defining the problem by focusing attention on the individual, albeit to help or change them. In Murrell's perspective, it is therefore preferable to attempt to intervene on more complex and systemic levels of the community. Nevertheless, intervention on the individual can also be valid if conceived within a broader plan, in which the individual-system relationship is analyzed as a reciprocal exchange, and stages of intervention on groups, organizational structures, or the environment are also included.

3. Population interventions. Here the strategy is to increase the resources of an "at-risk" population or group, such as through crisis preparedness programs or group training interventions. The difficulty here may be motivating and engaging the target population to participate, especially if preventive interventions are involved. A program to improve communication skills between parents and adolescent children can be very useful, but there is a risk that few parents will be interested in following it unless they have significant difficulties with their children. Therefore, interventions of this kind should also be implemented within a framework of raising awareness among community members and developing social participation and support.

4. Social system interventions. This involves making structural and functional changes in systems so that they make it easier for individuals to "manage problems." Examples of this strategy are counseling to change the behavior of key people in the system or, at an even more structural level, counseling to innovate the rules, constraints, distribution of tasks and responsibilities, and objectives of a given system. In this case, the main difficulties lie in defining clear and correct advisory relationships and in ascertaining which components – and with what motivations, alliances, and hegemonic capabilities – are pressing for changes in the system's structure. Otherwise, the risk is that of proposing changes that are problematic to manage or even incompatible and therefore likely to create sterile destabilization.

5. Cross-system interventions. Action directed at multiple systems involves efforts to create better coordination and more functional connections between various systems. This level of intervention is inherently complex, as those attempting it typically lack the influence or authority across all the systems involved. However, this strategy aligns closely with the ecological approach of community psychology, which emphasizes the interdependence between different sectors of society and underscores the exchange and cyclicality of resources. For example, addressing issues like youth unemployment or delinquency requires more than just interventions within the productive sector or correctional institutions. Instead, comprehensive projects that engage local authorities, educational institutions, social services, and community groups are essential to create meaningful and sustainable solutions.

6. Interventions on the whole social network. Sometimes the boundary between the intersystem level and the whole social network level is very blurred. Murrell is referring here to programs aimed at the community, such as through the mass media, or to cases of participation of community psychologists or other social researchers in the design of new community networks.

As De Piccoli et all (2021, p. 44) note, Murrell redefines: the ecological-systemic perspective [...] based only on the multiplicity of levels (such as the one proposed by Bronfenbrenner, for example), to a complexity analysis. The former focuses on the different constituent parts of the system, while the latter considers the dynamic processes between the different levels. That is, it is a matter of moving from the plane of contextual analysis, thus referring to a specific context (school, neighborhood, organization, etc.), to the plane of systemic analysis, which considers the interrelationships between the parts that make up a system and detect their mutual influence.

We were inspired by Murrell's model to develop some intervention strategies that we elaborated at the small group level (chapters 8-10) and especially for PMOA and empowering training (chapters 7 and 11).

Among the most recent innovations in ecological models within community psychology, Neal and Neal (2017) propose using *network analysis* to examine the connections between people, technologies, and the activity settings that unite them. In this approach, context can be described as "constellations of relationships and redistributed resources" (Johnson *et al.*, 2017, p. 101). This method opens opportunities to explore new levels of interaction. We believe that a truly ecological approach should also include two additional levels. The first level is the study of interactions between humans and the natural environment, often referred to by environmentalists as "Mother Earth." This encompasses the ecosystems where thousands of sentient beings live, all interconnected, and whose diversity humans have largely destroyed or damaged. Given the pressing issues of climate change and pollution that are harming the planet, we have dedicated a chapter to this topic (chapter 15). This chapter also presents examples of intervention research on potential solutions to environmental problems, such as energy communities, as well as new job opportunities for community psychologists – opportunities that are still limited but hold significant promise.

When it comes to relationships, smartphones and social media appear to have a dual impact: they can both isolate individuals and strengthen social ties. The virtual contexts created by these new online technologies seem to foster both well-being and malaise in individuals and groups. For instance, Santinello and collaborators (2018) highlight how cell phones have become integral to every aspect of our daily lives, sometimes with negative consequences, yet they have also been used in interventions aimed at promoting physical health (Bauer *et al.*, 2003; Park *et al.*, 2014). Additionally, new technologies have the potential to reach culturally or economically disadvantaged individuals. During the pandemic, many psychologists **were able to continue supporting their clients via video call platforms** accessible from smartphones, and some have chosen to continue this practice even after resuming face-to-face sessions, appreciating the ability to record and revisit sessions as needed. Furthermore, apps like Live Happy have been used to "promote happiness, strengthen social relationships, and foster the perception of living a life with meaning" (Santinello *et al.*, 2018, p. 203). Other apps, such as Mood Man, encourage gratitude, emotional awareness, and understanding of how different emotions influence our behaviors and relationships. Some apps help reduce stress and provide strategies for distancing oneself from negative thoughts. However, research on the effectiveness of these technologies has also shown that excessive use of smartphones and social media has contributed to a decline in average levels of empathy in recent decades (Konrath, O'Brien, Hsing, 2011). On the positive side, social media can help maintain distant relationships and alleviate social anxiety in adolescents. Thus, new technologies offer many potential applications for interventions aimed at promoting individual well-being and strengthening social relationships.

A very recent field of research and intervention examines how virtual communities can interact with geographic communities and activate the processes of positive interactions even among strangers. New technologies such as community networks have already been experimented with these networks also connect people online who live in the same neighborhood or geographic area. Sharing a virtual platform and a real space can foster the formation of new social capital, including making people aware of events in the neighborhood and allowing them to discuss a particular problem or desire for change that has arisen in that area, or to exchange information about available services. All these activities increase opportunities to connect with others interested in solving common problems. The first projects also arose in Italy with Social Street in Bologna,[5] created to counter the general impoverishment of neighborhood relations, foster good neighborly practices, and the exchange of help and resources (see chapter 11 for the importance of new technologies in empowering education).

A group of community psychologists at the Federico II University of Naples has for several years been conducting a series of compelling investigations (Gatti, Procentese, 2020, 2021; Gatti, Procentese, Schouten, 2022; Procentese *et al.*, 2023) carried out in Italy, but also in collaborations with various European countries, funded by European funds, using social media in various ways. Among the most promising is research aimed at ascertaining how certain constructs – such as a sense of community, attachment to place, levels of participation, civic engagement, and even the propensity to be together responsibly (*Sense of Responsible Togetherness*, sort) – can be promoted through new technologies. For example, some studies have investigated how specific social media (e.g., Instagram) can change the spatial experiences of a place using photos (Gatti, Procentese, 2021). The results suggest that some social media can change the experiences citizens have of their local community both by representing different ways of using urban spaces and the relational life that can be lived in these spaces and by bringing out more strongly the shortcomings and problem areas of the local community.

Other studies (Gatti, Procentese, Schouten, 2022) have investigated whether the use of *people-nearby* apps (PNAS) to meet people in one's local communities is influenced by the characteristics of the neighborhood in which one lives, in addition to individual characteristics. Three hundred and forty-three PNAS users residing in Italy and the Netherlands participated; the research aimed to examine the complex interactions between certain characteristics at two levels: personal (PNAS use, sense of community, *bridging* social capital and loneliness) and neighborhood (availability of social spaces, opportunities for socializing, feelings of trust, friendliness, safety and environmental support). Further research, also done in more qualitative and more narrative ways, may better clarify some of the conflicting findings of this and other surveys. For example, in terms of loneliness, it might be

5. http://www.socialstreet.it

useful to better examine the life histories and characteristics of people who report being lonelier and make more use of PNAS to increase their social networks, even though their neighborhoods offer ample space for dating and socializing in general.

For community psychologists like us, the dual use of PNAS and spatial features offers a range of new opportunities for interventions that design desired changes from below. By also using online tools, it is possible to get people who do not participate in presence to discuss and speak, as documented by young colleagues in Naples who have been able to get many men to express ideas and feelings on difficult topics such as domestic violence (see Marcella Autiero and Emanuele Esempio's testimony in chapter 17).

2.4.   The contribution of Lewinian theory Kurt Lewin's thought is too broad to be treated exhaustively here. We aim to point out the theoretical and practical insights most consistent with the community psychology approach.

The main contribution that community psychology derives from Lewin's work is field theory, which posits that every event is determined by the totality of factors present within the psychological field. By psychological field, Lewin refers to the complete set of factors – both individual and environmental – that are in an interdependent relationship at any given time. This field includes not only the subjective world, encompassing knowledge, emotions, representations, perceptions, aspirations, and projects but also the external environment.

Particularly endorsed by community psychology is the conception of behavior as a function of both the person and the environment, expressed in the well-known formula $C = f (P, A)$, which effectively encapsulates the ecological orientation of Lewinian theory. The relationship between P (Person) and A (Environment) is understood not as a linear cause-and-effect relationship but as a continuous and reciprocal transaction. In Lewinian theory, the articulation between the individual-subjective and the social-objective spheres is dynamic and active, materialized through people's actions.

As Amerio (2000) rightly points out, a particularly relevant aspect of Lewin's legacy for community psychology is the concept of the active subject. This concept emphasizes the importance of human transformative action and the power (though not omnipotence) to influence the biological, material, and social variables that shape one's existence. As Amerio suggests, action can be considered a true "psychosocial process," as it combines and integrates the subjective world of intentions and motivations with the objective world of environmental resources.

The ecological approach of Lewinian theory influences the way social phenomena are understood and, as we shall see later, the methodology of research (see chapters 13-14). Finally, a significant legacy of Lewinian theory is the conception and use of the small group, as well as the development of knowledge about small groups and the laws that govern them. The concept of the group, the theme of interdependence, and the transformative potential of small groups are central to community psychology and have been reinterpreted into specific intervention methodologies. These include the methodology of group work, as well as organizational analysis and community analysis, which use the small group as a fundamental tool.

2.5. Contributions of humanistic and cognitive-behavioral psychologies

Community psychology shares with the humanistic psychology of authors such as Rogers, Maslow, May, and Perls an emphasis on potential and positive resources to be enhanced and developed rather than focusing solely on dysfunctions and disorders to be treated. This perspective has led to strategies and methodologies that promote individuals' coping skills (active coping), strengthen the abilities of key people and non-professional practitioners, and support spontaneous and self-help groups within the community.

Inspired by this humanistic approach, we have applied social-affective education strategies in schools and communities across Italy (see chapter 11), achieving positive results, as documented in the new edition of *Star bene insieme a scuola* (Francescato, Putton, 2022). In recent years, positive psychology, which aims to develop positive individual qualities such as self-determination, self-efficacy, optimism, and both individual and social happiness, has also gained popularity. Seligman and Csíkszentmihályi (2000) identified five factors that contribute to optimal functioning, personal well-being, and the pursuit of happiness: positive emotions that enhance life satisfaction; engagement in life activities that reflect an active role in pursuing one's well-being; positive interpersonal relationships; the meaning individuals ascribe to their lives, or the perception that life has a purpose; and personal achievement, or the awareness of having accomplished something meaningful in life.

However, some community psychologists have criticized positive psychology for its focus on the individual, arguing that well-being, both individual and collective, is also significantly influenced by political and economic power dynamics. These critics emphasize that well-being can either increase or decrease depending on whether power is used to oppress or to promote justice and equality (Prilleltensky, Prilleltensky, 2006).

Behavioral psychology contributes significantly to community psychology by applying behavioral principles to the analysis and resolution of problems within community settings and by enhancing community capacity to achieve its goals (Fawcett, Mathews, Fletcher, 1980, p. 505). The cognitive orientation, particularly, becomes essential in understanding phenomena such as social learning (Bandura, 1969; 1978), the influence of mass media and social information campaigns, and the subjective ways individuals perceive their environment and quality of life. The cognitive-behavioral approach is often employed in planning large-scale interventions, such as health education initiatives, training programs for key individuals or at-risk groups, and efforts to change the goals and organization of a system. However, the behaviorist approach carries the risk of falling into a purely functionalist logic, focusing on the efficiency of individual interventions without considering a more profound conceptual framework. Recognizing and managing this limitation allows for the effective use of operational tools and techniques while still aligning with the broader conceptual principles of the system-ecological orientation.

3. The construct of empowerment as a bridge between radicals and moderates

In discussing the goals and theoretical foundations of community psychology, the construct of empowerment deserves special attention. The concept of empowerment serves as a bridge that can help reconcile some of the theoretical and practical divergences between the moderate and radical areas of the discipline.

Empowerment-centered programs embrace the moderate ideology of empowering individuals but do so from a perspective that is now political and emancipatory rather than merely therapeutic and reparative, as was common in the 1970s. These programs acknowledge the unequal distribution of resources and the disparities in access to sources of power – such as force, law, money, and knowledge – across different social and ethnic groups. At the same time, they recognize that individuals who perceive themselves as powerless may also be less able to identify and utilize the personal and environmental resources available to them. Thus, the development of the individual and the empowerment of the community to which they belong are seen as being in a circular, interdependent relationship. Empowerment-centered programs aim to enhance personal empowerment while also improving individuals' ability to analyze and navigate various social systems. This dual focus helps people understand the constraints and conditioning imposed by these systems, as well as the opportunities they present, thereby fostering action for change.

3.1. Meaning of the term Empowerment carries a specific connotation of power. Etymologically, the word is composed of a prefix, a noun, and a suffix. In English, the prefix *em* is added to words to create verbs that describe the process of placing someone in a particular state, condition, or position. *Power* is the noun at the core of the word, signifying "power" or "ability." Finally, the suffix *ment* is added to some verbs to form nouns that refer to actions, processes, or states. Together, *empowerment* refers to the process or state of being given power or authority, particularly in a way that enables individuals or groups to take control of their circumstances and improve their lives.

Empowerment has to do with the development, increase, and/or transfer of power. The power in question does not have a traditional meaning of coercion or manipulation, but it implies the idea that it can be extended and distributed. Empowerment deals not with power "over" someone but power "with," which can be developed to improve the condition of individuals, small groups, organizations, and communities in the participatory, decision-making, and expressive aspects. Rappaport (1981) introduced this concept in community psychology, defining it as a process that enables individuals, groups, and communities to increase their ability to *actively control their own lives*. To this must be added *critical awareness*, understanding the sociopolitical context and power structures; *collective action*, which refers to processes of participation to achieve shared and desirable goals; and the *mobilization of resources* that can promote change. Kieffer (1982) defines empowerment in terms of an individual's acquisition of political skills, competencies, and information that enable a person to become empowered: for Kieffer, someone who can control their own life by actively participating in associations and organizations gains power (becomes more *powerful*) by increasing their degree of political awareness and self-perception of their own competence is "empowered."

The participatory dimension is central to the construct of empowerment (Zimmerman, Rappaport, 1988), emphasizing its operational aspect – specifically, the essential activation required to achieve desired goals (Dallago, 2006). As some authors have argued (Cornwall, Brock, 2005; Pettit, 2012), empowerment and participation are deeply intertwined: without meaningful participation, empowerment remains "an empty, unfulfilled promise." However, while participation is often considered "the cornerstone of empowering strategies" (Wallerstein, 2006, p. 9), participation alone does not necessarily lead to empowerment. Participation can sometimes be manipulative or passive rather than truly engaging and empowering. Fostering the process of empowerment means promoting the conditions that enable people to recognize their resources and those of the environment, to develop personal skills and abilities for the resolution of difficulties, and to increase the equality of opportunity and responsibility of the individual in decision-making processes, sensitizing them to participation and sharing of goals and actions (Zani, 2012).

Francescato and Tomai (2021) consider empowerment a goal and a process. The process of empowerment, resulting from the interaction between individual empowerment, empowerment-enabled, and empowerment-fostered, helps to choose, and improves knowledge and skills toward a social and emancipatory dimension. From this perspective, proceeding with empowering interventions means activating resources and skills, increasing in subjects –whether individuals, groups, or communities – the ability to use their positive qualities and what the context offers to act on situations and change them.

In summary, within the field of community psychology, empowerment can be defined as a crucial path of change through which citizens, particularly those who are most disadvantaged, gain greater power and control over their lives. Empowered citizens, in turn, contribute to the development and growth of empowered communities, even during times of crisis or difficulty (Francescato, Zani, 2017). Perkins and Zimmerman (1995) highlight two common elements found in the many definitions of this construct over the years: first, the conception of empowerment as an intentional behavior that involves critical thinking, active interest, and group participation, leading to greater control and well-being in one's life; and second, the understanding that empowerment is distinct from, and goes beyond, related constructs such as self-esteem, self-efficacy, and locus of control.

3.2.   Diffusion of the concept of empowerment in different areas According to Piccardo (1995), the construct of empowerment has been present since the 1960s in at least four different areas: the policy, medical, psychotherapeutic, and managerial-organizational literature.

In the late 1960s, this concept was used in political language, starting with the civil rights movement and feminist movements. There, the concept of empowerment still has wide application, representing the goal of most community development programs implemented in third-world countries and backward areas of more advanced countries; it has gained relevance in various social movements such as those for the defense of the environment and the rights of disadvantaged groups.

In the medical and psychotherapeutic fields, the construct has been used to facilitate short and effective rehabilitative processes, helping the individual to break away more quickly from dependence on the medical figure and promoting the ability to self-diagnose and care for one's own health. In healthcare, the construct has an interesting application in the context of chronic diseases. The construct of patient *empowerment* refers to a patient's ability to independently and responsibly manage relevant aspects of their health condition. The World Health Organization (WHO, 2021) defines it as a process through which people gain greater control over decisions and actions that affect their health and involve both themselves and the community.

In adult pedagogy, the focus has been and continues to be, on promoting individuals' growth throughout the life cycle by enhancing their emancipatory aspects and encouraging active participation in defining educational goals. This approach also emphasizes establishing supportive educational relationships and respecting different learning paces.

Another field where the concept of empowerment has been particularly impactful, especially during the 1990s, is the managerial-organizational sphere. The rise of flexible markets and increased international competition has made the empowerment of human resources a primary objective for companies, which must foster a culture of participation and high commitment among their workers. As a result, successful organizations have become more horizontal, with leadership focusing not only on management but also on the growth and autonomy of employees (Piccardo, 1992). Since the 1990s, the literature on this topic has expanded significantly, exploring the role of empowerment in various aspects of organizational life. This includes the exercise of leadership (empowering leadership), the implementation of teamwork methodologies (empowered work groups), and the management of organizations (empowered and empowering organizations) (see, e.g., Cuneo, 1992; Shipper & Manz, 1992; Fisher, 1993; Morganti, 1998; Özaralli, 2003; Peterson & Zimmerman, 2004; Yukl & Becker, 2006; Francescato, Tomai, & Solimeno, 2008).

3.3.   The multidimensionality of the construct According to various authors, empowerment can be distinguished into psychological-individual, group, organizational, and community empowerment (Zimmerman, Rappaport, 1988; Piccardo, 1995; Zimmerman, 2000; Francescato, Tomai, Solimeno, 2008).

- *Psychological-individual empowerment.* Psychological-individual empowerment is grounded in well-established psychological constructs that span three interconnected domains: personality constructs, such as locus of control; cognitive constructs, like self-efficacy; and motivational constructs, including the desire and willingness to actively shape events (Zimmerman, Rappaport, 1988; Rappaport, Seidman, 2000). Zimmerman (1990) characterizes psychological empowerment as the result of a process that transitions an individual from learned helplessness – marked by passivity, distrust, and discouragement in problem-solving – to learned hopefulness, where one acquires problem-solving skills and perceives a sense of control. This transformation is achieved through gaining confidence in one's abilities via active participation and community engagement. Zimmerman (2000) further identifies three components of empowerment: perceived control, which relates to making choices and influencing life decisions; critical awareness, which involves understanding the social environment and its influencing factors; and participation, which encompasses the transformative actions

61

individualsundertake,oftenincollaborationwithlocalagencies,toeffect change in their communities. These components collectively enable individuals to navigate and positively influence their social contexts.

Massimo Bruscaglioni (1994a; 1994b; 2007) links the concept of empowerment to that of "possibility," which is the set of choices available to a person which is the increase in the person's opportunities for choice. The process that leads to a widening of the range of available alternatives is referred to by the author as "growth," while "regression" is the reverse path, leading to a narrowing of possible choices. Bruscaglioni summarizes the concept of psychological empowerment in two main factors: positive internal protagonism and external trust. The first factor encapsulates dimensions such as the tendency to view outcomes as produced by a person's actions like increasing one's competencies; the second factor involves the trust that people in general can affect positive changes.

- *Small group empowerment.* From the time we go to kindergarten through college, we attend groups that become the contexts where we can experience our power through participation, control, and critical awareness; then, when we get a job, play sports, or engage in volunteer work, we often operate with people who can offer us empowering and disempowering experiences, as Francescato and Putton describe in their book *Star bene insieme* (Francescato, Putton, 1986) and the new edition (Francescato, Putton, 2022). For these reasons, in this volume, we devote two chapters to small group work and self- and mutual-help groups, which are crucial in everyone's lives and in many of the intervention strategies we present (see chapters 8-9). An empowering leader, as Santinello and co-workers (2018, pp. 130-1) write, is "a coach of others' skills and democratic practices [...] who encourages their co-workers to share information, in a climate of experimentation, helps them see learning opportunities, tolerate mistakes, encourage discussions, and combine reflection and action." We hope that studying the two chapters on small work groups and mutual-help groups will help you become an empowering person.

- *Organizational empowerment.* Initially, community psychology did not focus much on organizational empowerment, as its primary interest lay in environmental settings that were more familiar, such as educational and social health services. However, in the 1980s, the field began to recognize that work settings are crucial determinants of the well-being and empowerment of employees and can serve as places for health promotion, similar to other settings like services, family, and school. With a preventive perspective and a concern for vulnerable groups, community psychology started to address issues related to unemployment and the prevention of work-related stress. By the 1990s, empowerment interventions were being promoted within corporate environments, adapting strategies to fit the world of business organizations and management (Francescato, Rosa, Pellegrini, 1997). The social changes and increased international competition characteristic of the globalized era have further heightened the need for empowered and empowering organizations capable of navigating numerous external changes (such as global competition, pressure to innovate, and mergers) as well as internal challenges (such as fears of job loss and difficulties in managing work-life balance) (Francescato, Aber, 2015). Empowering organizations contributes to personal well-being by enhancing knowledge of organizational functioning, increasing involvement in decision-making, improving the quality of relationships, and promoting collective well-being through greater awareness of political, economic, and social factors (Zimmerman, 2000).

An empowered organization develops successfully when it achieves its goals and influences the systems in which it is embedded (Perkins, Zimmerman, 1995), such as making environmentally friendly products that will contribute to the health and well-being of the community. At best, organizations can become empowering and empowered (Zimmerman, 2000; Francescato, Tomai, Solimeno, 2008; Nelson, Prilleltensky, 2005, pp. 23-44; Francescato, Aber, 2015). An organization is empowered when it can effectively read the economic, legal, local, regional, national, and international processes of change in the contexts in which it operates. Understanding the contexts in which they are embedded enables organizations both to seize growth opportunities and to prevent threats. Organizations are complex phenomena that cannot be ascribed to a single interpretive paradigm (Avallone, 1999; Bonazzi, 2002). Organizational Empowerment: is essential for reducing job stress and fostering a sense of capability in implementing appropriate behaviors across various life situations. It also involves valuing one's skills and nurturing a desire to actively participate in actions. Additionally, organizational empowerment includes building confidence in the belief that individuals, collectively, can effectively intervene in events, influence outcomes, and steer them toward better results. Francescato, in the late 1980s, introduced in the Italian context a theoretical-practical model of reading and empowering organizations from a community psychology perspective. This model, the PMOA already mentioned, is based on a complex and multidimensional conception of the organization (Francescato, Tomai, Solimeno, 2008) and will be discussed later (chapter 7).

- *Community empowerment.* At the community level, empowerment focuses on encouraging citizens' participation in social and political life while reducing individual and collective marginalization. Rappaport (1981) was one of the first to highlight that empowerment is a process through which disadvantaged citizens become more empowered. Perkins (2010, p. 207) describes it as "an intentional and ongoing community-centered process involving mutual respect, critical reflection, caring, and group participation, through which people with fewer resources gain greater access to and control over those resources" and "a process by which people gain control over their lives, democratic participation in their community, and a critical understanding of their environment." Interventions aimed at promoting empowerment at the community level seek to ensure that disadvantaged individuals are empowered to access resources more effectively, challenge societal power dynamics, and enhance social participation and inclusion. According to Christens (2012), empowered communities contribute to mental health and increase local power in decision-making processes. Several scholars emphasize the urgent need to promote community empowerment in our society, warning that without active and participatory local communities, the social fabric could fragment into isolated groups that harbor animosity toward one another. This, they argue, underscores the importance of fostering a greater sense of belonging, power, and democratic coexistence at the community level (Amerio, 2003; 2004). As Zani (2012) notes, social action and community development are among the most recognized approaches to fostering collective empowerment. Interventions rooted in social action arise from recognizing the unequal distribution of resources within communities. Social action aims to raise awareness among those affected by this disadvantage and to encourage a shift in the existing power dynamics. For example, citizens might organize cohesively to legally oppose injustice through nonviolent demonstrations. On the other hand, interventions based on the community development approach operate on the belief that communities have the knowledge, resources, leadership, and organizational potential needed to enact constructive change. By understanding the resources and issues within a specific area, communities can identify priorities and plan targeted interventions (*ibid.*). One of the most widely used empowerment techniques in this approach is the development of community profiles (Martini, Sequi, 1988; 1995), which map the context in which people live. Community analysis goes beyond merely identifying needs; it also aims to increase community members' awareness, promoting shared and collaborative change (Arcidiacono, Tuozzi, Procentese, 2015; see chapter 5).

3.4. Is empowerment measurable? As a multilevel relational construct, empowerment has sparked debate among critical community psychologists, some of whom argue against creating scales or measures for it. However, the positivist approach is prevalent in psychology departments, leading to the development of various scales to measure empowerment, many of which focus primarily on individual empowerment. One of the most well-known is

Spreitzer's Psychological Empowerment Scale (PES; Spreitzer, 1995), which includes four dimensions: meaning, competence, self-determination, and impact. Meaning assesses the alignment between work tasks and personal values and beliefs. Competence refers to an individual's confidence in their ability to perform work activities effectively. Self-determination measures the perceived control over one's work. Finally, impact gauges the extent to which an individual believes they can influence the strategic, administrative, and operational outcomes of their work. This scale has been validated in Italian with a sample of social and health workers (Pietrantoni, Prati, 2008). Zimmerman and coworkers (1992) developed a psychological empowerment scale that covers three dimensions: intrapersonal, relational, and behavioral. The first component measures the sense of control and self-efficacy; the second component measures the ability to understand sociopolitical factors that contribute to the different distribution of resources in the environment; and the third component measures participatory behaviors enacted in the community. In Italy, Francescato and colleagues (2007) developed and validated the EMPO scale, which measures personal and political empowerment across three components. The first component assesses the ability to set and achieve goals effectively. The second component measures feelings of hopelessness and lack of confidence. The third component gauges interest in sociopolitical issues, serving as an index of critical awareness regarding institutional power relations. These three dimensions can also be combined to provide an overall measure of personal and political empowerment.

3.5. Relationship between power, empowerment and health: critiques of the empowerment construct

Community psychology reframes personal and social problems by focusing on access to various types of power, enabling individuals and groups to expand or limit their opportunities for choice. It uses the concept of empowerment in a multidimensional sense, integrating individual and collective needs for power and control as a key element of health promotion. Power and empowerment are viewed as qualities with both individual and collective dimensions, shaped by the interaction between individuals and their contexts. Numerous interventions in psychiatric rehabilitation have demonstrated that increasing decision-making responsibility in a supportive environment enhances users' autonomy and sense of self-efficacy. In recent decades, patient empowerment has emerged as an effective self-care strategy for managing chronic conditions, significantly improving health and quality of life for patients and their families (WHO, 2006; McAllister *et al.*, 2012; Mora *et al.*, 2022), particularly when bolstered by strong organizational and community processes (Haldane *et al.*, 2019). Similarly, active involvement and engagement in communities is a central focus of many health improvement programs, with numerous studies highlighting the physical, psychological, and psychosocial benefits for those involved (Wallerstein, 2006; Attree *et al.* 2011).

Multiple research studies over the years have shown that the social outcomes of community engagement can also be particularly effective for at-risk population groups, such as residents of poor social and economic backgrounds (e.g., Cyril *et al.*, 2015) or fragile and/or isolation-prone groups such as the elderly (Woodall *et al.*, 2010; Francescato *et al.*, 2017).

The impact of the quality of the social fabric on health has been compellingly documented by Wilkinson and Pickett (2009) in their influential essay, which demonstrates that in more developed countries, there is a lower incidence of disease where the social fabric is stronger, and income disparities are less pronounced. They argue that beyond a certain income level, health improves not only through individual self-care but also through the effectiveness of the surrounding social fabric. Numerous epidemiological studies highlight the positive health effects of social networking: individuals who maintain more human contacts and actively participate in community life fare better than those who live in isolation. Attending a discussion or self-help group has a protective effect on health comparable to quitting a harmful habit like smoking. While behaviors can be changed individually, the availability of self-help groups depends on local opportunities, which are often influenced by regional or national policies that either support or obstruct the development of such groups.

Many older people see TV characters more frequently than they interact with relatives and friends, while many young people are seeking and forming social relationships online. As a result, community psychologists must engage with virtual and media communities, understanding both the needs these communities satisfy and the problems they may contribute to. The European Union, which aims to reconcile economic growth with maintaining social cohesion and increasing social capital, recognizes the importance of these issues. Therefore, it will become increasingly vital for European community psychologists to explore ways to enhance social capital not only in local and national communities but also within media and virtual communities.

However, some critiques (Zani, 2012) have highlighted that despite the multilevel nature of the empowerment construct, research on this topic often privileges individualistic conceptions, overlooking social, material, and political dimensions (Perkins, 1995; Prilleltensky, Nelson, 2000). This underscores the importance of placing greater emphasis on the collective dimension of empowerment. Empowering interventions risk being incomplete if they do not integrate the various levels through which empowerment is expressed. Community psychology offers several techniques (Francescato, Zani, 2017) that promote dialectical relationships between different levels of empowerment, such as community profiles. These techniques encourage pluralistic interpretations, value historical origins, employ a multidisciplinary approach, and give voice to minority narratives and local knowledge, thereby initiating potential processes of change. By bringing out latent resources and potentials – whether at the individual, group, organizational, or community level – and enhancing the ability to analyze contexts and critically understand political and social forces, these techniques foster the beginning of transformation processes, also telling minority narratives.

In the next chapter, we examine the constructs developed by critical community psychologists and by liberation and decolonization psychologists who have explored precisely these minority narratives.

3
Critical psychologies of community, liberation, and decoloniality viewpoints

1. Theoretical-political contributions of radical exponents of critical community, liberation, and decolonial psychologies

Adherents of critical community psychology, liberation and decolonial psychologies, and feminist movements highlight the severe distress experienced by much of the world's population, not only in countries with state-controlled economies under autocratic regimes like China, Russia and Iran but also in democracies dominated by globalized neoliberal capitalism. These approaches criticize moderate community psychology for focusing too heavily on individual or micro-level changes while neglecting the **inequitable economic and political structures of macrosystems**. Despite this shared critique, each approach delves into different issues and offers distinct theoretical constructs, suggesting various paths of action needed to bring about complex changes across diverse geopolitical contexts.

As mentioned in the introduction, leaders of the liberal economy have sought to influence public opinion, particularly in the United States, where it became possible for *super political action committees* to make unlimited contributions to parties and candidates. Large multinational corporations, driven by the pursuit of high profits for shareholders and substantial earnings for executives, have often been indifferent to ethical or environmental concerns, continuously seeking low-paid labor in the Third World. For years, the narrative of ever-increasing economic growth was portrayed as a triumph over scarcity, promising a better world for all. However, it has become evident that this growth is endangering the planet's future, destroying more than it creates (Raworth, 2017; Butera, 2021; see also Chapter 15).

In this new century, there has also been a decline in mass parties and the utopias of the 20th century, accompanied by an increase in citizens who disengage from political issues or abstain from voting, having lost faith in government parties and politicians. Many of these individuals focus only on the present, fearing ecological catastrophes in the future.

Conversely, there has been a rise in those who express their anger and hostility toward politicians, often through brief demonstrations – whether in response to factory closures, bereavements as in the Black Lives Matter movement in the United States, or solidarity with Afghan women barred from higher education or Iranian girls fighting for freedom. Every day, countless individuals voice their deep dissatisfaction with an unjust society on social media (Stevens, 2018; Ndlovu-Gatsheni, 2020). Radical psychologists argue that their profession should align with those who rebel against these injustices and strive to build a more equitable and inclusive society, although they differ on the best methods to achieve this goal.

1.1. Critical community psychology: the interactions between power, oppression, resistance, and liberation Critical community psychology is deeply inspired by the philosophy and critical sociology developed by the Frankfurt School. This school, which emerged from the Frankfurt Institute for Social Research in Germany during the 1920s and 1930s, comprised an interdisciplinary group of prominent philosophers, psychologists, sociologists, and political scientists, including Max Horkheimer, Erich Fromm, Theodor Adorno, Herbert Marcuse, and Jürgen Habermas. Critical philosophy, as articulated by these thinkers, examines how social structures and culture constrain human potential and actions, creating systems of oppression that are maintained through the perpetuation of social, psychological, political, and linguistic practices, customs, and norms that become internalized into subjective experiences.

Oppression can manifest both internally and externally. Internal oppression is often unconscious, leading individuals to blame themselves and become their own harshest critics without recognizing that their discomfort is also a result of the power that social structures exert over them. However, individuals can resist this oppression by challenging dominant ideas, leading to a new awareness that aids in their liberation. Resistance and liberation are, therefore, the ultimate goals, although various paths can be pursued to achieve them.

With the rise of Nazism and World War II, members of the Frankfurt School emigrated, and the group dispersed. In the 1960s and 1970s, critical philosophy experienced a resurgence in France with post-structuralist scholars such as Michel Foucault, Roland Barthes, Jean-François Lyotard, Jacques Derrida, Julia Kristeva, and Jean Baudrillard. These thinkers focused on exploring the role of language and social norms in sustaining oppression.

Particularly, post-structuralists criticized positivism and empiricism, arguing that a pluralism of theories, methods, and perspectives is essential for conducting research in the social sciences.

Critical community psychologists adhere to a pluralistic constructivist model, arguing that the constructs used in research and interventions are culturally shaped and do not necessarily represent an objective reality (Teo, 2012). They advocate for using a "transformative paradigm," which incorporates a variety of quantitative and qualitative participatory methods to analyze the social systems that produce inequality, injustice, and oppression. By doing so, critical community psychologists believe they can offer research participants opportunities for both individual and social transformation (Mertens, 2009).

Taking an ecological perspective, which posits that an individual's experience is shaped by the interaction between the person and their sociocultural context, Ora and Isaac Prilleltensky identify four interconnected levels of well-being: personal, interpersonal, organizational, and community (Prilleltensky, Prilleltensky, 2006). The personal level refers to an individual's satisfaction with their life and self, focusing on characteristics such as self-esteem, self-efficacy, and self-determination. To enhance well-being at this level, it is important for individuals to feel they have an active role in the decision-making processes within their community. At the interpersonal or relational level, well-being is influenced by the quality of relationships with others in one's life context, including relationships with family members, exchanges with partners, and emotional support from friends. The organizational level pertains to structured, often hierarchical contexts, such as the workplace, where well-being is promoted by social participation, collaboration, respect for diversity, and the recognition of one's work and abilities.

Finally, at the community or collective level, equity in the distribution of resources, duties, and rights among community members is essential to promoting social welfare. Social justice within the community is crucial so that everyone feels they are treated equally, regardless of the social group they belong to. According to Prilleltensky and Prilleltensky, the well-being of a community relies on its ability to emancipate itself from oppressive powers. Therefore, effective community intervention should not only provide resources but also enhance individuals' critical awareness of their ability to mobilize and effect change.

Prilleltensky and collaborators (2015) also constructed a multidimensional assessment model of well-being called ICOPPE, an acronym that refers to six specific dimensions of well-being: interpersonal (*Interpersonal*), community (*Community*), occupational (*Occupational*), physical (*Physical*), psychological (*Psychological*), and economic (*Economic*). Unfortunately, while these models can increase critical awareness by identifying various dimensions of well-being and highlighting the different forms of oppression that individuals must overcome, they often fall short of providing clear strategies for how the oppressed can effectively compel the oppressors to grant the resources needed to meet their needs. The practical mechanisms for achieving such empowerment and resource redistribution remain vague.

Aware of the significant challenges in achieving drastic socioeconomic and power shifts in society, critical community psychologists have largely directed their critique toward "mainstream" community psychologists, achieving notable successes. The critical wing of the discipline accuses the moderate wing of focusing too heavily on the individual as the primary level of analysis, thereby neglecting important constructs such as mutuality, community, and the influence of institutions on individuals. They charge moderates with supporting social institutions that directly or indirectly oppress and marginalize individuals and groups and with working within organizations like universities and research centers that uphold the dominant neoliberal ideology, which perpetuates inequality. Furthermore, they criticize moderates for failing to engage in *reflective practice* that examines how their values, biases, and professional norms contribute to the perpetuation of oppression. In contrast, critical psychologists see themselves as activists and promoters of social change, using their relatively privileged positions as academic researchers to devise actions that challenge mainstream psychology and advance society toward greater equity (Teo, 2005).

Critical community psychology began to attract a minority of radical community psychologists in Britain, the United States of America, and Canada as early as the 1970s, as we saw in chapter 1. But in recent years it has become the dominant theory in many countries in the Global South where colonialism oppressed and marginalized Indigenous Peoples, for example, in New Zealand, Australia, and South Africa, where the *psychology of decoloniality* has become widespread (Tebes, 2017; Stevens, Sonn, 2021), and even more so in Latin America, where *liberation psychology* has developed. In the next section, we examine together these two approaches, which have many points in common.

1.2. Psychologies of community decoloniality, and liberation in the Global South

Psychologists of decoloniality and liberation come from countries that were historically invaded and colonized by European imperialist nations, such as Britain, Portugal, Spain, and Holland, as well as, to a lesser extent, Germany and Italy. In recent decades, these countries have continued to be economically exploited by executives of large multinational corporations. With the complicity of local politicians, these corporations have devastated and polluted areas inhabited by poor peasants and indigenous communities. Mainstream psychologists have sometimes contributed to this exploitation by attributing issues like drug and alcohol abuse in these degraded contexts to mental problems rather than acknowledging the broader socioeconomic factors at play. Adherents of liberation and decolonial psychologies urge community psychologists to recognize how Euro-American scientific psychology has often relied on a universalistic view of humanity, using the "Euro-American man" as the hegemonic model in both research and professional practice. This conception of Euro-American superiority has been central to colonialism and coloniality, leading to the psychological construction of people from other cultures as inferior and not fully human – whether as slaves, Indigenous People, or women – who have been systematically silenced, ignored, or excluded (Teo, 2005; Montero, 2007; Adams *et al.*, 2015).

Decoloniality, as described by Mignolo and Walsh (2018), seeks to understand how coloniality influences our thoughts, feelings, and perceptions of ourselves, how we make sense of reality, and how we view our relationships with other humans, other species, and the natural environment. Decolonialists specifically focus on examining how Western psychology has contributed to sustaining the cultural dominance of Euro-American psychologists by damaging the psyches of individuals from groups deemed inferior, marginalized, or excluded.

Liberation psychologists aim to create pedagogical-political pathways that promote liberation from prejudice and support the full development of marginalized individuals and groups. Psychologists of liberation and decoloniality argue that colonial dominance by Euro-Americans persists today, sustained by neoliberal capitalism, which fuels the arms race and escalates wars and conflicts. This dominance is often legitimized as a necessary response to the perceived threats posed by Chinese and Russian autocracies to Western countries.

Carolissen and Duckett (2018) identify several key aspects of decolonial pedagogy: a critique of Euro-American and Western epistemologies, the refutation of the pathologization of marginalized people, the rediscovery of silenced historical events, the creation of new archives, the incorporation of indigenous knowledge into curricula, and the use of reflexive and ecological methods. Numerous authors (Fernández *et al.*, 2021) stress the importance of developing epistemic justice that embraces pluriversality, includes traditional indigenous knowledge, promotes ethical and historical awareness, critiques the subjective bases of power and privilege, and fosters mutual accountability. They emphasize that decolonial pedagogy must also address pressing global issues such as rising racism, gender and class inequalities, increasing violence, migration, and diasporas.

Maldonado-Torres (2016a, p. 23) emphasizes that the concepts of accompaniment and accountability are central to epistemic justice and the decolonial project: "building the world of you implies opposing the metaphysical catastrophe, the paradigm of war, and ontological separation" on which warlike neo-capitalism is based, a system that mythologizes individual freedom and glorifies competition against rivals. In contrast, the individualism promoted by capitalism is fundamentally opposed to the *ontology and epistemology of relationships* advocated by the decolonial perspective, indigenous values (Sonn *et al.*, 2022), and liberation psychology (Montero, 2007; Montero, Sonn, Burton, 2017).

Sloan (2016) argues that decolonialist theorists reveal how the rationality developed by Western culture has historically been used to justify and legitimize genocide, slavery, and exploitation and continues to oppress millions of people, whose behaviors are largely shaped by the contexts in which they are forced to live. To foster epistemic justice, it is essential to recover marginalized practices, processes, and knowledge-production activities within Euro-American psychology while also subjecting them to rigorous epistemic critique.

To achieve this, it is necessary to start from the grassroots, giving voice to the narratives of the oppressed and marginalized, particularly Indigenous Peoples, by developing a praxis that values their knowledge. It is crucial to disseminate indigenous paradigms based on relational ontologies, pluriversality and situated local knowledge, supported by ongoing critical dialogue, accompaniment, and relational deconstruction and reconstruction. The excluded and marginalized must be seen not as objects of research, but as bearers of valuable knowledge that can promote the radical changes needed to address the multiple interconnected global crises provoked by the neoliberal, virilist, masculinist, and warrior paradigm. Ecofeminists argue that we should learn from the survival techniques developed by indigenous women who care for the health of nature in various parts of the world (Guimarães, 2020; see also Chapter 15).

It is worth recalling what Martín-Baró, Aron, and Corne (1994, p. 46) wrote three decades ago about relational ethics and reflective practice:

"It is not a question of abandoning psychology; the question to be asked is whether psychological knowledge will be used to build a society where the welfare of the few is not built on the misery of the many, where the fulfillment of some does not require the deprivation of others, where the interests of the minority do not require the dehumanization of all."

Decolonialist psychologists, aligning with Ndlovu-Gatsheni (2020), assert that the decolonial project seeks global liberation by addressing the diminished quality of life for many and promoting the "re-humanization" of those historically dehumanized. We need to prioritize the cultivation of relationships, reconnect and reinvigorate bonds, and mobilize our global communities to build the "world of you" against ontological separation. There is a need to rehumanize oppressed people with a positive attitude of love and overcoming anger (Maldonado-Torres, 2007, p. 23).

Sonn and collaborators (2022) acknowledge that this is a challenging task, but they argue that community psychologists must take the lead in debunking "counter-catastrophic science" by developing solidarity through alliances and coalitions both inside and outside universities. We must engage with movements and groups dedicated to dismantling the colonial power matrix and creating futures that embrace conviviality, decolonized love, justice, and cooperation.

Liberation psychologists also emphasize the importance of freeing the oppressed from both the state and the process of oppression. The state of oppression involves the deprivation of essential material and psychological resources necessary for living, leading to suffering and discrimination. Oppression as a process involves institutional and informal actions aimed at diverting what rightfully belongs to the oppressed toward other individuals or groups, convincing the oppressed to internalize their inferiority. This self-legitimization of oppression leads the oppressed to believe in their negativity, feeling undeserving of resources and incapable of participating in political, social, and economic life. This negative self-view develops over time through prolonged material deprivation, reinforced by linguistic, cultural, emotional, and behavioral mechanisms that promote the humiliation, infantilization, and social disempowerment of the oppressed. Thus, power can maintain itself not just through violence and coercion but also by leading individuals to self-regulate by internalizing sociocultural prescriptions (Menegatto, Zamperini, 2018).

Overcoming oppression begins with psycho political education (Prilleltensky & Gonick, 1996). The first step toward liberation is becoming aware of one's condition – a process through which individuals and groups gain a clear understanding of the socioeconomic, political, cultural, and psychological factors shaping their lives and their capacity for change. This awareness is crucial for initiating the inner transformation necessary to liberate oneself from both external domination and psychological internalization, as emphasized by Ignacio Martín-Baró, a Latin American psychologist assassinated in 1989. The freedom envisioned by Latin American theorists differs from the consumerist neoliberal ideal; it is the freedom of a citizen who becomes aware of the oppressive system, understands the need for change, and believes that change is possible (Menegatto & Zamperini, 2018).

2. Toward a European theory of community psychology: a possible bridge between moderate and radical psychologists?

2.1. Major cultural and political differences between Europe and the United States of America As early as 1977 in the book introducing community psychology in Italy, and more prominently in the first university handbook (Francescato & Ghirelli, 1988), Francescato argued that while some patterns and approaches from U.S. community psychology offered valuable insights for psychologists interested in shifting focus from individual distress to promoting quality of life, it was essential to develop theoretical models and intervention methods tailored to the distinct European sociopolitical context. During conferences and informal gatherings organized by the European Network of Community Psychologists (ENCP), U.S. colleagues were often criticized for being overly pragmatic, action-focused, and resistant to engaging in complex theoretical reflections. They were perceived as present-centered, with little regard for the historical evolution of community contexts and were also accused of mythologizing the concept of community without acknowledging its potentially oppressive aspects.

Piero Amerio (2000) addressed these challenges by developing theoretical and practical foundations aligned with the European, particularly Italian, sociocultural, and political-institutional framework. According to Amerio, community psychology must integrate elements of traditional clinical psychology, such as the helping attitude aimed at change and attention to individual cases, with a focus on the political and historical dimensions of the social context, considering both its material and symbolic components. Amerio's concept of the subject differs from that proposed by Orford. In Amerio's view, community psychology sees the human person as an active subject in constant interaction with the world, with their psychobiological activity deeply intertwined with the social context. This subject is historically, culturally, and socially situated, and its active capabilities are realized within the resources and constraints of the social environment. Human problems are understood not only in their personal and subjective dimensions, as is traditional in psychology (focusing on discomfort, suffering, and psychological disturbance), but also in their objective and social dimensions, where both constraints and resources are located (*ibid.*, p. 190).

Using a multidisciplinary perspective that emphasizes the importance of history and politics, Amerio proposes interventions that address not only the subjective dimensions of a problem but also its objective aspects and the broader social dynamics at play. The community psychologist, according to Amerio,

conducts a comprehensive analysis of the material and symbolic relations that influence and shape the situation. This includes not only the geographic and human social aspects of the territory but also power relations, dominant values, forms of solidarity, and more. Given that this type of analysis requires perspectives, procedures, and tools often beyond the traditional psychological sphere, community psychology seeks appropriate collaborations both at a multidisciplinary level and with the knowledge that the local community itself produces. The goal is to uncover local knowledge through collaboration and co-participation with the people involved in the situation and engage the active participation of individuals by involving them in concrete social actions. Amerio emphasizes that since activities, resources, commitment, and action are inherently connected to perceptions, evaluations, and emotions, the intervention must continually address what is "constructed" at the interface between the individual and the social, the subjective and the objective (*ibid.*, pp. 191-2).

For Amerio, community psychology places at its center:

the recognition of the individual as a principle and value, understood within its social dimension; the importance of participation in the construction of the common good, which forms the foundation of the community as the repository of human values; and the constructive sense of action, seen as a process that integrates mental and practical activity, individual and social spheres, providing the individual not only with the ability to adapt to the context but also to work toward changing it (*ibid.*, p. 208).

European community psychologists have also delved more deeply into the negative aspects of past community movements than their American counterparts. While American psychologists often emphasize the positive ethical components of community building, aligning with communitarian thinkers like Taylor and Spencer (1989), who view the community as inherently embodying the common good, Europeans have critically examined the historical evolution of the concept of community. They have explored how communitarian ideals, across different historical periods, have been both sources of oppression and catalysts for emancipation and personal growth. Communitarian utopias have been employed to promote solidarity and individual emancipation but also to create totalitarian settings where many freedoms were curtailed.

Europeans tend to be more skeptical and cautious of collective ideals, perhaps because of the lingering memories of fascist and communist regimes that, in different ways, oppressed individuals by exalting the collective good over individual rights. Several European community psychologists criticize the naturalistic, individualistic perspective prevalent in psychology while also recognizing that the vision of an empowering community – where relationships are based on solidarity, dialogue, and participation – is still more of an ideal to aspire to than a current reality (Gelli, Mannarini, 1999; Amerio, 2000). Amerio observes that today's paradoxical situation is characterized by young people who take for granted the economic well-being and civil, human, and social rights they enjoy, yet many of them, focused solely on their individual development, show less interest in the common good and the dimension of the "polis."

However, the European history of the 20th century shows that. "forms of tyranny and perverse domination can lead to indifference, passivity, and closure in the private sphere in the face of public difficulties" (Amerio, 2000, p. 88).

In this changed sociopolitical context, it becomes all the more necessary to help foster, particularly in young people, an awareness of the historical link between the process of enhancing individual freedoms (individual empowerment) and the social struggles that granted European citizens the high civil, social, and human rights they enjoy today – rights that could be at risk in the future. Amerio (*ibid.*, p. 46) argues that:

community psychologists must examine the images of humanity and society that have been passed down to us and with which we live today. These are not just images but realities that are manifested in the forms of law, politics, and even collective sensibility, and which, at least in some respects, also serve as "models in adherence" to which – or in opposition to which – we can build projects for change.

Amerio shows how the ideas of freedom, dignity, and justice, which give substance to the conception of the individual as an end and as a value, have matured through "often hard and bloody struggles against tyranny, bullying, or even just against aging traditions and fears of change" (*ibid.*, p. 45). He also recalls that individual rights were not given to individuals by nature, so much so that even today many human beings, particularly too many women and minorities, do not enjoy the civil, human, and social rights that we Europeans take for granted. These rights were won through social and political pressures and the struggles of those groups who felt oppressed; that is, they are historical rights, born in certain circumstances marked by struggles to defend new freedoms against old powers:

"religious freedom is an effect of wars of religion, civil liberties of the struggles of parliaments against absolute rulers, political and social liberties are an effect of the birth, growth, and maturity of the movement of wage laborers, of peasants with little land and no property, of the poor who demanded from the public authorities not only the recognition of personal freedoms and negative liberties, but also the protection of labor against unemployment, and the first rudiments of education against illiteracy, and gradually assistance for disability and old age, all needs that the affluent owners could provide for themselves" (Bobbio, 1990, pp. xiii-xiv).

Amerio (2000) sees the local community as a system immersed in a network of larger systems with different missions and goals, including psychological ones. The expression "local community" now generally means a social system organized even at the political-administrative level, placed within a larger system such as the state. Thus, a region, a municipality, a small set of municipalities are defined by certain geographic but also human boundaries. A territory in this sense is not just a place: it is a culture that is expressed in the local dialect and cuisine, a set of conducts, places, and ways of life. Amerio also notes that the policies of political-administrative decentralization now underway throughout Europe involve a shift of powers from the state to the regions and municipalities. These policies arise not only to improve the administrative machine but also to bring citizens closer to institutions and meet psychological needs that the state cannot provide for. As early as 1960, Nisbet wrote:

"The state cannot provide for the need for security. No large-scale organization can address individual needs, by its very nature too large, complex, and bureaucratized, the state can excite popular enthusiasm, mobilize people for great causes such as wars, as a means of providing for human needs for identity, security, membership is inadequate "(Nisbet, 1960, p. 82).

The local community can become the place of interpersonal relationships that meet human needs for belonging, security, and identity. According to Amerio, (2000, p.45) "this this dimension of interpersonal relationship acquires particular value today as an element that can enable the preservation of a social fabric, increase social capital and a sense of community even in critical situations such as wars, disasters, or the entry of different ethnic groups: in the situations, now increasingly common in European countries, of the entry of individuals and immigrant groups into communities that have already been formed for some time, a good relational dimension can allow the old residents to face the inevitable difficult moments with greater serenity, fewer feelings of insecurity, and with a more open spirit toward the newcomers without fear of seeing their territory invaded and their identity compromised. The dimension of participation broadens the sense of relationship with the whole community as it leads individuals to discussion and dialogue.

Both American and European community psychologists have criticized the fundamentally individualistic and naturalistic outlook of traditional psychology, which sees psychological processes as purely internal to the individual and human beings and society as natural phenomena, that is, as "essences" considered, in their basic foundations, to be always the same, governed by precisely "natural" and unchanging laws, which psychology must discover. As Amerio (2000, p. 42) points out again:

considering human beings and society as natural phenomena means in practice, looking at the facts of history and social events as aspects that are added only later to the events of the already formed psychic life, imagining that human beings are born, grow and develop their mental skills in a kind of "vacuum," or at least, in an aseptic environment that does not affect them except as a kind of fixed and immutable datum which is the physical-biological environment.

The naturalist view according to Amerio produces two consequences:

1. impoverishes human beings because it reduces their actions (and even mental activity) to mere behavior based on cause-effect chains almost necessarily determined beyond any subjective choice and decision;

2. It stiffens the conception of the social context by leading one to see it as a fixed and unchanging datum, as if it were a physical environment rather than a product that humans have constructed throughout history and can, therefore, change and transform through their actions.

Europeans share different cultural values than Americans; this is especially true for authors from continental Europe (Germany, Spain, Portugal, and Italy). European scholars do not think that "men are born free," as the American constitution states. On the contrary, they recognize from a long historical memory – continuously made relevant to every European by the vestiges of the past that dominate city centers (not coincidentally called centri storici (historical centers) and not downtowns as in the United States) – that each person is born into a historically created hierarchical social context, which is therefore modifiable through human action. This context can facilitate, limit, or even oppress an individual. However, the individual can, in turn, influence and even modify the social settings with which they interact, depending on the social position they occupy.

European community psychologists therefore place more emphasis on the historical analysis of how different political ideologies, over the centuries and still today, have legitimized and continue to legitimize current hierarchical stratifications by making them appear "natural." Ornelas (1997; 2008) examines the history of social and health services in Portugal, linking it to the historical, ideological, and political changes that occurred in the country before and after the dictatorship. Similarly, Palmonari and Zani (1980) demonstrate an interest in the evolution of services in Italy in terms of historical, social, and political changes. Sánchez Vidal (2007) attempts to situate the values of community psychology within the broader history of the evolution of values in Spanish society before and after Franco.

Another important difference between European and U.S. culture concerns myths about how individuals can change their position within social hierarchical structures and the legitimization of these inequalities. Europeans do not believe in the myth of the self-made man, which is so prevalent in American culture and reflected in its linguistic metaphors. In the United

States of America, this myth tends to imply that everyone is fully responsible for their own life and that if they want to succeed, they must win the rat race. The *survival-of-the-fittest* myth emphasizes the inherent goodness of the race for survival, suggesting that only the best, most deserving, and fittest will succeed or survive. Those who do not win, can only blame themselves. In America, individuals who become wealthy on their own are celebrated as heroes. In Europe, however, the "new rich" do not enjoy the same level of social consideration; there is a general belief that a disempowered person can rarely become empowered through their efforts alone. Europeans are more aware that, from a historical perspective, individual freedom and empowerment were won through collective struggles for civil, human, and social rights, as Amerio (2000; 2009) also reminds us.

How internet and social media have changed perceptions of politics: This crucial difference in European and U.S. cultures has been fading in recent decades with the spread of the internet and social media, as a "global" culture inspired by that of the United States of America, centered on individual success, the pursuit of fun and strong emotions, becomes more widespread, especially among young people. In this value context, love, friendship, family, money, entertainment, and work come first, while politics is perceived as distant, corrupt, boring, or otherwise uninteresting (Amerio, 2000; Francescato, Putton, 2022).

The minority of young people who engage in movements, parties, unions, or volunteer work often see personal engagement to develop their personalities, build a network of ties – including emotional ones – and gain competence while addressing a specific, limited problem (Migliorini, Tartaglia, 2021). However, quite a few young people receive little information about politics because it is not discussed at home, or it is avoided as a topic that creates tension. Many also reported that they did not study civics in school. Among their friends, talking about politics is often considered unfashionable, and those who do so may face ostracism (Mebane, 2019). Yet, at this moment in history – following a period of overwhelming market economy and financial dominance – there is a great need for politicians who are interested in the common good and for a greater balance between political and economic powers. The significant changes occurring in the financial and labor worlds due to globalization make it even more crucial to recognize the importance of good politics in promoting a more equitable and inclusive economy.

As political scientist Zagrebelsky (1995, p. 4) has long noted, the decadence of politics is dangerous to the democratic growth of a society:

"If you "depoliticize" people through fictitious and elusive messages about the complexity of problems, endearing their laziness and passivity, which as a temptation exist in all of us, you are not eliminating politics, but you are dispossessing them in favor of small circles of power where real politics is made. Democracy, as the regime of maximum diffusion of political capabilities, requires quite the opposite: not that problems be fictitiously brought down to the level of those who understand nothing about them but, on the contrary, that everyone strives, as far as they can, to bring themselves down to the level of the difficulty of the problems. The extent and quality of democracy depend on this."

In this regard, we feel it is important to reiterate (see Introduction) Francescato and Putton's (2022) that, in the contemporary postmodern age, the antagonism between capitalist and communist ideology has contributed to emphasizing and prevailing certain Manichean divisions in the public and private spheres. Individual growth and social welfare, individual interests and collective needs, individual freedoms, and social equality are thus represented as opposing concepts, denying a complex and multidimensional reality that would require a balance between satisfying individual needs and achieving general goals. The twentieth century has seen the opposing extremes, in its various economic and political forms, of *anarchic liberalism*, which privileges the rights of the individual at the expense of duties to the community, and *totalitarianism*, which, as was the case under fascist and communist regimes, sacrifices individual demands. Today, we are in pursuit of a new model that respects the balance between the needs of the individual and the purposes of society, that values the diversity of each individual and protects it from the tyranny of the many while encouraging each individual to cooperate with others for common ends.

Positive signs in the growth of generative communities: Generative communities are formed by individuals and associations interested in the common good, as well as by organizations and services dedicated to caring for the generation after their own, striving to build equitable, sustainable communities underpinned by values of solidarity, valuing difference, and living together. The goal is to create communities that foster the development of trust, loyalty, empowerment, and hope (Senge, 2016).

The generativity of these communities influences not only the present but also the future. The bonds of trust, loyalty, and empowerment produced by social generativity not only improve life in the present but also promote cultural change within which people, systems, and structures are embedded: "The goal of this type of intervention is to create a fundamental paradigm shift toward social awareness, evaluation, action, and a culture of reflexivity, thus leading to innovative and adaptive self-renewing systems and structures" (Robinson *et al.*, 2017, p. 205). The main objectives can be summarized as:

1. Facilitation of critical consciousness;

2. development of skills aimed at self-sustaining action and change; and

3. promotion of reflexivity.

These goals are not far removed from the idea of empowerment, particularly community empowerment (for a discussion of the concept of generativity, see Marta, Pozzi, 2004; Marta, Scabini, 2007; Marta *et al.*, 2017).

2.2.   Bridging concepts between individual and social spheres and between moderate and radical community psychologists Community psychology addresses human and social problems at the interface between the individual and the social spheres, with community psychologists aiming to work concretely on the connections between the psychic and the social and between mental and practical activity. Consequently, a theory of technique must identify the central concepts that enable these connections, either in a broad sense or concerning specific problems. For example, Francescato, as early as 1992, demonstrated that a problem such as marital separation could be viewed from a community psychology perspective by exploring the interface between the individual and social, subjective, and objective aspects. Each couple tends to view their relationship as a private affair, often focusing – much like psychoanalysts and clinical psychologists – on explanations rooted in individual and interpersonal variables. Few couples fully recognize that the ways they try to meet each other's needs, define expectations, and perceive their union and separation are often influenced by a complex system of laws, religious beliefs, and social, economic, and cultural factors that shape relationships. A couple's relationship, from its inception to its development and eventual dissolution, is undoubtedly influenced by the historical period in which it occurs, including the economic, cultural, political, and legal context (e.g., employment opportunities for both sexes, culturally prevailing gender roles, current separation and divorce laws). The tasks attributed to the family institution encompass all the functions historically assigned to the family, which evolve: economic functions such as production, redistribution of income, and consumption; household management, including all aspects of housework; psychosocial functions related to the adult couple, including the socially approved satisfaction of needs for sexuality, intimacy, solidarity, and mutual support; and functions related to the reproduction and socialization of children, including decisions about the number of children to have, how to raise them, and the influence of the individual characteristics of the couple, their extended families, and social networks. These factors, along with variables related to basic personality and interactional dynamics, contribute to the unique biopsychic profile of everyone (Francescato, 1992b, pp. 10-11).

Again Francescato, analyzing the post-separation marital paths of adults and children, identified four "bridging concepts" that mediate between individual and collective, personal and social, namely crisis, social support, storytelling, and empowerment: The different views a given society has of separation as a catastrophic or potentially positive crisis, the various types of social support available, the way an individual and family story is transformed into a narrative that becomes a social myth and personal identity, the way one feels more powerful or powerless during and after separation are all factors that contribute to very different qualities of life after separation (Francescato, 1994, p. 280).

Several other community psychologists have similarly tried to find bridging concepts. In the 1990s, Orford gave special attention to the development of a general theory for community psychology that, assuming the circularity of the relationship between the individual and the environment, identifies key concepts *in the fields of person and environment*, i.e., central concepts that facilitate the connections between the two fields. In the field of the person, Orford (1998) points to identity, status, and feelings of self-worth; these are three constructs for defining the person, widely studied in clinical psychology, and complex enough in their conception to be able to bridge the personal and the social. Sense of identity and self-esteem, for example, are considered elements that are conditioned and affected by the social contexts to which the individual belongs. In the social field, Orford identifies four ways in which the environment can either facilitate or hinder individuals in developing their sense of identity, achieving and maintaining their social status, and fostering their sense of self-worth. These four factors are: holding a socially valued role, possessing or being able to develop a sense of control, enjoying social support, and having future life opportunities. Orford repeatedly emphasizes that power and support are two fundamental resources for the development and well-being of the "person in context," and he considers these elements central to any theory of community psychology. In Orford's theoretical framework, these four resources operate at three different levels of the social system:

- **Micro-level**: includes systems of belonging such as family, work, services (health, educational, recreational, public and private), political or religious groups, etc;

- **Local community**: more difficult to define in a geographic sense, it can range from the condominium to the town to which it belongs; some theorists are also currently reflecting on the existence of "communities without territory" (for a more in-depth discussion of the topic, see chapter 10); and

- **Culture**: includes political, legal, sociocultural, and religious norms and structures at state or nation levels.

As early as the early 1980s, Francescato developed models of multidimensional organizational analysis, exploring the interactions between objective and subjective variables, and introduced the use of narratives and film scripts as strategies for understanding the past and designing organizational change. She also proposed a multidimensional analysis model of the functioning of work groups (Francescato, Ghirelli, 1988; Tancredi, Francescato, 1989; see also chapters 7, 8).

Narratives that support traditional or innovative interpretations of existing social hierarchies can influence empowerment and some of the person (self-esteem, status, sense of identity) and context variables identified by Orford (social roles, social support, future possibilities). For instance, consider the impact of the narrative proposed by the women's liberation movement on women's self-esteem and identity and their evolving social roles; or the different influences of the "Uncle Tom's Cabin" or "black is beautiful" narratives on the self-esteem identity, and conceivable social roles for African Americans in the United States of America. Constructionists have documented how it is possible to offer multiple interpretations of what mainstream sociology and psychology have presented as a singular social reality. Community psychologists and social constructionists, therefore, not only share some ethical values – but also demonstrate the limitations of a traditional clinical and psychiatric approach.

Inspired by the theoretical contributions of moderate and radical psychologists discussed so far, Francescato and Zani (2013; 2017) and Francescato and Aber (2015) sought to schematize the theoretical premises guiding the intervention strategies developed by Italian community psychologists (Francescato, Tomai, 2001; Francescato *et al.*, 2009) who have adopted a multidisciplinary perspective. The methodologies of community analysis and participatory multidimensional organizational analysis – presented in chapters 5 and 7 – are clear examples of this emphasis on the integration of disciplines.

2.3. Growth of attempts to integrate intervention constructs and methodologies

Some recent Italian manuals on community psychology, involving a large proportion of the discipline's faculty and researchers (Arcidiacono, 2017a; Arcidiacono *et al.*, 2021), have integrated a fundamental ethical stance that promotes the principle of justice in human coexistence and the development of professionalism characterized by trust, reflexivity, positionality, and intersectionality (Arcidiacono, 2017a). These books also draw on constructs developed by critical and liberation psychologists, and particularly from feminist perspectives. The selection of intervention tools and the problems addressed – such as social marginality, gender-based violence, addictions in deprived contexts, the complexity and diversity in multicultural societies, and the environmental and social sustainability of current lifestyles – reveal a creative integration between the theoretical-political assumptions of traditional community psychology and the constructs and goals of social change proposed by community psychologists from the Global South.

In this chapter, we have discussed the contrasting theoretical contributions of moderate, critical, and more radical (of decolonization and liberation) community psychologists. We believe that knowing and appreciating different points of view for "reading" social problems can enrich the theoretical-political and methodological heritage of a community psychologist who wants to accompany people, groups, organizations, territorial and virtual communities in their paths of change. We have summarized the main principles discussed in the chapter in table 1, also indicating some characteristics that should be present in the intervention strategies of a community psychologist.

Therefore, intervention strategies in community psychology should:

1. **Encourage pluralistic interpretations of a social problem** that bring together and integrate different types of knowledge of both objective and subjective types and increase the points of view from which a situation can be considered. Promote the use of perspectives, procedures, and tools from different disciplines and bring out, by activating forms of collaboration and participation, the knowledge produced locally by the people involved in the target social problem.

2. **Examine the historical origins of the social problem** and the unequal distribution of power to access resources in the social context. Critically reflect on how dominant narratives legitimize this unequal distribution of power.

3. **Giving voice to other existing minority narratives** that break the tacit consensus by which social actors accept the systems of convention in which they are immersed. Promote the production of new metaphors and/or new narratives that make thinkable new scripts and new roles for individuals and social groups and create new bases for legitimizing change. Pursue epistemic justice and the re-humanization of the oppressed.

4. **Promote and implement empowerment projects** that create bonds between people who share a problem and increase the social capital of a community.

5. **Identify strengths** that can be leveraged to achieve the desired changes.

6. **Identify which problems**, among those identified as priorities, **can be solved** at the level of the **group involved in** the empowerment project and which require interventions at other levels (organizations, networks, local community, region, state, European Community, global level, other international bodies, etc.).

TABLE 1 Guiding principles for a theory of technique in community psychology

1. Placement of problems	Community psychology addresses human and social problems at the interface between the individual and collective spheres, between the psychological and social spheres. Human problems always have an individual side, in that it is essentially the individual who suffers and must cope with them; they also have a social side, in that they arise from social situations, and it is often in the social that they find the material and/or psychological tools to be addressed. It pursues epistemic justice and the re-humanization of the oppressed.
2. Concept of individual	Community psychology sees the human person as an active, historically, culturally, and socially situated subject whose competencies find their possibility of implementation in a specific environmental context, which poses constraints and impediments, and offers opportunities and resources unequally for individuals. Humans are seen as social agents who construct meaning in their interaction with others. It advocates a relational ontology.
3. Concept of environment	The social environment is a hierarchical and historically created context; inequalities of power and access to resources among individuals are not natural but historical and modifiable.
4. Relationship between individual and environment	The hierarchical social context can facilitate or limit the individual, who can in turn influence the social settings with which they interact depending on the position they occupy there and the available interpretations of the origin and legitimization of existing social stratifications.
5. Complexity of the social system	Transactions between individuals and hierarchical social context occur at multiple and multidirectional levels (individuals, small groups, organizations, local communities, macro-communities, virtual communities). Empowerment can be developed at various levels.
6. Levels of intervention	Community psychologists examine problems not only in their personal-subjective dimension, as is traditional in psychology, but also in their objective-social dimension, where constraints and resources are situated. Concepts like pluriversality, accompaniment, and responsibility play key roles in this broader analysis, ensuring that the complexities of both individual experiences and social contexts are fully considered.
7. Conscientization clarifies the link between individual empowerment and social struggles	There is a historical link between processes of enhancing the freedoms of individuals and struggles for human, civil, and social rights. Conscientization is a cognitive and emotional process that leads people to be aware of the circumstances that affect their living conditions and leads them to act for change through active participation and psycho political education.
8. Role of narratives	Personal, community, cultural, and political narratives connect the individual and collective spheres, psychological and social aspects, because they provide both traditional and innovative interpretations of social hierarchies that influence individuals' identity, self-esteem and status, their empowerment, socially valued roles, and thinkable future life opportunities. They promote generative communities.
9. Integration between positivist and constructionist models	It uses different kinds of knowledge, derived from both traditional scientific approaches and modern theories of social constructionism, using the paradigmatic model to search for invariances in individual-environmental context relationships and the narrative model to understand and facilitate personal, organizational and social change.
10. Use of resources and critical issues	Community psychology emphasizes positive experiences (strengths) as well as stressors (problems, discomforts, negative experiences). It values trust and hope.
11. Constructive role of action	Community psychology emphasizes the constructive meaning of action as a process that articulates mental and practical activity, the individual and social spheres, providing the individual with the ability not only to adapt to the context, but also to change it.

4
Clinical and community psychology: where do we stand?

1.   Introduction

The emergence of a new discipline often results from processes of detachment and differentiation from its original matrix. Scholars from various disciplines – whether pedagogical (e.g., Fraisse, Piaget, Reuchlin, 1990), psychological (e.g., Kerr, Bowen, 1990), or sociological (e.g., Gallino, 1987) – agree that such an evolution of thought is driven by the ability to shed outdated perspectives, pose new questions, and develop new methods of inquiry. We believe that the rise of community psychology followed a similar path, influencing its relationship with clinical psychology from which it originated. The connection between community psychology and clinical psychology has always been a subject of intense debate, given the significant differences in areas such as the causes of psychological distress, levels and strategies of intervention, and the planning, implementation, and evaluation of services (Rappaport, 1977; Bostock, 1998).

As discussed in Chapter 1, the term "community psychology" first appeared in the 1960s and quickly spread globally, signaling a shift in psychology's focus toward societal issues during a period marked by intense civil rights struggles and challenges to the status quo (Wandersman, Florin, 2003; Reich *et al.*, 2007). The innovation of community psychology lay in its opposition to the medicalized, individual-centered model of traditional clinical psychology at the time.

Clinical psychologists gathered at the Swampscott (1965) conference intended to publicly disapprove of the idea of "mental illness" decontextualized by cultural factors such as inequality in accessibility to resources, poverty, or racism (Kelly, 1990; Gergen, 1999; Goodman *et al.*, 2004).

Criticisms of the dominant paradigm of clinical psychology emerged from a growing recognition of social justice and inclusion principles, highlighting the threats posed to these values by Western healthcare industry norms (e.g., Page, 1998; Nelson, Prilleltensky, 2005, pp. 23-44) and by hegemonic epistemologies that perpetuated social inequalities (Gone, 2011; Allen, Mohatt, 2014; Avissar, 2016). Psychotherapeutic interventions were also critiqued for being predominantly deficit-oriented and reductionist, in contrast to approaches emphasizing empowerment and resource development (Liang, Tummala-Narra, West, 2011; Jason *et al.*, 2019).

Historically, community psychology broke away from traditional clinical psychology driven by reformist ideals, consistent with the well-known slogan "give the psychology away" and the quest for progressive independence (Orford, 1992). Community psychologists argued the importance of turning attention to preventive interventions and promoted participation, empowerment, mutual respect, and advocacy (Gregory, 2001).

In this cultural-historical setting, how has the relationship between clinical and community psychology been organized?

2.   The relationship with psychodynamic matrices

Psychodynamic theories have been particularly opposed by community psychologists within mainstream clinical psychology (Liang, Tummala-Narra, West, 2011). By the 1950s, these theories had gained substantial influence in the psychiatric establishment (Haley, 1969; Francescato, 1977a), and during both World Wars, American psychology was largely dominated by the psychodynamic approach. The community mental health movement rejected this intrapsychic focus and the notion that individuals are dominated by forces hindering change (Gergen, 1999; Plante, 1999), positioning itself as an alternative discipline. Feminist and transcultural psychology also criticized the psychodynamic approach as conformist and misogynistic (Marecek, Kravetz, 1998; McLellan, 1999). Overall, psychodynamic treatment was seen as time-consuming, expensive (and thus inaccessible to the less affluent), unsuitable for minorities with low levels of verbalization, and ineffective for real-life change (Schwartz, 1997; Druiff, 2001).

The need to distance itself from conservative positions and define itself from a new perspective has driven community psychology to seek its identity by emphasizing its differences from mainstream approaches.

In this attempt, community psychologists seem to have set aside attention to the role of emotions and experiences, both individual and collective, in understanding social phenomena. The scarcity of discussions on emotion in community psychology is probably due to how the more traditional models of clinical psychology conceptualize people's emotional processes. Many of these situate emotions in intrapsychic dimensions and describe them as individual and private phenomena (Rappaport, 1977; 1981); this greatly hinders the integration of theories of emotion within the theoretical framework of community psychology.

We believe that the new discipline's need to "distance itself," particularly from a psychodynamic tradition, has prompted an emphasis on cognitive and pragmatic dimensions over emotional ones in community processes, leading progressively to a rejection of the notion of the unconscious (Rimé, 1993; King, Shelley, 2008) and a neglect of the role of affect (Leon, Montenegro, 1998; Gibson, Swartz, 2008; Caputo, Tomai, 2020).

3. The place of emotions in the community perspective

Emotions are a socially constructed phenomenon shaped through interaction with the environment, making them a psychosocial phenomenon. While the affective component has always been present in community processes, emotional dimensions are often overlooked in the scientific work of community psychologists. There is limited literature offering a systematic theoretical explanation of the emotional processes that influence both communities and the psychologists working within them (Gibson, Swartz, 2008). According to Rimé (1993), this oversight reflects a "resistance" toward clinical psychology – particularly psychodynamic psychology – and a broader tendency in social psychology to prioritize cognitive and rational aspects of psychosocial processes over emotional ones (Caputo, Tomai, 2020).

Some authors have acknowledged the importance of emotions in community work, emphasizing their connection to consciousness and action (Lane, Sawaia, 1991; Leon, Montenegro, 1998). Emotions, in this perspective, not only initiate and sustain consciousness and action but also guide and are shaped by them, contributing to environmental change.

The literature in community psychology reflects different stances on the role of emotions (*ibid.*). Initially, emotions were largely omitted from theoretical explanations of experiences, a myopic approach that characterized the field, particularly in its early decades. A second position involves the use of constructs that integrate emotion with cognition and perception. However, in these cases, emotions are often addressed superficially through constructs like "learned helplessness," "shared emotional connectedness," and "sense of community," which some authors find vaguely or inadequately defined, limiting their applicability (Gibson, Swartz, 2008). In constructs like emotional connection and a sense of community, the individual's relationship with others is often depicted more as an objective experience than as a subjectively perceived one (Koh, Twemlow, 2016).

Finally, a third, evolving perspective emphasizes the importance of emotions in community work and their link to consciousness and action. This view has gained traction over the past 15 years. For example, some researchers have highlighted how emotions and unconscious factors influence political power (Mannarini, Fedi, Trippetti, 2010; Gelli, Mannarini, 2014), while others have explored the role of emotional experiences and the individual-community connection in forming community identity (Yang, Xin, 2016).

Overall, these critical or ambivalent positions on the emotional world seem to clash with the practice of reflexivity, which is widely recognized as a core competency in community psychology and is deeply intertwined with introspective awareness of one's values and biases (Arcidiacono, 2017b; Case, 2017; Akhurst, 2020).

Regarding the role of emotions within the community perspective, it's important to highlight authors who have adopted positions more attuned to the complexity of reality, particularly when their contributions involved designing models and tools for intervention in various contexts. Francescato (Francescato, Putton, Cudini, 1986; Francescato, Tomai, 2001; Francescato *et al.*, 2009; Francescato, Aber, 2015; Francescato, Zani, 2017), consistent with the principles of the technical theory of community psychology (see chapter 2), has proposed a multidisciplinary perspective that integrates positivist paradigms with constructivist ones. She employed the former to identify regularities in the interactions between individuals and their social environments and the latter to capture the conscious and unconscious emotional dimensions – such as emotions, values, fears, and hopes – of the groups within the investigated settings. Her methodologies for research and intervention include the analysis of "objective" and rational variables alongside cultural and psychodynamic dimensions, utilizing both numerical and quantitative instruments as well as projective narrative or graphic tools. According to Francescato, postmodern paradigms and social constructionist theories allow for the exploration of the world of meanings and the "sense-making" processes through which individuals organize their interactions with both material and relational environments. The complexity of these models facilitates the exploration of contexts – whether community, organization, or group – in both their diachronic and synchronic dimensions and along a continuum from maximum objectivity to maximum subjectivity.

Recently, the importance of emotion in community work has gained recognition from authors who, within community and ecological perspectives, advocate for a psychodynamic approach that emphasizes the affective aspects of sense-making in community interventions (Caputo, 2015; Langher *et al.*, 2019b; Mannarini, Salvatore, 2019).

In conclusion, interventions designed to promote social change through sociocultural processes have long focused on cognitive rather than affective sense-making processes, creating a significant gap that needs urgent attention (Mannarini *et al.*, 2012). This review underscores that overlooking the role of emotions in community psychosocial research and intervention is no longer acceptable. However, community psychology still seems to lack the language and theories needed to fully address and integrate the communal nature of emotion.

4. The contribution of the psychodynamic approach

The study of social processes has led clinical psychology to move away from the traditional "individual-centric" orientation and progressively shift the focus from individual intrapsychic dimensions to the relationship between individual and context, seen as circular and interdependent.[6] Consistently, clinical intervention has become increasingly conceived as anchored in (organizational and cultural) context and serving individuals, groups, and organizations. In this regard, a cultural approach to clinical psychology has been conceptualized as addressing issues of social coexistence rather than individual distress (Carli, Giovagnoli, 2011).

Compared to the past, clinical psychologists are more concerned with politics and society, and their training increasingly incorporates community-related knowledge to develop more differentiated intervention skills (Jenkins, 2016; Kloos, Johnson, 2017). Integrating culture and community contexts into clinical practice becomes a necessary step to remain relevant in an increasingly diverse 21st century (Nagayama Hall, 2005).

Among the various foundations of clinical psychology, psychoanalysis has a longstanding tradition of focusing on emotions as a social phenomenon and views groups (or communities) as settings where an individual's identity and emotional experiences are shaped (Parker, 1997). This psychoanalytic perspective aligns with community psychology in the belief that the "individual" is a product of their social environment. Relationships – both present and past – are filled with emotionality that influences not only dyadic interactions but also extends to the broader social networks that individuals are part of. Modern dynamic psychology no longer supports the notion of a "given," objective, and linear reality over time; instead, it views reality as co-constructed, relational, and multidimensional (Arcidiacono, Lavanco, Novara, 2021).

Various branches of clinical psychology, such as humanistic psychology, cognitive-behavioral approaches, and systemic approaches, have significantly

6. Carli, Paniccia (2003); Nagayama Hall (2005); Maddux, Tangney (2011); Jordan, Lovett, Sweeton (2012); Grasso, Cordella, Pennella (2016); Carli *et al.* (2016); Langher *et al.* (2019a; 2019b); Bucci *et al.* (2021); Carbone *et al.* (2021).

shaped the conceptual framework of community psychology (see Chapter 2). In contrast, the psychodynamic perspective has struggled for broader recognition. Over the years, some authors have advocated for a deeper examination of how psychodynamic theories can contribute to understanding community phenomena and shaping interventions (e.g., Borg, Lynch, 2005; Gibson, Swartz, 2008; Borg, 2010; Liang, Tummala-Narra, West, 2011).

A systematic review by Caputo and Tomai (2020) identified several potential contributions of the psychodynamic approach to community psychology. The review highlighted four main psychodynamic/ psychoanalytic frameworks: interpersonal psychoanalysis, the Adlerian approach, the Tavistock model, and Lacanian psychoanalysis. Among these, interpersonal psychoanalysis emerged as the most frequently used. The review found several points of synergy between community psychology and psychoanalysis. All psychodynamic perspectives examined emphasized intersubjectivity and the reciprocal interactions between individuals and their community environments, viewing psychological distress as closely linked to dysfunctional adaptation at the sociocultural level. Notably, interpersonal psychoanalysis aligns with theories of primary prevention and empowerment (Rappaport, 1981; Zimmerman, 2000) by focusing on enhancing the individual's agency to navigate daily complexities and anxieties. This approach supports exploring and addressing tensions between conflicting positions rather than avoiding them.

In all the psychodynamic theories discussed in this paper, the community is portrayed as a complex entity, not merely the sum of its individuals, but rather as one characterized by its unique identity, shaped by repetitive patterns and fundamental meanings established throughout its history. These elements govern the relationships among its members and their interactions with the outside world. From this perspective, emotion is viewed as a shared sociopolitical experience that highlights the dialectical tension between the desire for change and the fear of the unknown. The community's function, for instance, is to provide a sense of belonging, cohesion, identification, security, and stability. However, when faced with conditions of risk or uncertainty, the continuity of the community is threatened, leading to feelings of anxiety.

The unconscious dimension of a community may manifest through the enactment of defensive mechanisms that serve to obscure genuine concerns or desires. These defensive stances can cause the community to rely on limited meaning-making processes, adopt rigid or repetitive modes of functioning, and display inconsistent or ambivalent behaviors (Arcidiacono, Lavanco, Novara, 2021).

In this way, the apparent irrationality of community life, with which community psychologists frequently struggle, is read as the community's tendency to avoid its painful realities or reenact its powerless position.

5.   Implications for community interventions

To recap, several psychodynamic insights may be fruitful for community and psychosocial interventions. First, the distinction between desire and demand, proposed by Lara Junior and Ribeiro (2009), allows us to rethink the role of community practitioners in consultative work. The exclusive focus on the explicit demands of communities can lead professionals to overlook the unmet needs and desires of social groups and disregard the symbolic component inherent in their demands. Avoiding reflection on these symbolic contents exposes community practitioners to the risk of colluding with client fantasies and meanings and, consequently, limiting their ability to make decisions (Carli, Paniccia, 2003). Interventions and change processes, instead of differentiating meanings and creating new shared narratives, could thus result in coercive or homogenizing practices (Borg, 2004). Reducing the potential for individual choice and depriving community members of the opportunity to understand their painful realities, therefore, can have several iatrogenic effects: reinforcing dependency and "victim-blaming" processes, perpetuating hegemonic epistemologies, fostering power imbalance, and hindering the development of community cultural competence (Cherniss, 2002). Another potential contribution of psychodynamic theories lies in analyzing transference and, particularly, countertransference within counselor-community interactions. Countertransference reflects the psychologist's emotional involvement and engagement in the helping relationship (Lavanco, Novara, 2013).

Community workers often encounter perplexing phenomena, such as social groups or organizations expressing conflicting positions or obstructing progress while seeking help (Gibson, Swartz, 2008). These behaviors are often driven by shared anxieties, leading to community practices and beliefs aimed at self-protection, even when they conflict with rational goals (Hinshelwood, Skogstad, 2000). Although these experiences are common, they are seldom recognized as a starting point for understanding community dynamics. Psychodynamic theories can shed light on these emotional reactions and enhance introspective awareness of one's biases and assumptions through reflexivity (Case, 2017; Fernández, 2018).

Considering the role of emotions can help community psychologists acquire more accurate contextual knowledge and create more coherent interventions. Psychodynamic concepts offer access to the often-neglected emotional domain, leading to a deeper understanding of both the communities served and the psychologists' responses.

6.   Clinical and community psychology or clinical community psychology?

From a historical perspective, community psychology broke away from traditional clinical psychology as a "rebellious daughter" in search of progressive independence and its ideals. To this end, it has gradually positioned itself on the fringes of influential contemporary movements, leaving the field open to reductionist and pathologizing paradigms of community mental health (Martin, Lounsbury, Davidson, 2004; Timimi, 2010; Hartmann, St. Louis, 2010; St. Paul, 2010). Arnault, Gone, 2018), and emphasized pragmatic and cognitive aspects in the pursuit of sociopolitical change, devoting less attention to subjective dimensions (King, Shelley, 2008; Koh, Twemlow, 2016; Hartmann, St. Arnault, Gone, 2018).

Against this backdrop, over the decades, as we have seen, several authors have argued for the need to develop a dialogue between clinic and community to stimulate transformative change in community mental health (Hartmann, St. Arnault, Gone, 2018) and to recover the role of affect and its connection to action (Lane, Sawaia, 1991; Rimé, 1993; Leon, Montenegro, 1998; King, Shelley, 2008). As early as 1990, more than 30 years ago, James Kelly proposed that the two disciplines should be considered complementary rather than opposing, pointing to the preventive approach as a fruitful field of integration (Kelly, 1990; Jenkins, 2016; Kloos, Johnson, 2017).

Compared to the past, as discussed in the previous section, clinical psychologists are more attentive to political and social issues, and their training increasingly includes knowledge and skills related to intervention in community settings (Jenkins, 2016; Kloos, Johnson, 2017). Moreover, within the psychodynamic tradition of clinical psychology, some emerging paradigms provide useful insights for community work (Swartz, Gibson, Gelman, 2002; Borg, 2010; Koh, Twemlow, 2016; Bermudez, 2019). What is the current relationship between the clinic and the community? Does the synergy between the two matrices of psychology allow the development of an integrated perspective? Does it authorize talking about "clinical community psychology"?

Some authors (Caputo *et al.*, 2020) have recently tried to answer these questions by polling the existing literature. From their examination, the clinical community approach appears to be mostly defined as deriving from clinical psychology and mostly aimed at expanding the provision of services and related interventions (Jason, Aase, 2016). Without going into the specifics of the study, to which we refer, it seems interesting to point out that several of the publications considered dealt with the training paths of clinical psychologists. Professional training in clinical psychology shows its limitations when young psychologists are confronted with the demands of life contexts, particularly community issues such as social violence, inequality, power dynamics, and minority integration. Work with social marginality also involves restructuring the intervention setting, moving from private practice or psychotherapy services to services in communities, to home-based intervention settings. These settings represent new frontiers for clinical psychological work that require exploration. On one hand, there is a need to rethink traditional intervention categories; on the other, it is essential to offer services where the psychological role focuses on rethinking relationships and activating processes of exchange between the family and its surrounding contexts (Roberti, 2017).

Indeed, the shift in the clinical psychologist's role from that of a technician to a consultant demands greater responsibility, a deeper understanding of community organization and resources, and the development of additional professional skills for effective clinical psychological intervention (Jason & Aase, 2016). Particularly in cases where the goals involve prevention and/or health promotion, focusing on community issues reveals the inadequacies of the reductionist and deficit-oriented paradigm typical of traditional clinical psychology. Instead, skills specific to community psychologists – such as the ability to work with social groups and organizations and to implement participatory interventions that emphasize resource development and the individual-context relationship – become indispensable (Ockene *et al.*, 2007).

Regarding the core professional competencies of the clinical community psychologist, the literature highlights four main areas: group processes, project planning and evaluation, interprofessional collaboration, and participatory methods. These competencies involve soft skills and require multi-level interventions that account for the complexity of community problems (Trickett, 2009). The study cited above identifies many synergies, at both theoretical and practical levels, between community psychology and clinical psychology. However, these synergies are not yet fully supported by systematic theoretical reflection on the potential integration between these disciplines.

Working in community contexts of varying breadth and complexity – whether group, organization, or community – is particularly challenging for psychological intervention because it frequently exposes gaps in theoretical models and practical approaches. No single psychological theory seems fully capable of accounting for the multiplicity of factors at play (Gibson, Sandenbergh, Swartz, 2001). Calls for psychological intervention from groups, organizations, and communities often arise from the complexity and multifaceted nature of problems; responses to such problems tend to integrate community dimensions and clinical aspects in practice more than in academic psychology. Addressing cultural, social, and linguistic differences in community work requires a high capacity to constantly navigate new sets of rules and meanings (*ibid.*), which could be better understood through a cultural approach to clinical psychology. At the same time, when faced with challenging situations such as intervening in disadvantaged and insecure settings, managing social conflict, poverty, or environmental degradation, the principles of community psychology may prove more beneficial than conventional clinical practice (*ibid.*; Jason & Aase, 2016). Effective professional responses require action–research methods that are as participatory as possible, providing solutions at the local level and integrating idiographic and ecological paradigms (Caputo, 2013; Francescato & Zani, 2017; Francescato *et al.*, 2017; Langher *et al.*, 2019a, 2019b; Mannarini & Salvatore, 2019).

In conclusion, how can we address the question of the relationship between clinical and community psychology? Some authors have called community psychology an area of intervention (Koh Yah & Castillo León, 2014) or a specialization (Jason & Aase, 2016) within clinical psychology. Despite this historical perspective, we believe that community psychology has now acquired its own identity, preserving it from the risk of "blurring" with its origins while allowing it to have closer relations with even the most neglected members (such as psychodynamic theories) of its family of origin. Signs of this are evident in the recent scholarly contributions that specifically reflect on the relationship between clinical psychology (including psycho-dynamically oriented psychology) and community psychology (e.g., Lavanco & Novara, 2013; Koh Yah & Castillo León, 2014; Jason & Aase, 2016; Hartmann, St. Arnault, Gone, 2018; Caputo & Tomai, 2020; Caputo *et al.*, 2020). More significantly, the most recent handbook of community psychology in the Italian language (Arcidiacono *et al.*, 2021) includes a chapter dedicated to the "psychodynamic perspective" of community psychology. It states that "the community approach positively values the legacy of psychodynamic approaches to understand and explain individual-context interactions, particularly in their group and macro social aspects" (Arcidiacono, Lavanco, Novara, 2021, p. 48).

While the time may not yet be ripe to speak of a true community clinical psychology perspective, the emergence of a new and independent field of knowledge may not be necessary or desirable if clinical and community psychology can find ways to operate in a functional, integrated manner that respects each other's identities.

A new and independent field of knowledge could originate if, borrowing Stark's (2011) definition, a *linking* science, i.e., a science (and practice) concerned with synergistically and integrally connecting constructs, models, and practices from different fields of knowledge for a more comprehensive understanding of the relationship between individuals and social contexts, were to be created.

Part 2
Methods

5
Community development

1.   Introduction: brief reflections on the concept of community

The term "community" has the same root as "common" and "communication": according to some etymologists, the root is derived from *cum-munia* ("common duties"), according to others, from *cum moenia* ("walls," "common fortifications"). In any case, the prefix cum *emphasizes* the aspect of a relationship, of shared context, of globality of the interactive system.

The first scientific definition of "community" is traced back to Tönnies (1963). This author, at the end of the nineteenth century, distinguishes between community and society and places them at the extremes of a *continuum* within which he positions social aggregates of different breadth, complexity, and characteristics. Synthetically, a community is conceived as a "living organism" based on languages, meanings, spaces, shared experiences, and feelings of belonging and participatory dimensions. In contrast, society is understood as a social artifact, based on rationality and exchange, aimed at achieving concrete, mostly individual goals.

The definition of the term "community" also appears complex because it has its roots in different sciences and perspectives, from the psychological study of small groups to systems theory, from cultural anthropology to sociology. A recurring theme of interest in the various definitions of community in the social sciences of the twentieth century is the relationship between relational and spatial dimensions. That is, different authors oscillate between valuing relational aspects and social relations or the sharing of physical spaces, connoted by constraints and resources, that influence the choices and actions of individuals. The development of means of transportation and mass communication tends to bring citizens of the "global village" closer psychologically and physically, allowing the opportunity for a representation of space/time not necessarily tied to narrow geographic areas. Territorial rootedness for certain communities or certain social groups remains a significant element, while for others the sense of belonging to a community seems to be based solely on a cultural or psychological identity. The territorial dimension of community, therefore, is often but not always present. Throughout the evolution of studies on this concept, most authors have acknowledged the significant role of the territorial dimension. For instance, scholars from the Chicago School emphasized that the concept of community revolves around three central ideas: rootedness in the territory, the presence of social organization, and interdependence among members (Park, 1952). Similarly, Amerio (2000) identifies territory, relationship, and participation as the three most appropriate and usable dimensions for community psychology.

The ecological perspective of community psychology (Levine, Perkins, 1987) has traditionally leaned toward valuing the spatial aspects of communities while emphasizing the impossibility of disregarding the relational ones.

Emphasis on the territorial dimension of the community, also understood in a cultural, organizational, and sociopolitical sense, makes it possible to define it as the concrete context in which people's daily lives are actualized, social problems take specific forms and where it becomes possible to grasp the continuous transaction between objective factors (geographic, economic, urban aspects, etc.) and psychological-individual factors (perceptions, representations, needs, expectations, etc.).

Embracing a concept of community that is complex and anchored in the territorial context allows us to investigate the processes through which the relationship between the individual and the community is articulated and the ties that connect people and to the territory they share, a feeling commonly referred to as a "sense of community."

2.   A sense of community

A sense of community is a foundational construct of community psychology and explicitly marks its departure from traditional clinical psychology and biomedical, individualistic models of distress management. Community psychologists seek to move beyond analyzing and intervening at the individual level, incorporating an ecological-cultural dimension (Levine & Perkins, 1987). In this framework, the social environment contributes to shaping the individual – his psyche, goals, hopes, as well as his problems and disorders. Life contexts and community dimensions are seen as spaces where discomfort is expressed and addressed by community psychology.

The concept of a sense of community was introduced by Sarason (1974), who defined it as "the perception of similarity with others, a recognized interdependence with others, a willingness to maintain this interdependence by offering or doing for others what is expected of them, a sense of being part of a fully reliable and stable structure" (p. 174). This comprehensive definition highlights the subjective and relational nature of the sense of community. It is a perception, subject to constant verification, and is not fixed but influenced by the experiences of individuals. This sense is rooted in ties of interdependence, which require active maintenance, and in feelings of security and stability toward the community context.

Sarason's sense of community can be viewed both as a subjective experience, indicating a supportive network, and as a cohesive and motivating force that promotes the well-being of the community. However, the loss of this sense is evident in devalued settings where fragmentation and subgroup formation occur. A limitation of Sarason's contribution is its predominantly theoretical nature, lacking in operational approaches that would enable the use of tools for survey and measurement.

Ten years later, McMillan and Chavis (1984) took up Sarason's definition and proposed an operational development of it, still dominant today, from which it was later possible to construct a measurement scale. According to the authors, the constituent elements of the sense of community are as follows:.

- The feeling of belonging and sense of general connectedness (*membership*): the feeling of belonging is defined in order of the level of identification developed (the certainty of being accepted in the group, of having a role, the willingness to sacrifice oneself for the common good). Related to the development of belonging is the practice of group rituals (appearance, customs, language) that ensure the establishment of the boundary of the community, the line of demarcation through which it is established who is "in" and who is "out." The evidence and sharing of symbols and languages are thus necessary by the very definition of the group;

- influence and power (*influence*): the ability to count and directly influence the fates of the structure is undoubtedly an incentive factor for group participation. The sense of influence is thus directly related to the physical size of the structure. Organizations that are too large tend to develop anonymity and conformity and may influence the emergence of intermediate structures (volunteerism, parties) between the individual and the macro collectivity (Warren, 1981);

- *integration and fulfillment* of *needs*: real or imagined needs met through the group are a real reinforcement of membership. A successful group thus increases its cohesive strength. It also helps determine the scale of needs that is similar among members;

- *shared emotional connection*: a set of common values, beliefs, and expectations. Such elements as contact, success, the pursuit of group goals, the sharing of emotionally important events, even if dramatic, common investment, the sharing of rewards and punishments, and bonds of a spiritual nature develop a greater sense of belonging and consequently strengthen the community itself. It is the shared emotional connection then that marks the quality shift from a collection of people to a true community.

McMillan and Chavis produced a synthesis aimed more directly at guiding social workers in fostering a sense of community. Their approach encourages social workers to promote local leadership (as opposed to relying solely on externally programmed interventions), to respect the languages and norms intrinsic to community culture, and to employ research-intervention methodologies along with participatory modes of awareness and problem-solving. Mannarini and collaborators (2021) highlight the promotion of value dimensions as an effective strategy for enhancing the sense of belonging, responsibility, and cohesion within communities. According to the authors, values shape the social, cultural, and normative environment and are embodied in social practices across all sectors of the community, such as schools, the health system, and policies. As a result, community-based participatory intervention programs could include activities specifically designed to activate values among participants.

Moreover, civic engagement actions implemented within one's community are instrumental in fostering greater trust and higher levels of a sense of community.

2.1.   Correlates of the sense of community A sense of community has immediately attracted considerable interest among scholars and professionals. Research has shown that it is highest in the elderly while it decreases significantly in the adolescent period (Prezza *et al.*, 1999; Zani, Cicognani, Albanesi, 2001). This indicates how the stages of life of greatest fragility and dependence are those in which the territory and attachment to the place of residence are most relevant to the individual. The sense of community, moreover, seems to vary with the size of the reference territory (higher in small towns than in cities; Puddifoot, 1996; Prezza, Costantini, 1998), age, income, and the number of years lived in the neighborhood (Ross, Talmage, Searle, 2019). Community is an essential component of the human experience and has an impact on residents' well-being. The perception of belonging to a community can improve the well-being of individuals and the quality of their social life (Prati, Cicognani, Albanesi, 2018; Sohi, Singh, Bopanna, 2018).

A growing literature over the past two decades has documented an association between a sense of community and well-being (e.g., Prezza *et al.*, 2001; Farrell, Aubry, Coulombe, 2004; Albanesi, Cicognani, Zani, 2007; Prati, Albanesi, Pietrantoni, 2016). Each of the dimensions of well-being (emotional, psychological, and social, as well as anxiety and depression) is significantly related to the sense of community (Coulombe, Krzesni, 2019), which, for that reason, is becoming an increasingly valued construct in health promotion fields (Ross, Searle, 2019).

Numerous studies, moreover, have found sense of community positively correlated with social and political participation (e.g., Prezza *et al.*, 2001; Obst, Smith, Zinkiewicz, 2002). However, although the sense of community and participation are closely related, the direction of this relationship is not yet fully clarified. In the conception proposed by Levine and Perkins (1987), a sense of community and participation have a mutually reinforcing relationship: participation fosters a greater sense of community, which, in turn, leads to greater participation. Consistently, in numerous studies, a sense of community has been found to make one more attentive and responsive to problems and difficulties present in one's home context; foster a sense of social responsibility and civic engagement (e.g., Talò, Mannarini, Rochira, 2014; Chandra *et al*, 2016; Procentese, De Carlo, Gatti, 2019; Procentese, Gatti, 2021); strengthen the sense of connectedness and the building of better neighborhood relationships; and improve the sense of control and empowerment of an area (Speer *et al.*, 2013; Cicognani *et al.*, 2015).

A sense of community, therefore, is commonly understood as a force that attracts individuals to the community and activates their participatory and socially engaged behaviors by fostering the conditions of their well-being.

2.2.   Some reconsiderations of the construct Over the years, several authors have proposed a revisiting of the notion of a sense of community by hypothesizing its distribution along an axis from a positive to a negative pole. Following a study carried out in 1996, Brodsky and colleagues conceptualized the existence of a negative sense of community, understood as a centrifugal force that symbolically pulls individuals away from the community to which they belong when the context is perceived as dangerous and/or deprived; in this case, emotional and physical distancing from the community takes on a protective value.

Some scholars (Mannarini, Rochira, Talò, 2014), mirroring McMillan and Chavis, have proposed a model of a negative sense of community composed of four dimensions:

1. distinctiveness: the need to stand apart and differentiate oneself from other members of the community;

2. abstention: an attitude of passivity and indifference toward the community;

3. frustration: perception of the community as an obstacle to the satisfaction of the individual's needs; and

4. alienation: the feeling of estrangement from symbols, traditions, and values of the community.

A further revisiting of the sense of community has been proposed by Nowell and Boyd (2010). Their thesis is that the sense of community was initially measured and understood exclusively through a "need theory perspective", according to which the community would represent a resource that individuals draw on to satisfy a variety of material and psychological needs. This key to understanding allows for an adequate explanation of why, as several research studies claim, feeling part of a community is generally associated with higher levels of psychological well-being and active engagement. However, this perspective does not consider the possibility that a sense of community may be rooted in values and social responsibility (Perry, 2000). Taking this broader view also allows us to account for those behaviors enacted on behalf of others motivated not by self-interest but by ideals, values, or a feeling of responsibility.

Based on these considerations, Nowell and Boyd propose the Community Experience Model, a conceptual model according to which individuals may perceive their experience in the community in terms of two dimensions: a sense of community as a resource and a sense of community as a responsibility. These two aspects of a community experience are distinct but complementary and respond to a different logic. The former emphasizes the extent to which the collective well-being of a community constitutes a resource for the physical and psychological needs of individuals; the latter refers to a feeling of personal responsibility in protecting or enhancing the individual and collective well-being of the community that is independent of any expectation of personal profit (*ibid.*; Nowell Boyd, 2014; Boyd *et al.*, 2018).

2.3.   Organizations as communities Macmillan and Chavis' model has been used in territorial communities but also in organizations where community spaces are built, in schools and work settings especially in the last two decades (Zani. Cicognani 2012; Chioneso, Brookins 2013). Several authors have explored why work settings are considering sense of community an important component of organizational culture, that adds positive relational benefits (Brytting,Trollestad 2000; de Vries 2001; McBride, 2006; Bryan *et al.* 2007; Peterson *et al.*,2008). Emotional connections and reciprocal support are variables that foster intrapersonal empowerment (Hughes *et al.* 2008); workers' wellbeing (Boyd, Nowell, 2014) and the prosperity of enterprises (Mintzberg 2010).

School settings are developing sense of community and researchers examine schools as communities, Goodenow (1993) introduced the construct, Prati, Cicognani, Albanesi 2017 define it:" sense of belonging to a school as a community, emotional connection and ties with other students, and satisfaction of personal needs through this belonging". Sense of belonging to a school community predicts both students' wellbeing and school success (Vieno er al.2013; Prati, Cicognani, Albanesi 2017, Francescato, Putton 2022). Recently two Italian Community psychologists have started promoting sense of community among inhabitants of the same building creating "The Circle Of Good Inhabitance" both face to face and online during the Covid pandemic. These pilot projects are important to diminish the isolation and indifferences that prevail in many residential buildings of middle and big cities where neighbors barely know each other, and often have condominiums conficts-

3.   Community development as an expression of the radical wing of community psychology

Community development can be defined as a process that aims to create conditions for social and economic progress through active community participation (Rothman, 1974; Heller *et al.*, 1984) by providing meaningful experiences in which spontaneous groups and organizations in the community can be involved (e.g., Kelly, 1987). The goal of *community development* is to create a supportive and integrated social network, based more on citizen mobilization and participation than on professional expert advice, that enables citizens to self-determine their transformative processes. Indeed, the strategy of community development moves from the assumption that social change can take place more effectively and in a direction of greater freedom and equality if citizens' initiative and involvement in defining the goals and practices of transformation can be developed. Participation and sharing, for community psychology, are both guiding principles and goals to be pursued, resources that enable the implementation of social programs. In them, the assumptions of the radical wing of the discipline, which prioritizes a focus on growth and development over a "restorative" perspective, are most clearly expressed. Indeed, participation, in addition to being a resource, constitutes a way of guiding and controlling programs, which can correct and reduce the gap that often exists between professionals and the recipients of interventions.

As we discuss in chapter 1, in the 1970s, "radical" scholars and practitioners harshly criticized the traditional psychiatric care approach and the way community-based interventions were implemented. They highlighted the manipulative nature of psychotherapy, the stigmatizing role of psychiatric services, and the predominantly restorative – rather than emancipatory – approach even in interventions from the moderate wing of community psychology (Ryan, 1971; Francescato, 1977a). These critiques gained traction in the 1980s, especially within community development practices.

Cooperative strategies aimed at integrating various forces and components for social change through group mobilization were being explored in North America as early as the 1970s.

In Italy, the past three decades have seen social and legislative developments that fostered a focus on local communities and service participation. This shift has been driven by the crisis of the welfare state, the emergence of more complex societal needs, and the demand for improved service quality. Laws promoting transparency, such as Law No. 241 of August 7, 1990, Law No. 15 of February 11, 2005, and Legislative Decree No. 97 of May 25, 2016, as well as the introduction of the Service Charter, have further supported this transition.

The crisis of the welfare state has highlighted the potential of local communities as valuable resources and areas for investigating needs. Without active and participatory communities, societal structures risk fragmentation, making it essential to develop a sense of belonging and promote democratic coexistence processes (Amerio, 2003; 2004).

3.1. Community development and social action "Community development" is a wide-ranging expression that can encompass different types of interventions (Heller *et al.*, 1984). Let us try to list, referring to Clinard (1970) and Levine, Perkins (1987), the main modes of community development:

- create a sense of social cohesion, improve interpersonal relationships, and develop belongingness at the neighborhood and neighborhood level;

- support and stimulate self-help, volunteer, and other spontaneous gathering experiences;

- raise awareness and inform citizens about the most relevant issues in the community and propose common goals for action;

- identify and promote the capabilities of local leaders;

- develop civic awareness, respect, and communicative exchange among the different cultures and ethnicities in the community;

- utilize the skills of professionals and the *know-how* of researchers to support the mobilization of pressure groups and social change;

- provide training in conflict management, *decision-making,* and problem-solving techniques;

- contribute to the coordination between the actions of different departments and the thrust of opinion and social action movements.

Some of these functions can be encapsulated in the term *advocacy*, which means support, defense, and action to protect the rights of certain social groups. As can be easily guessed, in some cases community development presupposes the (perhaps latent) presence of resources, knowledge, and leadership skills capable of implementing changes with cooperative strategies through consensus building. In other cases, the deficient and/or unequal distribution of resources and the discrepancy between the goals of different social groups create contradictions that are not easily healed and exclude the presence of common motivations for all. In these situations, rather than community development, we speak of social action (Heller *et al.*, 1984; Levine, Perkins, 1987), that is action that aims, through conflicting strategies, to rebalance an unequal distribution of resources and modify power relations. Social action thus presupposes that the interests of the conflicting parties are not easily reconciled. Therefore, change will be supported by those components that can find advantage in the transformation process.

While it is theoretically possible to distinguish between community development and social action, in practice, the differences are often less clear. Interventions tend to alternate between cooperative and conflict strategies to broaden consensus while achieving tangible results. For this reason, we use the term "community development" in a broad sense, encompassing all the various intervention and social action strategies that can be implemented at the *community* level.

Scholars widely agree that participation, empowerment, and social responsibility are key factors in fostering community development. Participation and a sense of responsibility are closely linked to the sense of empowerment and the exercise of power experienced by residents (Higgins, 1999; Cuthill, Fien, 2005; Nikkhah, Redzuan, 2009).

Orford (1996; 1998) considers power one of the social resources necessary for the well-being "of the person in context"; according to this author, participation and empowerment require the perception of an adequate sense of power. Of the same opinion are Martini and Sequi (1995, p. 24), who argue that "a high sense of responsibility cannot last for a long time in a condition where they think they can do nothing to change the situation.

As we will also see in the following sections, community development projects aim to grow a sense of responsibility, power, and competence by promoting the development of intermediary structures (associations, neighborhoods, and self-help groups) and encouraging voluntary participation, as this facilitates the promotion of a sense of belonging and social empowerment and supports a sense of responsibility.

4.   Understanding the community: community analysis

In the process of community development, the goal of the psychologist is to act as a "consultant" who intervenes in a convergent and complementary manner with both other practitioners and community members. The specifics of the community worker's intervention, that is, to promote community competence, can only be achieved through careful and precise knowledge of the reality in which he or she wishes to operate. To know a community, it is necessary to identify the many factors that interact in it simultaneously and that give rise to a complex and changing system. To facilitate its understanding, Martini and Sequi (1988) devised a tool, "community analysis," now widely used and tested in its effectiveness, which allows the multiplicity of variables in a territory to be assessed (territorial, demographic, productive activities, services, institutional, anthropological, psychological profile; see Box 1) and their mutual interdependence, and to draw, therefore, a "profile of the community" under examination.

BOX 1 The community analysis

- Spatial profile: includes all data related to the territory (extent, physical composition, climate, natural resources, environmental degradation, infrastructure, etc.) and their usability.

- Demographic profile: covers the number of inhabitants, divided by age group, gender, schooling, population increase/decrease, migration flows and mobility.

- Profile of productive activities: presence and development of primary, secondary and tertiary activities; rate of environmental harmfulness.

- Profile of services: social-health, social-educational, recreational-cultural public and private.

- Institutional profile: concerns political-administrative organization, ideological references, presence of particular institutions.

- Anthropological profile: history of the community, its values, social attitudes, degree of cohesion.

- Psychological profile: affective dynamics, sense of belonging, collective identification, degree of openness/closure of subgroups, level of participation, cooperation, affective security.

The process of knowing, as is well known, is subject to distortions of various kinds depending on the objectives, the cultural filter, and the level of awareness of the examiner; therefore, however accurate it may be, the reading of community remains approximate. It is based on these reflections that the most appropriate research tool consistent with the principles of community psychology remains, again, participatory research, where participatory means not so much, or not only, the participation of the practitioner in the life of the community but the active participation of the community in the process of knowing and designing interventions for it.

Community profiles analysis is a method that enables researchers and community members to identify the needs, resources, and gaps within local communities, institutions, and services. The approach gathers three types of data: objective (such as demographic information and economic indicators), subjective (mainly derived from interviews with key informants from diverse backgrounds), and symbolic (e.g., through dramatization and drawing). This method helps to pinpoint both the strengths and weaknesses of a community and prioritize critical areas for action plans and interventions. It has been frequently applied in community development projects in Italy (e.g., Arcidiacono, Tuozzi, Procentese, 2015).

Community analysis, while initially conceived as a tool for knowledge and "diagnosis," gains additional value and potential when carried out through participatory methods (Martini, 1996). In this context, it is not just an analysis of needs to inform the planning of appropriate services, though that may be one outcome. Instead, it becomes a fundamental moment of change. It allows the community to become aware of their conditions, needs, potentials, resources, limitations, values, and desires. The importance lies not in the data itself but in the meaning the various social actors attribute to it through a process of "collective negotiation" (Martini, Sequi, 1995, p. 57).

In analyzing each profile, both objective, numerical data and more subjective data, such as individual perceptions and group narratives, are examined. Based on the type of data collected, the profiles are categorized into hard profiles (e.g., territorial, demographic, economic, institutional, and service) and soft profiles (e.g., anthropological-cultural, psychological, and future-oriented, as introduced by Francescato, see Section 4.1).

The investigation of soft profiles, in addition to questionnaires and interviews, makes use of specially designed techniques and the use of focus groups. Focus groups are small discussion groups led by one or more moderators and focused on a topic to be investigated in depth (Corrao, 2000). In the focus group, the participatory atmosphere encourages the exchange of ideas, opinions, and experiences among its members. Unlike the individual interview, in which the respondent relates only to the interviewer, in the focus group, the answers are "contaminated" by the reactions of others, and it is in this process of mutual influence that modes of co-construction of descriptions and meanings are activated.

The focus group instrument, modified *ad hoc* (Francescato, Tomai, Solimeno, 2008), has proved useful not only in the investigation of soft profiles but also in the assessment of what we might call "soft aspects" of hard variables, i.e., the social perceptions that different groups in an area have of certain structural features of their community (e.g., one can ask a group of parents what they think of school services in their area).

In Box 3 at the end of the chapter, we provide a detailed description of the profiles, indicating for each: the aspects of the community it considers, the sources where to find the information, and the tools that can be used.

4.1. Changes made to the survey instrument

This community survey model has been used extensively in our work; the repeated and diverse (in terms of objectives and contexts of application) use of the instrument together with the instrumental input of some Austrian community psychologists

(Ehmayer, Reinfeldt, Gtotter, 2000) led us to introduce changes in the number of profiles, mode and survey instruments.

An eighth profile was added by Francescato to the seven profiles proposed by the authors: the *profile of the future*. This profile assesses how the community experiences the relationship between the present and the future and what events it fears and/or hopes for. It is explored through interviews with key people and focus group discussions on three stimulus questions, "What do you think this city will look like in ten years?", "What do you fear most for the future of this city?", "What do you wish most for the future of this city?"

Other changes introduced by Francescato concern the indicators considered for the evaluation of each profile and some research tools (Francescato, Tomai, Ghirelli, 2002). In the *institutional profile*, an assessment of the type of network relationships the municipality has with other municipalities, with the region, with the province, and with the European Community has been introduced. In the *demographic profile*, types of immigration are explored by noting the flow of changes of residence and trying to grasp, overall, the elements capable of activating social mobility. In the *profile of productive activities*, the implication that globalization processes may have on the economic development or decline of local communities is also considered; for example, whether organizations providing jobs in the area are particularly subject to international competition is assessed. Finally, in the *anthropological* and *psychological* profiles, more space is given to identifying group emotional experiences and dominant and marginal narratives.

In terms of innovations in survey instruments, innovative techniques were introduced in the spatial profile and the anthropological and psychological profiles. For spatial profile analysis, collaboration with the Austrian group mentioned above led to the development of two techniques borrowed from the experiences of environmentalists:

- The *walk*: two people from outside the community are asked to walk two hours through the streets of the area under consideration and note and/or photograph aspects of the fixed environment that strikes them. They must also make a list of the positive and negative aspects they find by describing the first impression the place makes on an outsider;

- *Neighborhood photographs*: in this case, people from the community are asked to photograph the places they think represent their city and neighborhood and to illustrate what negative and positive aspects of the area the photos portray.

- For the assessment of the anthropological and psychological profile, Martini and Sequi (1988) proposed the analysis of documents and interviews on the history, festivals, special events of the area, social support, values, sense of community, degree of belonging and identification with the social context; the tools they favor are questionnaires, interviews, and focus groups. In addition to these methodologies, Francescato introduced the use of two other techniques (Francescato, Tomai, Ghirelli, 2002):

- *Neighborhood drawing*: participants are asked to draw an element or scene that represents the neighborhood. All drawings are then hung on a wall, and the group of participants expresses the emotions that each drawing arouses; then, the group makes free associations by looking at all the drawings. The associations are then distinguished into positive, negative, and neutral, paying particular attention to those repeated several times;

BOX 2 The movie script play technique.

The movie screen play technique (Arcidiacono, Tuccillo, 2017; Francescato *et al.*, 2002), involves asking participants in a group to develop a film script inherent to their community. The group is asked to choose the genre (documentary, fantasy, detective, historical, comedy, drama, fantasy, etc.) and title, define the main characters, develop the plot, and construct a definite ending in the present and, if desired, in the future. At the end of the work, participants narrate or perform their screenplay. After verbalization, the scripts are commented on and discussed by the group to identify, share, and discuss the cultural patterns that emerged, the language used, the interactions, and the projective and introjective dynamics of the organization and the various subgroups. Mebane and Benedetti (2022) developed a variety of indicators to interpret the multiple meanings of the scripts, which help participants better understand the psychodynamic processes of their communities

---

- The *screenplay*: in this case, the group is asked to invent a film representing life in the community (see Box 2). The screenplay technique can also be used at the beginning and end of the community survey to compare the change that has occurred in the knowledge and perception of one's local area. Content analysis techniques are used for comparison, often showing a climate of resignation and general devaluation of one's community in the scripts produced at the beginning, and reevaluation or deeper knowledge of the resources in the area in the scripts produced at the end of the survey.
- These animative techniques serve to explore the attitudes and experiences that the various subgroups have toward the community and the type of emotional response that the environment can evoke in them; the environment can activate different emotional responses and, therefore, different behaviors depending on the physical and social characteristics that connote it. The proposed techniques are used to explore emotions that various subgroups experience toward their community, values, fears, hopes, problem-solving and coping techniques, perceived difficulties and obstacles, and tools and support networks used.
- A multitude of indications can emerge from this type of projective

data; however, we simply ask the groups that participated in this work to identify the weaknesses and strengths that emerged, which can help us understand the psychological and anthropological profiles of their area. Using these more emotionally engaging techniques also allows people taking part in participatory research to engage with groups and individuals with whom there is often little or no opportunity to meet and exchange views.

4.2.   How to use the instrument We will now present how we use the eight profiles to learn about a community and assess its degree of empowerment. Usually, community profiles have been investigated by us at the request of a particular client:

- Environmental groups who wanted to increase citizen participation in the construction of *Agenda 21*, an EU-sponsored project to foster sustainable development;

- Social and health services that wanted to design their interventions with the needs of the local area in mind;

- Social cooperatives and other entities that wished to participate in projects for the elderly or youth, or other at-risk groups, promoted by regional or national laws or European projects (e.g., *Equal, Youthstart,* and *Leader*) that require the involvement of various social actors in an area, or more rarely by mayors who wished to launch territorial development projects.

First an interdisciplinary research group is habitually formed consisting of at least two local experts for each profile (Francescato, Tomai, 2002; Francescato, Tomai, Mebane, 2004) coordinated by one or more community psychologists from outside the community. This group carries out the community analysis through the following stages of investigation.

- *Step 1. Preliminary analysis*. Preliminary analysis is an activity routinely carried out by the interdisciplinary group to draw an initial description (based on the knowledge and opinions of the group) of the community under consideration. This initial "diagnosis" allows for the emergence and comparison of strengths and problem areas perceived as priorities by the group members. For this reason, the initial research group must consist of people and key witnesses from the community who are diverse in terms of age, *status*, role, profession, knowledge of the community, etc. The community diagnosis made through the preliminary analysis will be compared with the data progressively collected during the analysis of the different profiles. This will allow us to constantly compare subjective data with objective data, soft data with hard data and to arrive at a more shared and integrated description of the community reality.

To perform the preliminary analysis with the initial research group, the community psychologist uses the brainstorming technique. In the first associative phase of brainstorming, group members are asked to list the weaknesses and strengths of their town or neighborhood. The group can choose indifferently to start from the brainstorming on strengths or from the brainstorming on weaknesses; the important thing is to have, at the end of the associative phase, two separate lists. Having finished the associations, in the convergent thinking phase, the group leader must:

a) for each proposed strength and/or weakness, ask how many others consider that aspect to be a strength or weakness in their community;

b) Present the outline of community profiles to the group;

c) help the group classify the strengths and weaknesses that emerged in the eight profiles (e.g., "environmental degradation" is framed in the spatial profile, "good schools" in the service profile, etc.).

Adding up the weaknesses and strengths placed in each profile will provide both an initial assessment of the community as a whole and a balance of the individual profiles. A prevalence of weaknesses accompanied by a paucity of resources will suggest an overall under-empowered community. In addition, this first analysis allows for an understanding of which profiles the interdisciplinary research team perceives as most salient, richest, or most critical.

The preliminary analysis allows for clearer planning of the subsequent research work, for defining which profiles to delve into depth and which key people in the community to contact. At the discretion of the interdisciplinary team, the preliminary analysis can be repeated with other representative groups in the community and/or key people who are contacted in the continuation of the research work. It is very important to involve both dominant and marginal groups; especially in multiethnic communities, it is desirable to do focus groups with each ethnic group (Benedetti, Mebane, Onacea, 2010). From the results of these various preliminary analyses, an overview of perceptions of the various profiles by different groups can be obtained.

- *Step 2. Profiles analysis.* The goal of participatory surveys carried out by community psychology is to encourage the comparison of different descriptions and to activate change as early as during data collection. Data collection and reading, if done in a participatory and shared way, becomes a key moment in a process of change. At this point, the interdisciplinary group, following its interest and expertise, gathers the empirical data available for analysis of each profile and/or activates processes to find the unavailable information. The group members, therefore, evaluate and decide on the need to organize focus groups with different social groups to explore specific aspects and/or questions. For each profile, the interdisciplinary group reflects on the data and ascribes meaning to it, indicating strengths, weaknesses, and weight using objective indicators (increase in environmental degradation, decrease in crime, decrease in school *drop-out*, increase in public transportation services, etc.). Finally, the group compares these "objective" analyses with social perceptions from the various preliminary analyses obtained in the various focus groups and/or interviews conducted.

- *Step 3. Identification of priorities for change.* The final stage of the work represents a particularly significant moment in the research. It involves an assembly meeting in which all the people who, in different capacities, were involved in the data collection are invited to participate. This is where all the posters and all the material that can document the research process carried out (e.g., the contents of the discussions that took place in the various focus groups) are displayed. The participants are then divided into subgroups, which are given the task of pointing out the priority problems and resources in the area that emerge most clearly from the data summarized in the posters. The work of the subgroups leads to the drafting of a poster of priority strengths and problem areas based on which the groups develop concrete proposals for change, including identifying those who have the power to implement the desired change. The technique of community profiles encourages the integration of different knowledge by bringing out the plurality of viewpoints; it is a tool that can be profitably used to explore the territorial context through specific ways of involving citizens. The purposes of the intervention vary relative to the clientele, most of which consists of social and health services, social cooperatives, voluntary associations, and schools.

4.3. Community consultation The research mode (participatory research) proposed to investigate the community more adequately makes use of both traditional tools, such as the questionnaire and interview, and less traditional ones, such as organizing group discussions and focus groups. The role of the psychologist, however, remains that of process and resource activator and community consultant. In performing this role, the practitioner must follow a methodology that contemplates successive stages of a counseling relationship, established, however, not with a single client, but with social forces or organized groups that promote change. The stages of community consultation are, for example, described by Redman, Cullari, and Fabris (1985) as follows:

- Preliminary analysis;

- Input and definition of counseling;

- Diagnosis;

- Setting goals and procedures;
- Assignment of operational roles;
- Implementation of the intervention;
- Evaluation;
- Maintenance;
- Exit;
- Follow-up.

As can be seen, these are common stages in all processes in which the consultant contributes to planning and evaluation; in the case of community consultation the stakeholders who relate to the consultant (or team of practitioners) are quite numerous and active. In the entry stage, for example, there is a meeting of representatives from all levels of the client community who are interested in change. The needs assessment directly involves population and community representatives, both in data collection and data processing. In the planning of interventions, moreover, the community psychologist often plays the simple role of facilitation and advocacy of the community's self-determination processes. Finally, in the final maintenance, exit, and follow-up stages, the expert ensures that the community is completely autonomous in managing the innovative processes initiated.

Community intervention is the practice most consonant with the community approach because it is most consistent with the systemic and ecological orientation of the discipline and is aimed not at individual behavior modification but at the development of resources and skills in the community, the extended involvement of members (seen as resource and *know-how* holders), and the use of strategies that contemplate the management of social dynamics among groups and the activation of partnerships and networks among institutions, formal and informal agencies in the area (Tricket, 2009b; Trickett *et al.*, 2011).

But what are the groups, the collective stakeholders with which the community psychologist relates? Two types of aggregations can channel community participation: organizations promoted by state and local governments and groups created spontaneously by citizens. Institutional groups (called *government-mandated* in the United States) are organizations promoted by the authorities themselves to contribute to the action of state bodies or public services and to operate a "bottom-up" control of the implementation of laws and social problems. In Italy, this participatory orientation began in the 1970s, targeting only certain segments of the population (e.g., to encourage parental participation in school governing bodies). In recent decades this orientation, as previously mentioned, has undergone considerable expansion (e.g., the establishment of councils: council of people with disabilities, youth council, neighborhood committees, etc.). The action of institutional groups can have, on the one hand, a function of stimulation and verification by the community and, on the other hand, the effect of creating consensus and adherence to official initiatives. The community psychologist can thus act as a consultant, an expert embedded in or in contact with institutional groups.

We believe that for a community psychologist, the area of social policy should be a non-secondary part of their sphere of activity, while the politician and administrator might receive a non-ideological through the attentive contribution to collective processes. Spontaneous grassroots groups, which include neighborhood associations, ecological organizations, minority rights, and freedom groups, service and volunteer groups, are also included in community consultation.

5.   The volunteer movement

We mention here an issue that is an increasingly relevant element of the social fabric and cannot be overlooked in community development interventions. Volunteer groups are non-public organizations that offer a service and/or act to protect social rights to respond to needs and problems present in the community. They are spontaneous, nonprofit, self-managed aggregation groups that are financed with funds from predominantly private sources, although collaboration with and the possibility of funding from public systems are not precluded. Depending on the case, the workers in these groups may be all volunteers or part volunteers and part salaried professional workers. The areas of expertise and the areas to which the work of voluntary groups is directed are extremely varied: home care for the elderly and disabled, nursing care, health emergencies, prevention of child abuse, protection of the rights of the sick, youth work, mental health, drug and alcohol states, hospital care, etc.

Both in Italy and in many Western countries, people who feel personally and socially responsible and who creatively integrate the two traditional ideological orientations of liberalism, based on individual freedom, and socialism, centered on common welfare, are increasing. In volunteering, people can test their abilities to provide help, establish social relationships, and form primary groups that facilitate the development of a high sense of self and new personal and social empowerment. What we want to emphasize, from the perspective of community psychology and "community development," is the possibility of conceiving of volunteer groups as an important resource, a mode of spontaneous aggregation whose spread and integration in the territory is often an indicator of the degree of participation and sense of responsibility of the population. It should be understood as an autonomous resource but not separate from institutional services, capable of enabling citizens to actively intervene and find more ductile and "humane" social responses. The key point is to identify ways of linking different levels of intervention, as appropriate, either by seeking integration in programs or by leaving full autonomy to different forms of action. The strategy of community development can thus offer a comprehensive perspective to integrate articulated social realities into an overall vision of promoting quality of life and social change in a democratic and progressive sense.

6. Critical evaluation of community development techniques

Used together, the techniques described in this chapter--from community profiles to community development strategies--almost all adhere to the guiding principles of community psychology. Certainly, profiles encourage a pluralistic view of community, integrating different types of knowledge of both objective and subjective types. To do a community analysis with this method, procedures, and tools from various disciplines are used, and locally produced knowledge from community residents can be brought out by activating forms of collaboration and participation. Through the anthropological profile, the historical roots of the community's strengths and problem areas can be explored, and through the profile of productive activities, the unequal distributions of power to access resources in the social context can be examined. Through focus groups with members of minority or marginalized groups, and especially through the movie script technique, minority narratives that challenge dominant ones or otherwise give another perspective on the community can emerge. Certainly, community development techniques facilitate the coming together of people who share a problem or situation and bring out the link between individual empowerment and social struggles. They also promote the implementation of change projects and identify strengths that can be leveraged to achieve desired changes. Finally, these techniques make it possible to identify which problems among those identified as priorities can be solved at the local level and which require interventions at other levels (regional, state, European, etc.).

6
Social networks and social support

1. Social networks

The concepts of *social network* and social *support* have gradually emerged as constructs of remarkable fertility and effectiveness for describing the structure of interpersonal relationships that characterize daily life, and the interweaving of social, institutional, and spontaneous resources present in the community. The epistemological innovation consists of overcoming a sectoral view of the community, habitually described in its distinct social and therapeutic agencies (the family, the school, the office, the counseling center, the group of friends, or the civic committee, etc.). Rather, the attempt aims to examine the overall social field in which individuals, groups, and organizations are immersed and analyze its positive and negative influences. The relevance of such constructs to the development of integrated therapeutic interventions and large-scale preventive actions is easily understood.

1.1. The origins of social network studies The systematic study of networks has developed through the contributions of three strands of study: the Manchester School of Anthropology, the Gestalt School, and the Harvard School.

The first systematic analyses of social networks were conducted by English social anthropologists, with Barnes (1954) playing a key role through his study of a Norwegian fishing village. He attempted to map the web of relationships surrounding everyone, identifying key features such as breadth, density, and types of interpersonal exchange. Initially, "personal network" was used to describe the specific configuration of ties around an individual, while "social network" referred to the connections among all members of a population. *Gestalt's* main contributions came from Moreno, the founder of sociometry, who emphasized the study of relationships, and Lewin, who focused on the social field and spatial representation of psychological phenomena. In the 1970s, the Harvard School further advanced the study of social networks by developing *network analysis*, which modeled social structures and their properties.

1.2.   Studies on the characteristics of social networks The essential aspect of the study of social networks is the description of the various distinctive characteristics and properties. Among the various formulations, the most comprehensive and concise is that of Marsella and Snyder (1981), who argue that social networks can be characterized by four dimensions, each of which includes some interrelated variables. The two authors point out that distinguishing the dimensions of social networks is important for conducting research and interventions on specific situations and different supportive effects. The dimensions identified by Marsella and Snyder are as follows:

- *Structure*: this dimension includes morphological variables such as amplitude, density, frequency of interaction, and the individual's position in the network;

- *Interaction*: this dimension is composed of variables that describe the relationship between various actors in the network, reciprocity, symmetry, directionality, and multiplicity;

- *Quality*: variables describing the affective quality of ties are included in this dimension. That is, networks can be represented in terms of friendship, intimacy, affective closeness;

- *Function*: this dimension describes the specific function performed by network members. Indeed, networks can provide information and feedback, emotional support, material help, problem-solving advice, etc.

A fruitful construct that originated in the Harvard School is that of "weak bond strength" (Granovetter, 1974). Weak ties are occasional, short-lived ties of minimal emotional contact with unfamiliar people, but they can foster contact with other social networks and access to a whole range of resources (information, social situations, etc.) that would otherwise be hard to reach. Granovetter (1973) observed, moreover, that strong ties tend to concentrate interactions within the groups a person belongs to, while weaker ties may facilitate the integration of members of different groups. Based on this observation, he hypothesized that community organizing was easier in neighborhoods where ties between family and friends are less close and therefore there is less distrust of outsiders.

1.3.   Functionality and value of networks Through social networks, information, values, and modes of behavior are spread, and role socialization is accomplished. They are recognized as having preventive and rehabilitative potential. Since the 1970s, numerous research works have been concerned with studying the connections between characteristics of networks, the degree of adaptation, and the well-being of individuals. The conclusion progressively reached by these studies is that the value and functionality of networks are related to the life stage and/or type of event to be faced (Stokes, 1983). For example, in the event of a parent's death, a young child may receive the best support within a network of high density and intimacy. Rather, in situations of mobility, job change, or other types of transitions, an extended low-density network marked by weak ties will be desirable. A study by Hirsch (1980) verified, for example, that in a sample of women who were going through delicate transitions, the presence of higher-density networks was associated with lower self-esteem, more obvious physical symptomatology, and worse adjustment. A small, dense network can trap the individual within a limited set of expectations, norms, and social contacts, hindering the transition to new social roles (Walker, MacBride, Vachon, 1977).

Another line of research initiated during those years investigated changes in the size and composition of networks concerning age and life events (Wrzus *et al.*, 2013). Concerning age groups, cross-sectional and longitudinal studies have shown that the social network increases until young adult age and then steadily decreases; in contrast, the family network remains stable in size from adolescence to old age. Notably, more peripheral and marginal relationships continuously decrease in number during adulthood and are also more quickly damaged by non-normative events (Lang, 2004).

Transition to parenthood, widowhood, and entry into the workforce are among the most investigated life events. Entry into parenthood tends to reduce social networks (Bost *et al.*, 2002), due to both decreased free time and changes in daily routines and increased focus on the child and partner (Bernardi, 2003; Rözer, Poortman, Mollenhorst, 2017). A similar phenomenon occurs following the loss of a spouse; in fact, the temporary social withdrawal that frequently accompanies the mournful event causes a major depletion of one's social network (Antonucci *et al.*, 2001; Zettel, Rook, 2004). In contrast, as one enters the workforce, the overall social network tends to increase due to the inclusion of new colleagues (Morrison, 2002; Back, Schmukle, Egloff, 2008).

Finally, a strand of investigation, also initiated during this period, has examined the social networks of psychotic or neurotic patients. Several researchers (Tolsdorf, 1976; Cohen, Sokolovsky, 1978; Sokolovsky *et al.*, 1978; Vaughn, Leff, 1981) have agreed in finding that when compared with "normal" control groups, the social networks of schizophrenics are characterized by less complexity and quantity of intimate relationships and especially by more asymmetrical and dependent relationships. It was found that schizophrenics are less able to maintain the reciprocity necessary for the formation of friendships: the primary deficiency appears not so much in receiving as in offering support. In a study of the networks of neurotic patients (Silberfeld, 1978) it was found that, compared with the control group, people with neurosis had less numerous and articulated social networks, with a lower proportion of work-related interpersonal relationships and less time devoted to interpersonal relationships.

Social networks, due to their characteristics and the resources they contain, are increasingly being conceived of as an effective and usable tool for intervention on the individual and in contexts, as we will see more fully in chapter 10 on the methodology of networking.

1.4.   Social networks and health From a combination of the rich literature on social integration, attachment, and social networks, a belief has gradually emerged that the nature of human relationships – the degree to which an individual is interconnected and embedded in a community – is vital to individual health and well-being, as well as to the health and vitality of entire populations (Berkman, Glass, 2000, p. 137). A celebrated and pioneering study in this regard was the longitudinal epidemiological research conducted in California's Alameda County. This study explored the relationship between sociodemographic conditions, lifestyle, well-being status, and health. *The Alameda County Study* aimed to identify risk factors for poor health and mortality within a community. Since 1965, it has examined the behavior of 6,928 people over twenty years (Berkman, Syme, 1979). This extensive body of data allowed for the identification of seven health habits, now known as the "Alameda 7," which are strongly associated with physical health status and long-term mortality.[7] From 1979 to 2004, continuous analyses performed on the data collected in the Alameda study have continued to confirm the hypothesis that lifestyles are important factors in long-term health (see, for example, Wiley, Camacho, 1980; Kaplan *et al.*, 1987; Schoenborn, 1986).

Although initial studies focused primarily on health behaviors, later studies have also begun to assess the impact of social relationships on health conditions. Social relationships, especially intimate ones, are important determinants of health, particularly associated with the development of chronic diseases (Housman, Dorman, 2005). For example, marital relationships are an important protective factor, especially for the male gender (e.g., Kotler, Wingard, 1989). Low-quality social ties, in contrast, are more frequently associated with

---

7. Never have smoked; drink less than five drinks at one time; sleep seven-eight hours a night; exercise; maintain weight appropriate for height; avoid snacks; eat breakfast regularly.

cancer prognosis (Reynolds, Kaplan, 1990) and increased risk of death (Yen, Kaplan, 1999).

Studies over these decades have reinforced the idea that existing social networks in the community contain potential supportive resources. Extensive studies, epidemiological and otherwise, conducted in the intervening years have provided clear and convincing evidence that the presence and quality of close relationships are among the most reliable and robust predictors of illness and lifespan (Holt-Lunstad, Smith, Layton, 2010; Sbarra, Law, Portley, 2011; Robles *et al*, 2014; Shor, Roelfs, 2015; Holt-Lunstad, Robles, Sbarra, 2017; Vila, 2021), comparable to biomedical and behavioral risk factors (Holt-Lunstad *et al.*, 2015; Schetter, 2017).

These scientific findings have, over time, also been taken up by major health organizations. The OMS, for example, has listed social support networks among the most important determinants of health (Commission on Social Determinants of Health, 2008). Nevertheless, some authors believe that the promotion of social connectedness is not yet sufficiently recognized as a priority goal by public health systems (Holt-Lunstad, Robles, Sbarra, 2017).

2.   Social support

The resources inherent in social networks have been formalized in the construct of social support. This is a complex construct that refers to the various forms of help that a person can exchange within the network of relationships in which he or she participates (Orford, 1992); it leads individuals to perceive that they are cared for or receive assistance and comfort from others when they need it (MacGeorge, Feng, Burleson, 2011).

Two types of support systems are recognized in the literature: the informal network system and the formal network system (Maguire, 1994; Thoits, 2011). The informal system refers to the relationships present in primary groups in spontaneous aggregations of various kinds, that is, intimate and close ties with family members, friends, people with whom one has a good degree of confidence or with whom one shares affection, interests, and goals. The formal system is, on the other hand, composed of institutional structures and professionals working in contexts of care, rehabilitation, education, and psychosocial prevention. It is from the action of these two types of systems, which are often interdependent, although not always integrated, that social support arises that can promote healthy individual development and strengthen stress-coping skills.

2.1.   Types of social support But what are the dimensions of interactive behavior likely to be perceived as social support and to have a positive influence? In response to these questions, there are several interesting conceptualizations of the dimensions that qualify social support. We aim here to offer a summary of the categories or dimensions identified by leading authors (Caplan, 1979; Gottlieb, 1981; House, 1981; Moos, Mitchell, 1982; Barrera, Ainlay, 1983; Cohen, Wills, 1985).

The most frequently mentioned theoretical and operational dimensions of social support are emotional, informational, and instrumental (House, Kahn, 1985). Following a careful review of the literature, Langford and coworkers (1997) also add esteem support to these three dimensions.

Emotional support (also called affective or expressive) is considered by many authors to be a key element in defining the construct (Tolsdorf, 1976; Leavy, 1983). It includes listening behaviors that express interest and understanding. Through this form of expressive and confidential support, the person receiving help feels considered and accepted despite their difficulties. Indeed, their self-esteem is strengthened precisely because he or she feels attention and support for their own problematic experiences or experiences

Informational support helps in defining, understanding, and dealing with problematic events. It can also be understood as cognitive guidance, offering guidelines and advice, and support in evaluating the event. Feedback related to interpersonal perception is particularly important.

Instrumental support refers to the provision of services and/or concrete actions, performance of tasks, and financial help. This mode can reduce stress, either by directly solving the problem through the additional material resources provided or by lessening the physical and psychological burden on those who are managing the difficult situation.

The last dimension that can be identified is estimative or evaluative support; it involves the communication of information relevant to self-evaluation, rather than problem-solving (House, 1981). It refers to expressions that affirm the appropriateness of acts or statements made by another (Langford *et al.*, 1997). Hirsch (1980) found that knowing how one is viewed by others is a supportive factor that promotes cognitive restructuring processes.

2.2. Effects of social support Over the past three decades, research has ascertained the impact of social support in a truly vast number of life situations and domains: for example, in relieving stress and depression (Cohen, Wills, 1985; Wang *et al.*, 2014), in improving psychological adjustment and feelings of control (Hoekstra-Weonebers *et al.*, 2001; Barrera, Fleming, Khan, 2004), in facilitating *adherence* to medical prescriptions (DiMatteo, 2004) and disease recovery (e.g., Epplein *et al.*, 2011; Høyer *et al.*, 2011).

Highly focused on by research are adaptation to chronic diseases, social integration processes, and processes of change and/or transition.

1. In chronic disease adaptation processes, research has consistently

demonstrated numerous benefits that patients receive, especially from their informal networks. In people with diabetes, for example, social support improves emotion regulation, coping, glycemic control, and quality of life (Van Dam *et al.*, 2005; Strom, Egede, 2012; HillBriggs *et al.*, 2021). In heart failure, support networks appear to influence behaviors related to disease management and the patient's functional status (Bucholz *et al.*, 2014; Graven and Grant, 2014). Finally, in cancer patients, social support plays a crucial role in addressing psychological problems resulting from poor adjustment to cancer and its treatment (Rizalar *et al.*, 2014). By fostering the patient's acceptance, positive *reframing*, and sense of humor, social support networks help the patient to be more determined in the fight against the disease and counter feelings of helplessness and hopelessness (e.g., Kawa, 2017; Lauriola, Tomai, 2019; Tomai, Lauriola, Caputo, 2019). Finally, significant positive correlations have been reported between social support and organic health *outcomes* (cancer onset, cell proliferation, lifespan; e.g., Ikeda *et al.*, 2013; Yağmur, Duman, 2016).

2. Within social integration processes, social support emerges as a key factor in the community integration of individuals or groups with frailty. For example, social support plays an important role in the community integration of individuals with severe mental illness (Terry, Townley, 2019), as it provides them with opportunities to experience mutual relationships and reduce isolation. Similarly, with refugee populations, the presence of a viable support network can provide aid in terms of employment and housing and make long-term settlement possible (Hanley *et al.*, 2018). Moreover, in these populations, increased perceived social support significantly reduces feelings of hopelessness, loss of motivation, and lack of hope for the future (Yildirim *et al.*, 2020).

3. In processes of change and transition, social support facilitates adaptation to the new condition by improving coping strategies, reducing stress, and promoting the development of conditions of well-being. For example, social support from family members and friends can predict a successful transition from primary to secondary school (Martínez *et al.*, 2011) or improve the autonomy and personal growth of college students in the transition between their first and second years of study (Malkoç, Mutlu, 2019; Cobo-Rendón *et al.*, 2020). Also, on the topic of transition, the construct of social support has been much studied in relation to a relevant event such as the birth of a child. In this period of life, the presence of social support seems to facilitate adaptation to the new condition at many moments of the parenting experience. For example: it reduces fear of childbirth (e.g., Calpbinici, Terzioglu, Koc, 2021), protects against the risk of premature birth (e.g., Hetherington *et al.*, 2015), predicts birth weight and fetal growth (Feldman *et al.*, 2000), and prevents *postpartum* depression (Tani, Castagna, 2017; Leahy-Warren *et al.*, 2020).

2.3. How to explain the effectiveness of social support The mechanisms that explain the effectiveness of social support generally refer to two study perspectives. The first perspective hypothesizes a linear and direct connection, i.e., a "primary effect" (Main Effect Model; Cohen, Wills, 1985) between social support and well-being. A continuity of supportive action promotes, as various research seems to confirm, personal development, the acquisition of appropriate coping modalities, and the maintenance of mental and physical health. Conversely, insufficient support may have a pathogenic action and result in increased vulnerability. According to the primary effect model, social support performs a function independent of stress levels; its health-promoting action occurs both in the presence and absence of stressful events (Orford, 1992). This line of research examines the effects of support throughout the individual life span, aiming to capture how people find embeddedness in their social environment. The research of Berkman and Syme (1979) can be considered among investigations of this kind. Or, again, one can cite an investigation by Marmot and Syme (1976) in which it is hypothesized that the risk of coronary heart disease may increase during emigration and uprooting from one's original culture and social ties. These data are also confirmed by more recent studies reporting the significant psychosocial costs of emigration, and damage to organic and mental health (hypertension, diabetes, cancer, and depression; see for example Lu, 2012; Rosenthal, 2018). Surveys conducted according to the "primary effect" perspective thus have the merit of highlighting how being relatively isolated or, conversely, integrated into one's social network greatly influences health status; however, in showing connections between very broad social phenomena, they have the limitation of offering little insight into the interrelationship processes between different variables.

More analytical and complex appears to be the second line of studies on social support. This approach moves from the *buffering* hypothesis, that is, the hypothesis that social support acts as a buffer or protective cushion against stress and moderates its consequences. It is thus examined how social resources support individuals facing critical moments and stressful events in their existence. The *buffering* effect of social support on stress has been confirmed across various life events. For instance, the studies on social support during pregnancy mentioned earlier align with the *buffering hypothesis*. When it comes to spousal loss, individuals who perceive high levels of social support tend to experience lower levels of anxiety and depression (Somhlaba, Wait, 2008). Other life events that have garnered significant interest include securing employment and managing the stress related to job loss. Social support in these contexts serves as a protective factor, bolstering self-esteem and self-identity and promoting more effective behaviors in finding new employment (e.g., Gore, 1978; Mazerolle, Singh, 2002).

2.4. When social support intervenes As can be seen, research associates the action of social support in various ways and with different types of stressful events. According to Cohen and Wills' (1985) analysis, it can exert its *health-protective* role at different points or moments in the sequence from the potentially stressful event to the stress reaction. First, support can intervene between the stressful event (or the expectation of such an event) and the stress reaction by attenuating or preventing the stress appraisal process. In other words, the perception that others can and will provide needed resources can redefine the potential harm attributed to a situation and/or support the perceived ability to cope with environmental demands and thereby prevent a particular situation from being rated as highly stressful. Second, appropriate support can intervene between the experience of stress and the onset of pathological effects by reducing or eliminating the stress reaction or directly influencing physiological processes. Support can alleviate the subjective impact of stress by offering a solution to the problem, reducing its perceived importance, acting as a regulator of the neuroendocrine system so that people are less reactive to perceived stress, or facilitating healthy behaviors.

Figure 1 The effects of social support as a moderator of stress, at different stages of the sequence Linking potentially stressful stimuli to the stress reaction

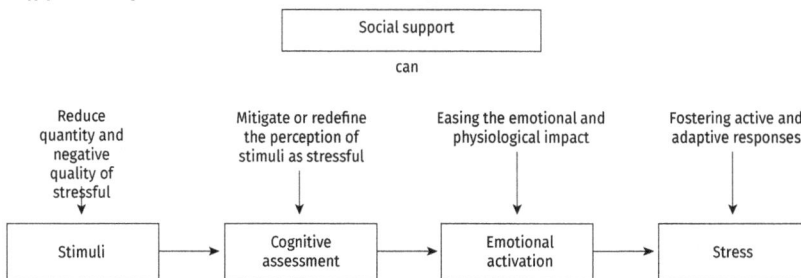

According to the theory of social support as a "stress moderator" (which we have represented in graphic form in Fig. 1), social support is thus able to:

- Reduce the quantity and negative quality of stressful stimuli;

- Mitigate or redefine the perception of stimuli as stressful;

- Alleviate the emotional and psychological impact of such stimuli;

- Promote active and adaptive responses.

2.5. The concept of "perceived" social support In 1983, Heller and Swindler published a lashing and, in our view, reasoned critique of the survey approach.

According to the two authors, the validity of much social support research is challenged by methodological shortcomings and conceptual ambiguities. Measurement scales are often not calibrated, the variables of support and reactions to stress are not adequately distinguished, or support is assessed through an overly broad range of personal characteristics. The authors argue, however, that the major limitation is not so much methodological as conceptual. They note that in many surveys, the concept of social support comes to be confusingly identified with that of social networks. There is thus a risk of using the two terms indiscriminately or examining relationships between vague or overly general variables. For these reasons, the two authors propose to make the distinction between network and support more evident by attaching importance to the assessment process; they, therefore, suggest using the concepts of perceived social support (*perceived* support) and support *seeking (support seeking)*. According to this terminology, "social networks" describe the social connections available in the environment that vary in structure and function; "perceived support" refers to the perceived valuation of being supported; where "support seeking" takes over in response to a threat and because of the need to receive help or information. The richness and quality of ties present in the social network can, therefore, be considered the "objective" condition that enters a dialectical relationship with "subjective" variables such as perceived support and support seeking.

The relationship between network and social support is thus to be understood in an interactive and bidirectional sense: not only because the supportive effects of networks are cognitively mediated by individual perception, but because the quality of the social network itself is not an a priori condition but also the result of personal ability and motivation to establish and maintain meaningful ties. Overall, it seems to us that the alternative between a simplistically environmentalist approach and an attempt to develop more complex models that account for the multiple interactions (or transactions, as Murrell would say) between different social levels and between objective and subjective variables also emerges in this field of community psychology.

Figure 2 A three-dimensional model of the interdependent relationships between social networks, social support, and individual capabilities

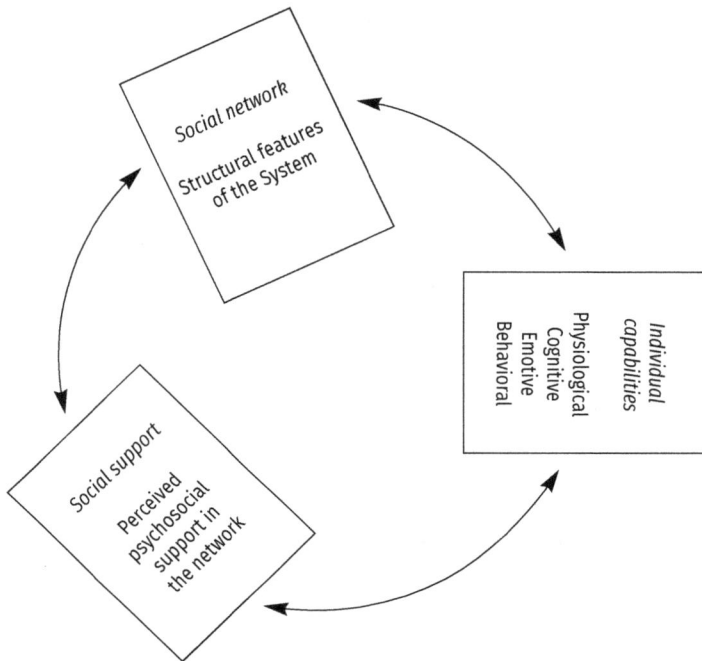

We thus find Heller and Swindler's call to keep social networks distinct from perceived support very clear and useful, while nevertheless studying their dynamic relationships. In Figure 2, we have attempted to represent a three-dimensional model that interdependently connects social network (i.e., the structural features of the system), social support (understood as perceived psychosocial support in the network), and individual capabilities (understood as ways of reacting to events and seeking social support).

Among the individual variables that influence the search for social support, attachment style has attracted much interest. Attachment theory has been used to understand the development of individual differences in the seeking, provision, and perception of available social support. Attachment theory suggests that early relational experiences give rise to an ingrained pattern of expectations, needs, emotions, and behaviors in interpersonal situations. Solid and effective interactions with attachment figures reinforce the use of social support as a distress regulation strategy and encourage optimistic beliefs about the ability to receive support from significant others. Conversely, disappointing or frustrating interactions with attachment figures erode a person's confidence in seeking sources of help in their context. Numerous studies have found that people with a more secure attachment style have a more favorable expectation of the possibility of receiving help, communicate more openly about their distress, react more empathetically and compassionately to the suffering of others, and are more likely to provide effective help to partners in need (e.g., Mikulincer, Shaver, 2009; Collins, Ford, Feeney, 2010).

2.6.   The intersystem link between formal and informal support systems We can consider the formal and informal networks as two systems of help present and grafted into the social fabric. Theoretical formulations and survey results on social support networks and systems over the years have made clear the desirability of making optimal and coordinated use of resources from formal and informal systems. Working toward the integration of the two types of systems has useful spin-offs for both professional services and citizens. Indeed, informal systems increase social participation in the management of health and quality of life while formal systems are monitored and stimulated to improve their organizational effectiveness. Likewise, the social costs of professional interventions themselves can decrease using spontaneous resources (Croce, 1995; Ridolfi, 2013a). Coordinating professional interventions with the natural resources that each community possesses (Ridolfi, 2013b) is referred to as *community care*, and has been proposed as a possible solution to cope with the deep crisis that welfare systems have gone through in recent decades.

2.7.   Hints on measures of social support The instruments used to measure social support are numerous and very different from each other due mainly to the different ways used to operationalize the construct. Usually, this process refers to one of the two components of social support identified by Cohen (1988): structure and function. Structure pertains to quantitative aspects, i.e., the presence and number of social relationships (e.g., number of friends, presence of a spouse, group membership), the frequency of interactions, and the amount of interconnectedness (e.g., the degree to which a person's friends know one another) (Sherbourne, Stewart, 1991). Using this type of approach, the focus is on describing the structure of the social support network in which the individual is embedded to have a tangible measure of how likely he or she is to receive support when in need and to assess the person's participation and integration into their network. Useful tools for this purpose are, for example, the *Arizona Social Support Interview Schedule* (Barrera, 1980) or the *Arizona Life History Battery* (Barrera, Sandler, Ramsay, 1981).

Measurement instruments that refer to a functional approach of social support aim to capture the type of resources present in the social network.

Among the most widely investigated theoretical and operational dimensions of social support are emotional, informational, and instrumental support (e.g., *Social Support Questionnaire*, Sarason, *et al.*, 1983; *Social Provision Scale*, Cutrona, Russell, 1987).

Among the most widely used instruments is the *Multidimensional Scale of Perceived Social Support* (MSPSS, Zimet *et al.*, 1988). This scale, besides being short (12 items), provides both a total score of perceived social support and a score related to three different sources of social support (family, friends, and significant others). Together with the *Interpersonal Support Evaluation List* (ISEL, Cohen, Hoberman, 1983), the MSPSS is one of the few instruments validated on the Italian population (Prezza, Pacilli, 2002). The ISEL is a self-report questionnaire consisting of 40 items divided into four subscales, also widely used internationally and validated on the Italian population by Moretti and collaborators (2012).

Undoubtedly, social support from informal networks has been the most studied type of support. However, especially in the past two decades, research has shifted its focus toward studying the potential benefits of social support provided by secondary groups and/or formal networks (Brucker, McKenry, 2004; Reblin, Uchino, 2008) and on the construction of related measurement instruments. Particular attention has been paid to the effect of social support provided by professionals in the healthcare setting. Indeed, healthcare professionals, during their profession, have opportunities to provide numerous types of social support to their patients, thus coming to constitute a formal social support network for the sick individual. It has been seen, for example, that the social support of the physician can reduce stress related to disease management and improve glycemic control in diabetic patients (Venkatesh, Weatherspoon, 2013; Wardian, Sun, 2014), as well as contain the fear of disease progression in cancer patients (Ban *et al.*, 2021); the informational and emotional support of the nurse can help the patient reduce pain during screening mammography (Fernández-Feito *et al.*, 2015).

Measuring the support provided by healthcare workers is a challenging area of research that can have considerable use in areas as diverse as, for example, healthcare quality assessment (Malley, Fernández, 2010). The support of healthcare providers has been operationalized and measured in a variety of ways (Remblin, Uchino, 2008), resulting in a great variability of instruments, making it difficult to make comparisons between data. For example, despite the multidimensionality of this construct, many studies either examined formal social support as a global construct (e.g., Hipkins *et al.*, 2004) or considered multiple aspects of support but measured it as a single, aggregate, global score (e.g., Reynolds, Perrin, 2004). Other authors have focused on only one aspect of the construct, such as emotional support (Kuuppelomäki, 2003; Ansmann *et al.*, 2012) or informational support (Rutten *et al.*, 2005). Finally, measurements of formal social support can also be traced in the literature using qualitative data (open-ended questions, interviews, focus groups; Venkatesh, Weatherspoon, 2013).

A more recent study tested the factor structure and psychometric properties of a new measure of social support for healthcare providers that reflects the quadruple model of the construct (Tomai, Lauriola, 2022). The *Healthcare Provider Social Support* (HPSS) assesses four dimensions of social support: emotional, informational, instrumental, and esteem. These domains were inspired by the studies of Cutrona and Russell (1987). The authors excluded the dimensions of *Opportunity for Nurturance* (the sense of being necessary to others for their well-being) and *Social Integration* (the importance of belonging to a group that shares similar interests) because they were considered forms of social support primarily provided within informal networks (*ibid.*).

7
Participatory multidimensional organizational analysis (PMOA)

1. Theoretical premise

Organizations are complex phenomena that cannot be ascribed to a single interpretive paradigm (Avallone, 1999; Avallone, Borgogni, 2007). Over the years, numerous theories have developed that have tried to read organizations from different points of view.

Functional and pragmatic aspects have been emphasized to describe the logical-rational design of the organizational system (Butera, 2008). "Organic" reading models have been used to look at organizations as systems that adapt, evolve, and disappear (Borgogni, 1994; Weick, 1997). Other models, finally, have valued the irrational and unconscious aspects by considering each organization as a system of symbolic translation of the collusive dynamics that characterize the context (Carli, Paniccia, 2003; Carli, Paniccia, Giovagnoli, 2010; Bisogni, Dolcetti, Pirrotta, 2021).

No model, approach, theory, or metaphor used in isolation can be considered exhaustive (cf. Bruscaglioni, 1982); for this reason, we need multidimensional approaches to the analysis of organizations, capable of directing our observations to multiple aspects and sides of the organizational phenomenon. This "kaleidoscopic" way of approaching the study of organizations has been developing since the early 1980s simultaneously and in parallel in the United States and Europe by several authors (Bruscaglioni, 1982; Francescato, 1982; Morgan, 1986; Jackson, 1987; Flood, Jackson, 1991).

From the United States comes the idea (Burrell, Morgan, 1979) that theories of the social world should be conceived as four key paradigms depending on different basic assumptions about the nature of social science and society.

Indeed, the social sciences assert that social reality is either "objective or subjective"; similarly, the nature of society is read by some as "regular change" and by others as "radical change." Combining the "objective-subjective" and "regular-radical change" dimensions results in a matrix that defines four key sociological problems: functionalist, interpretive, radical structuralist, and radical humanist. This Burrell and Morgan network allows systems methodologies to be related to different sociological paradigms and to assess which assumptions underlying a particular systems methodology are appropriate in some problem situations and not in others.

Morgan (1986; 1994; 1996) also proposes reading organizations using metaphors, that is, through multiple keys that, by creating even irrational connections, allow for a broader and more complex view of the reality we are examining. Some of the "images of an organization" used by Morgan are machine, organism, brain, culture, political system, psychic prison, flow and transformation, an instrument of domination, etc. Often, a theoretical approach is connected to only one of these metaphors to a single interpretive model; choosing to view the organization according to only one of these metaphors risks losing the interpretive richness that the use of multiple metaphors, on the other hand, would ensure. It is also because of this reflection that in the 1990s, building on the philosophy of "critical systems thinking" and using the concept of metaphors, Flood and Jackson (1991) developed the methodology of TSI (total systems intervention).

This method refers to several systems metaphors to foster creative thinking about the organization and its problems; the metaphors are linked by a framework--the System of Systems Methodologies--to the various theoretical approaches on systems, so that once it has been determined which metaphor is most relevant to the organization's issues and problems, the appropriate system-based intervention methodology or group of methodologies can be chosen and then move on to implementation.

The IST, in line with a complementary perspective, thus combines work on metaphors with the System of Systems Methodologies and with knowledge of individual systems approaches.

Flood and Jackson (1988) and Jackson and Keys (1984) sought to integrate the Anglo-Saxon managerial approach with the European structural school by promoting, precisely, a complementary perspective. Jackson (1987) argues, in fact, for the complementary role that various systems methodologies can play in the overall task of decision-making and problem management in management, and that diversity should, therefore, be recognized as a sign of strength and not weakness in the systems movement. Even in Italy, in the 1980s, Bruscaglioni (1982) argued how different organizational theories are not competing but that each paradigm sees only a portion of the "truth" of the problem or system. Bruscaglioni hypothesized, then, that we were in a phase of isolation of approaches in which each author made his theory a general theory of organizational reality and pointed out the need to go through a phase of mechanical complementarity of approaches, in which at least partially complementary addresses borrowed from other disciplines as well as one's own are used, for an integration of the different approaches following two paths:

- Develop complementary theories and perspectives focusing on organizational phenomena whose complexity requires the use of concepts derived from different theoretical approaches (e.g., role analysis);

- Study phenomena traditionally addressed by one approach using variables traditionally peculiar to another (e.g., relationships between psychosocial variables, such as work motivation, and functional variables, such as organizing a task and distributing functional responsibilities). Francescato (1982), at the same time as Bruscaglioni, began to develop the hypothesis of a multidimensional approach to organization that would consider both structural and functional, psycho-environmental and psychodynamic aspects of a system. The basic assumptions are that:

- it is possible to identify variables common to all organizational realities;

- different organizational theories are centered on different organizational phenomena;

- each theory is accompanied by a set of tools and techniques that enable a certain "reading" of organizational functioning;

- none of these "readings" is more "true" than another but they offer different views of the same reality.

Francescato believes that using more than one reading increases people's ability to identify their organization's problem areas and strengths and thus make a multiple diagnosis of organizational functioning. She also argues that this multidimensional view increases the awareness of people involved in the organizational analysis of the interconnections that need to be considered if changes are to be promoted that address both the productive needs of the organization and those of the individuals in it (organizational empowerment).

2. The model of the PMOA

Organizational reality can be represented as a kaleidoscope that shows different images depending on how it is moved: none of the very many possible combinations can be considered "truer" than the other. Similarly, even in the analysis of the same organization, it is possible to have dozens of different readings, depending on the criteria of analysis and the orientations we choose.

In work practice, however, it is hardly appropriate to make such a complex and comprehensive reading, nor is it necessary for the objectives one is pursuing to try to know everything possible. Therefore, it is useful to follow an outline that can guide one in identifying salient, non-sectoral aspects.

What should be the characteristics of such a scheme? The most pragmatically effective organizational analysis should, in our opinion, detect strengths and problem areas, the variables to which these problems are related, and the factors on which it is easiest to affect to achieve the desired changes.

Donata Francescato has developed a PMOA scheme (Francescato, Ghirelli, 1988; Francescato, Aber, 2015) in which participants are invited to consider four different keys present in every organization: the strategic-structural, functional, psychodynamic, and psycho-environmental dimensions. The strategic and functional dimensions are referred to as **hard variables**, as they include the most objective and measurable aspects of the organization; they are grounded in a realistic ontology and consider the organization as an ordered, unified, and cohesive whole in which consensus on the goals to be achieved and rationality in the behavior of individuals prevail. The epistemology of reference is positivist, deterministic, and nomothetic. The psychodynamic and psycho-environmental dimensions are considered **soft variables** because they are related to affective and relational aspects, both conscious and unconscious, often conflicting or divergent, created by the individuals who are part of the organization and are grounded in a constructionist epistemology and ontology. To take both hard and soft aspects into account is to explore organizational reality in its conscious and unconscious aspects and to enable the promotion of change. Not to imply hard dimensions in the reading of the organization and the consequent intervention means to implement actions that are not anchored, in the specific context, to the target market, to the available resources. Conversely, not considering soft dimensions leads to triggering change processes without adequate reflection accompanying the development, achieving only the increase of system conflict or "orthopedic changes" (Carli, Paniccia, 2003).

The PMOA is an action–research that requires the presence and active involvement of all levels of the hierarchy: if it is a small organization (less than twenty people), all members participate, or otherwise, one works with a group in which the different components are represented; for example, in a school, students, teachers, non-teaching staff, managers, parents, or managers, executives, and employees will be involved, in a company blue-collar workers or president, members, operators, a cooperative. In this, the PMOA differs from most current intervention models, in which a consultant tends primarily to follow directions and pursue the goals of the organization's leaders. In contrast, the PMOA aims to increase people's ability to know and interpret the organizational phenomena of the different work realities in which they participate to facilitate them in governing critical issues and desired changes. The client of the community psychologist is not the direct formal client but all the people who belong to the organizational unit under consideration (Morganti, 1998; Francescato, Tomai, Solimeno, 2008).

The community psychologist plays in the PMOA pathway the role of facilitator and connoisseur of this integrated organizational method to help different professionals diagnose organizational functioning. The participatory mode stems from an emancipatory-liberatory interest in all individuals who make up the organization, even those in the humblest roles, providing everyone with the opportunity to systematically confront different views of the organization and giving each view equal epistemological dignity. In this way, individuals belonging to the organization are seen as actors in an intentionally intended change (Borgogni, 1994). The **use of multiple readings enhances** people's abilities to formulate multiple diagnoses that enable the promotion of changes capable of responding to both the productive needs of the organization and those of the individuals in it. This makes an organization empowered and empowering (Francescato, Tomai, Solimeno, 2008).

The PMOA is a tool that promotes a pluralistic view of the organization because it integrates subjective and objective variables, theoretical knowledge, and professional knowledge from different disciplines; this allows all members of the organization to have a voice, even those who, by hierarchical level or functional role habitually have less power to break the tacit consensus on norms and behaviors (such as sexual harassment, prevarication of the fragile) undesirable to members with less power.

Because of these characteristics, we have ascertained that the methodology is most effective in less top-down organizations, in which the grassroots elect its top management, such as in cooperatives, voluntary associations, trade unions, in small family businesses, schools, and more rarely in large companies where a human resources director, or top manager, wants to do something innovative.

2.1.   The strategic–structural dimension It is the dimension that examines an organization's goals and the resources needed to achieve them. It begins by exploring the strategic history. The original strategic objectives are clarified, and whether and how they have changed over time; budgets, balance sheets, corporate forms, and physical and legal structures are then analyzed. The sociological-structural approach has traditionally focused on the phenomena of power and wealth distribution within organizations and the quantity and quality of human performance. For the aspects we consider, the analysis is enriched by examining the opportunities and constraints that come from the territory in which the organization operates. It may be useful to carry out not only an analysis of what lies within the company but also a study of the complex and articulated boundary zones and modes of interfacing.

The organization is then examined for factors that concern the distribution of power and wealth: contracts, shareholdings, budgets, laws, deeds, and documents that testify to the presence and use of material and human resources and illustrate the purposes and constraints of the organization. Together, this information gives us an initial structural reading of the organization studied. Key aspects, such as the type of internal social stratification and the level of success achieved by the structure, are photographed (or seen in their historical development). Already, with this initial approach, it is possible to identify some areas characterized by problematic aspects and some strengths on which to set organizational development can emerge.

Information on the structural variables we are interested in can be found through:

- Sources of a legislative nature: knowledge of regional and sectoral regulations allows us to detect the social mandate assigned to public bodies, the legal and economic constraints within which different organizations operate, and the opportunities provided (subsidies, tax breaks, protected markets). More generally, the analysis of legislative sources provides valuable and often overlooked elements for assessing the degree of power formally exercised by the political system over the organization;

- Statistical, economic, and social sources (like chambers of commerce, etc.) that provide information regarding the characteristics of the environment, development stages, and prospects of the sector we intend to analyze;

- Notarial and associational acts: the bylaws and the amendments that are made to them. These sources outline for us the framework of formal characteristics and expected participation, statutory purposes, methods of access to offices, formal requirements due, geographical spread, and degree of influence of other organizations;

- Contracts and resolutions, from the examination of which we can detect the autonomy that exists in the management of labor relations, including in terms of any agreements or consultancies, scheduled deadlines, arrangements for renewal or termination, scheduling of incentives and penalties.

- Minutes of boards of directors and similar bodies, and analysis of budgets, budgets, and final accounts from which we acquire information on the organization's financial status, development prospects and resource utilization, and relationship with the market.

Respecting the participatory perspective, the retrieval of information useful for knowledge of this organizational dimension is entrusted to the members of the organization, while the critical reading of the collected documents is carried out through group interviews. The set of legal-economic-political variables comes to characterize the structural conditions of the organization. These conditions are not only significant in themselves but also contribute to the creation of the organization's culture and values and either help or hinder the satisfaction of the needs and expectations of the individuals within it. For example, working in a company that is a leader in its industry, or in a company with significant liabilities and forced to place groups of workers on layoff, has immediate implications not only on the perception of one's job security but also on one's motivation, degree of involvement, and self-image and future.

Structural conditions also affect the programmatic directions and decision-making mechanisms of organizations.

2.2. The functional dimension The functional dimension is related to operational management; it considers the articulation of roles, flows, and functions to be performed and answers the questions, "Who does what?" and "In what time frame and with what tools?"

The functional model we initially referred to was the one developed by Tancredi (1981), chosen for its characteristics of simplicity and completeness. According to Tancredi, in functional terms, every organization can be seen as an organism embedded in the environment and consisting of three systems interacting with each other and with the context:

1. management control system is responsible for planning operational activities, consistent with environmental constraints and resources and internal objectives, and controlling their effectiveness and efficiency. It examines the degree of development of the functions of *planning* objectives, *organizing* processes for implementing established plans (such as delineating organizational levels and units, assigning tasks and responsibilities, specifying rules and procedures), and *controlling* operational *efficiency*;

2. the operating system comprises a set of functions related to the process of producing and/or delivering services. This system thus includes the various aspects and different phases that enable the acquisition, transformation, and placement of resources specifically 'worked on' by the organization. The framework was initially conceived for product organizations, but Tancredi tested the adaptability of his framework by participating in a workshop facilitated by Francescato in a drug addiction treatment center (SERT), where the resources to be 'worked on' were individuals with substance use disorders.;

3. the information system, which includes information on operation and

results; typical management information in the information system concerns the effects, the *outputs of* activities, and makes it possible to collect data on how and how much resources have been used, but also how management data are acquired, entered and stored, to provide data that can be evaluated by the management control system. According to Tancredi, business problems can be reviewed systemically by identifying both what dysfunctions occur (and where) and what consequences decisions made to correct those dysfunctions may have in all subsystems. The scheme is general and usable for any type of company. Companies are, in fact, different in terms of organizations and products produced, but the functions, as constituent elements of logical operational development, remain the same, although with different emphasis for each in relation to the type. Some subsystems may have less importance or disappear altogether, but the pattern remains the same whether in the case of a manufacturing or service company or the case of a political party or a social welfare center. Subsequently, this model was integrated with some aspects of Quaglino's (1985; 1996) and Butera's (1990; 1995) analysis of the organization, attempting to examine the existing relationship between the activities carried out over a certain period and the objectives formulated by the organization, to check for possible existing congruencies or inconsistencies. With this type of analysis, it is possible to identify both which functions are not being performed well and which objectives are not being achieved or are only partially achieved.

2.3. The psychodynamic dimension The structural and functional approaches described so far assume that consensus on the goals to be achieved and rationality in the behavior of individuals prevail in the organization: the organization is seen as a mode of interaction between individuals with common goals related to the process of resource transformation. The psychodynamic approach offers us the possibility of understanding an organization on the level of irrational experiences, that is, it considers the organization as it is subjectively experienced on an unconscious level.

According to the psychodynamic approach (Cotinaud, 1975; Lapassade, 1975; Enriquez, 1980; Carli, Paniccia, 1981; Carli, 2020), certain forms of behavior are the manifestation of a logic other than that of rationality and consensus. The irrational has its roots in another reality, however closely interrelated with the known one: it is a primary logic that sometimes prevails over the secondary one.

The individual works for an organization whose goals he often did not choose, and he perceives his own experience as alienating, shattering, and contradictory to the need for unity and an overall coherent identity. Moreover, the organization offers the individual only an extremely partial role identity: everyone is recognized by others in an organization not for who he or she is as a person but as a role, as an anonymous unit in a structure. The frustration of the need for wholeness, identity, and coherence leads to feelings of aggression that are poured out on the organization, which, as an alienating object, stands as the recipient of all man's negative and destructive impulses: the organization is *bad*. But the organization also provides the opportunity to support oneself with a job and, by offering a role, also provides social recognition and *status* in society. Thus, the organization is also rewarding and stands as the recipient of man's positive impulses: it is *good*. As Muti (1986, p. 61) writes, "It is thus a contradictory kind of relationship that binds man to the organization, and in its profound ambivalence, it calls to mind man's original relationship with a single object, the mother."

This logic of the unconscious tends to be reactivated by specific current situations in the environment that evoke the primary mother-child dyadic relationship. For example, the evaluative processes used in many organizations activate numerous and complex unconscious processes related to the friend-enemy schema that characterizes the unconscious fantasy of social order (Carli, Paniccia, 2003; Biggio, 2008). In formulating judgments, not only merits and abilities are often considered, but also numerous less rational criteria (Francescato, Tomai, Solimeno, 2008).

Among the elements that are favored as fundamental to the analysis of an organization is the examination of the boss-employee relationship. For example, affective and power dynamics, conflicts among employees, and between bosses and employees are evaluated. Belonging to an organization and holding a position evidently poses the individual with the problem of whether to assume the responsibility that comes with their role and *status*. On the one hand, the individual tends to take responsibility, and on the other hand, to reject it since taking responsibility involves risks and anxieties. This results in an anxiogenic intrapsychic conflict, which can be enacted at the organizational level. The result of this process may be to transform the conflict between responsible and irresponsible parts of oneself into an opposition between responsible and irresponsible people in the organization. This mechanism operating between different roles is also often present in the relationship between the individual boss and the individual employee. In this case, the conflict occurs because the superior tends to project the more irresponsible parts of himself onto his subordinate while the other is inclined to project the more responsible parts of himself onto the boss.

According to Bruscaglioni (1982), the consequence may be unconscious collusion between boss and employees that results in the empowerment ofmanagers shifting responsability to lower levels and delegation upwardemployees who delegate problems they could solve to their boss. The boss and employees enact various defensive mechanisms in the face of this situation. Boss conflicts can be of three types:

- A conflict between phantasmatic-type tendencies related to the omnipotence-impotence problematic;
- An ambivalence between the desire for domination and the guilt associated with the desire for power;
- An unpleasant feeling of dependence on one's collaborators, as the execution of one's will is in the hands of those who execute.

The defensive behaviors assumed by the leader could be as follows:

- The relinquishment of one's leadership role;
- The delegation to their subordinates;
- The performance of tasks that do not involve the immediate exercise of authority;
- All those behaviors in which ambivalence is denied through a despotic use of authority;

Among employees, intrapsychic conflicts may arise:

- By peer rivalry and feelings of envy and jealousy for the attention of the leader experienced as a parent;
- By an alliance with peers to override the leader and the resulting guilt and fears about the succession struggle; and
- By the desire to receive protection from the leader and thus to be attractive in his eyes.

While the defensive strategies of the leader, according to Jaques (1966), are primarily aimed at eliminating the exercise of authority, the defensive strategies of employees are focused on shifting the sources of tension in the leader-collaborator and colleague-colleague relationships towards external objects. This results in the phenomenon of scapegoating, confusion of responsibilities, lack of clarity in roles, and a victimized adherence to the authority relationship.

In contrast to the investigation of organizational culture (Avallone, Farnese, 2005), the instruments are exclusively qualitative, subjective in nature, and essentially include narrative and projective techniques. Among the **narrative techniques** are the collection of mottos and jokes about the organization and the individual work novel (each participant recounts the main events of their work history). The latter allows us to understand the values and meanings of lived experiences that guide the actions of people and the subgroups to which they belong. Francescato proposed the movie script, which is also used in community profiles and empowering training (Francescato, Tomai, Mebane, 2004; Solimeno *et al.*, 2008; see chapters 5, 7, 11).

Among **projective techniques**, we particularly use drawings and free associations to facilitate participants' expression of meanings and emotions. The drawing technique consists of several steps. First, participants are asked to draw a picture with their own organization in mind. The directions are generic so that people feel free to use objects, subjects, metaphors they prefer. The community psychologist provides the group with pencils, pens, colors, etc. The drawings are processed individually, usually within 15 minutes, then displayed on a wall. Each drawing is commented on with words, emotions and images, which are affixed next to the drawing, by the subgroup members, making associative chains that let what is not conscious surface, as happens in group analysis (Di Maria, Lavanco, 1993; Neri, 2021). The members of each subgroup are then asked to produce associations, which will be noted on a poster board, on all the drawings made by the group. Finally, members of the two subgroups are asked to observe the pictures made by the other group and, if they wish, to write the associations on these drawings using different colors, to keep self and hetero-produced associations distinct. The drawings can be discussed using graphic, content, relational, metaphorical and chromatic indicators. For analyses of chromatic aspects, we refer to the indicators proposed by Biasi and Bonaiuto (1992; 1997) and Bonaiuto and collaborators (1996), who associate negative emotions with the colors purple, black, and gray and positive emotions with colors such as pink, white, yellow, light blue, and light green. We also identify stress and comfort designs. The former are usually made in black and white, or with colors that convey negative emotions, while the latter with colors that express positive emotions and harmoniously distributed. In addition, associations are classified as positive, negative and neutral activating comparisons and discussions on the emotional aspects of the shared organizational reality (Francescato, Tomai, Solimeno, 2008; Francescato, Putton, 2022).

2.4. The psycho-environmental dimension In this dimension we include both the traditional psychosocial approach to the study of organization and the new contributions made by community psychology and environmental   psychology.

Such approaches start precisely from human needs and analyze the networks of relationships inside and outside the organization, considering the relationships between individuals, groups, and the environmental context on the level of conscious perceptions. The variables used in the analysis of organizational phenomena are generally:

- *group phenomena*, understood as a set of people whose members are in a relationship of interdependence, that is, they influence each other. In particular, the group's influence on the individual and their behavior, cohesion, conformity and deviance, change and resistance to change, role distribution, leadership, and affectivity have been studied;

- *leadership* and management *styles* aimed at identifying the optimal behavior of one who performs the functions of a leader in both achieving goals and promoting social integration;

- *communication* (structure of exchanges, two-way or uni-communication, roles and attitudes in communications are some of the themes of this broad phenomenon among the most studied in this approach);

- *needs* and *motivations*, *attitudes*, etc;

- The *degree of psychosocial agreement* between environmental pressures and individual expectations.

The instruments used for this dimension are varied: initially Berne's psychological position self-diagnosis questionnaire was used for the study of interpersonal communication; for the study of roles and power various tests and scales; and for the study of organizational climate such instruments as Spaltro's (1977) organizational check-up questionnaire. More recently there has been use of questionnaires measuring achievement in work, the need for external recognition, and the need for affiliation. Especially in recent years we have used climate questionnaires such as M_DOQ (D'Amato Majer, 2005) and OCS (Borgogni *et al.*, 2005), which pay attention to group characteristics.

Other tools we use are semi structured interviews to detect psychosocial agreement, that is, the close coincidence between the demands of the organization and the expectations and resources of individuals and functional subgroups. In particular, the following are analyzed:

1. positively perceived congruencies between individual and organizational *desiderata*, which promote feelings of pleasantness, adequacy, and satisfaction in performing the tasks inherent in one's function;

2. existing discrepancies between individual and organizational *desiderata*, which create dissatisfaction, discontent, and a sense of inadequacy, conflicts, etc. An attempt is then made to find feasible solutions, to improve psychosocial agreement, through brainstorming in functional subgroups or interviews with experienced colleagues.

We have also developed exercises aimed at increasing collaboration and decreasing conflict by giving assignments to individuals who have pleasure in performing a certain function and adequate skills to perform it to the best of their ability. In fact, in many organizations, we have found among problem areas that often some people would enjoy performing a function but do not have sufficient skills, while other people have the skills but no longer take pleasure in performing their function or are tired of performing it for too long, and perhaps would like to try their hand at new tasks. We usually ask members of the organization (or functional subgroups) to indicate, for each function, who performs it, the degree of pleasure in performing it, and the degree to which they possess specific skills. Then we invite people who would enjoy performing a specific function, but do not have the appropriate skills, to ask the colleague who said he or she no longer enjoys performing that function if he or she is willing to pass on certain skills. Sometimes the solution found is to shadow the competent person for a period or undertake specific training courses. These ways can be used for task planning with respect to new activities or with respect to problem-areas identified in the PMOA's journey, paying attention to psychosocial agreement, to encourage, in the assignment of functions, the co-presence of "pleasure" and "competence."

3.   The path of the PMOA

To carry out a PMOA, a heterogeneous group of representatives from different professional groups and hierarchical levels must be created and met at least six times to diagnose the strengths and problem areas of their organization.

The PMOA's course of work begins with the establishment of a group composed of representatives from each subgroup in the organization. This group works coordinated by the professional psychologist following certain stages.

Group Constitution

- Step 1: Preliminary analysis. Preliminary analysis is, as also in community analysis, an activity that aims to draw an initial description (based on the knowledge and opinions of the group) of the context under consideration (in this case, an organization) using the brainstorming technique. This initial "diagnosis" allows for the emergence and comparison of strengths and problem areas perceived as priorities by the group members. The diagnosis made through the preliminary analysis will be compared with the data progressively collected during the analysis of the different organizational dimensions. This will allow us to constantly compare subjective data with objective data soft data with hard data, and to co-construct a new and shared representation of organizational reality. The brainstorming associations will be placed by the group into the four organizational dimensions. This initial analysis allows the group to grasp which aspects of the context it perceives as most valuable or most critical.

- Step 2: Analysis of the four dimensions. The group explores one dimension at a time using appropriate methods and tools, reflects on the data, and assigns meaning to it, pointing out the strengths and weaknesses of each organizational aspect. The work is done in groups using clipboards that should be kept for use later in the inquiry process.

- Step 3: Identification of interdependencies among the dimensions and priorities for change. The final step for the group involves finding the interdependencies among the four dimensions; critical points and areas of strength in one dimension will likely have an influence on the critical issues and resources in the others. The consultant helps the group of participants identify the connections among the various dimensions to have a complex, systemic view of their organization based on which to establish priorities for change and decide from which dimension it is appropriate to begin implementing it.

This scheme of analysis is an *in-progress* tool that is being modified in relation to the different organizational realities investigated in recent years to place itself more and more in the stage of "integration of different approaches" rather than "mechanical complementarity of approaches."[8]

---

8. For the evaluation of POAM done in small firms, schools, volunteer organizations, municipalities, sport centers, etc., see Francescato, Tomai, and Solimeno, 2008.

8

The group as a *behavior* setting and as an intervention tool

1. Critical premise

Individuals carry out their existence by transiting through formal and informal groups. The quality and quantity of relationships experienced in these settings determine the richness and potential of each person's network and personal conditions of well-being. For these reasons, for community psychology, the group understood as an environmental setting, is one of the privileged places of intervention.

All the intervention strategies of this discipline presented in this volume are implemented in groups. Organizational analysis, research intervention, self-help groups, and social support intervene on existing groups within specific organizations and/or the broader community or promote the creation of new groups and new aggregations of people with common interests. The literature on small groups has been growing in recent decades and is currently vast, ranging from cultural anthropology to social psychology, from occupational psychology to psychoanalysis, and from political science to psychiatry.

This growth, as well as the increase in interest in research on groups and their functioning, has also been facilitated by the evolution of the organization of work, which has seen a multiplication of intellectual functions and thus moments dedicated to meetings and teamwork in every work context (health, educational, corporate, third sector, etc.).

Given the richness and heterogeneity of groups operating in different contexts, it becomes necessary to identify what is specific about the way groups are conceived and used in community psychology.

2.   The small group for community psychology

Trying to delineate boundaries, the small groups that community psychology deals with are, first and foremost, hetero-centered groups, work groups, that is, groups centered on a common work, on an objective purpose outside the group.

Community psychology, therefore, is not concerned with therapeutic groups, groups that are self-centered since they are exclusively focused on emotions and relationships existing within them, nor with dynamics groups or family groups.

2.1.   What groups does community psychology use? Community psychology makes its conception and use of the small group proposed by Lewin. The scholar takes a systemic view of the group and considers it a dynamic whole, qualitatively different from the set of individual elements that compose it. With respect to the interaction among members, Lewin emphasizes its dimension of interdependence and the condition of mutual influence whereby any change in one of the parts transforms the structure and dynamics of the group, and conversely, any change in the psychological field produces effects on each component. In his work, Lewin, therefore, greatly enhances the transformative potential of the group, both at the individual level (as in work on T-groups) and at the social level (techniques for modifying attitudes). Finally, recall the well-known studies on leadership styles in groups (Lewin, Lippit, White, 1939) and the discovery and use of the feedback mechanism.

Among the various group types, T-groups (training-groups, "training group") of the Lewisian matrix are the groups (in terms of objectives and methodologies) most akin to those used in community psychology.

The T-group can be defined as "a learning experience by direct implication, through which participants acquire a greater sensitivity to group phenomena and a more accurate perception of themselves and others" (Badolato, Di Iullo, 1979, p. 143). T-groups were born in 1946 when Lewin, working with a learning group, discovered the importance of feedback: he realized that providing group members with information about their attitudes and modes of interaction allowed people to enact more impactful learning because it was emotional as well as cognitive.

Originally, T-groups involved two phases: a first in which some problem was discussed (hetero-centered group) and a second in which an observer communicated their observations of the group interaction and process. Thus, the first phase was task-oriented, and the group dynamic analyzed in the second part was that of a discussion group. Progressively, the evolution of discussion groups became clinically oriented, and the T-group became, like the therapy group, a self-centered group but, unlike the therapy group, aimed at healthy people who wish to know themselves better through confrontation with others.

What are the groups that community psychology deals with and uses? They are groups mainly centered on a task, a goal to be achieved, and the conscious aspects of relationships between people, and vary from a minimum of group sense (number of people coming together for short meetings) to a maximum (a team that also does group work together).

A common goal and operational interdependence: The main element of relevance in the group concept of community psychology is the experimentation with the operational interdependence that is created among members to achieve a common goal. The presence of a common goal is what creates life for the group. Interesting in this regard is the distinction made by Muti (1986) between an identical goal and a common goal. A common goal responds to even very different individual needs, desires, and motivations but is shared by all group members and is attainable only through the contribution of each. An identical goal, it could be said, is "more than the same," it is something that unites but does not connect since it does not require coordinated intervention for its realization and meets individual, often overlapping needs. Examples of this are people waiting for the bus or standing in line at the post office: their goals are identical, and the presence of the other is irrelevant if not in the way. It is only the presence of a common goal that can activate a relationship of interdependence among group participants as they engage in its achievement. The connection with others becomes evident and acquires meaning when members must find ways so that each one's operational contribution can be productively integrated with that of others, identifying competencies, assigning tasks, assigning roles, and scheduling times and ways to verify effectiveness.

Muti (1986) also distinguishes between:

- workgroup: understood as a formal group of people who constitute a small organizational unit with a certain degree of operational management autonomy to achieve the objective;

- teamwork: understood as a method that implies the existence of an operational goal to be achieved by coordinating the actions of people with interdependent purposes, needs, and desires.

According to Muti (1986), this distinction makes it possible to clarify some terminological difficulties: a team is a working group, and a meeting can already be considered a working group. Where the conditions mentioned above are met, these working groups do group work.

It seems to us that this distinction also helps to understand why formal groups, although they have common goals, often fail to do teamwork because there is no real need (or willingness) for people to coordinate their actions.

For example, in a counseling center, teams have a mandate to respond to users' health questions but may accomplish this mandate in different ways of operation. They may decide to work on individual cases by thinking of themselves as separate professions (the gynecologist may do their work alone, as may the sex counselor or pediatrician) or by coordinating their professionalism and setting common operational goals. The lack of need for integration causes the existence of a common operational goal to fade away and can lead to group meetings where people "stick together" for affective reasons, with no need and/or ability to do teamwork. When the needs and desires of the various practitioners do not coincide, it becomes difficult to do group work, even if we get together.

What, according to this definition, are the groups with which the community psychologist works? All contexts in which some people organize their activities to pursue shared goals. So, within organizational contexts such as health (e.g., multidisciplinary teams), educational (e.g., classroom groups but also groups of teachers or parents), corporate (e.g., teamwork), third sector (social cooperatives, associations, volunteer groups). But also, all those working groups that are formed "in the territory," such as citizens' committees, social movements, inter-organizational networks (see network work, chapter 10), partnerships, and all working settings that require knowledge and care of group processes.

2.2.   The intervention of community psychology in work groups The basic assumption of community psychologists is that work groups can be positive but also negative settings for the people in them. All of us have had experiences of working group meetings from which we have come out tired, tense, bored, feeling that we have wasted time and that we have accomplished nothing.

Some may have experienced real psychological violence, humiliation, and disconfirmation and developed a real aversion to group work. This phenomenon, called *group hate* in the literature (Sorensen, 1981), is defined as the fear and/or revulsion that many people feel about working in groups or teams; it correlates negatively with cohesion, consensus, and satisfaction (Myers, Goodboy, 2005).

The goal of interventions in group settings is to improve the quality of life of the people in them. Improvement can be achieved by enhancing the positive characteristics of the environmental setting so that the pressure it places on participants provides growth opportunities and makes demands that are more congruent with participants' needs. Or interventions aimed at enhancing the sense of power (empowerment) of group participants can be implemented, avoiding the double trap of feeling omnipotent or powerless by helping members and the group to set realistic goals whose achievement can increase the individual's and the group's sense of competence.

We believe that work groups can enhance the quality of both work and life, but only under certain conditions. First, the group's goals must be appropriate and truly require collaborative effort. This means that the contributions of several people are essential to achieve the goal, rather than meeting in a group purely for formal, legal, or habitual reasons. Second, participants must possess the knowledge and skills to positively influence the group by offering their expertise and acting as facilitators for the tasks at hand.

Working in teams is not a spontaneous skill but is also, above all, an acquired skill. To work in teams, one must be trained to communicate, and have both the ability to express oneself and the ability to listen, and be trained to cooperate, organize work, and know and manage certain group dynamics. Mucchielli (1986, p. 90) believes that as early as elementary school, children should be educated in group work, which they now do without any preparation: "There would be less to fight against individualism, rivalry, and the will to power if training in group work began at school."

In every work setting in which we have operated, we have found that while people are trained for various tasks, it is taken for granted that they know how to conduct work meetings and how to manage or participate effectively in work groups. That is, it is not considered that the group situation creates a specific environmental setting. Only in the presence of common goals congruent with the group work method and with trained participants and conductors can discrete functioning be achieved (the group and the people in it and the group itself can experience empowerment). In the absence of these prerequisites, groups as environmental settings are often a source of discomfort rather than well-being for participants.

Therefore, community psychologists intervene in work groups, broadly understood, seeking to improve the group setting through multiple strategies. They also intervene by creating specific *on-the-job* training programs devised according to the various environmental settings to teach individual members and the group as a whole that knowledge and skills (knowing how to do and knowing how to be) related to group functioning that will help them become more competent and thus more "powerful," in the sense of more aware of the limitations but also of the opportunities that this mode of work can offer.

3. The literature on working groups and teamwork

Several authors have developed real manuals for promoting group work by offering a variety of practical exercises for the reader to self-diagnose their behavior in work groups and practice improving their functioning (Johnson, Johnson, 1975; Mucchielli, 1980; Hanson, 1981; Gordon, 1987; Maxwell, 2002; Lencioni, 2011; Sinek, 2014; Bartholomew, Sama, 2020).

Italian authors who have dealt with the topic of work groups have usually talked about small groups in general, with few practical references to the phenomena that occur predominantly in work groups; almost none offer practical exercises. Instead, they have favored theory about groups, and while U.S. authors rely mainly on social psychology, European authors in general, and Italians in particular, also give much space to the psychoanalytic orientation (Novaga, Borsatti, 1979; Bruscaglioni, 1982; Spaltro, 1985; Muti, 1986).

An influential scholar in the Italian production scene who seems to us to be particularly in line with the orientation of community psychology is Spaltro, whose contribution we find interesting to analyze briefly. During his studies, this author has always devoted special attention to the phenomenon of the small group (Spaltro, 1969; 1985; 1990), its insertion in productive contexts, and the radical changes that its use would and has brought about.

As in the "Italian tradition," he too favors a theoretical-methodological discourse on small groups, viewing them as bearers of a new culture:

Indeed, despite the suspicion and resistance that small groups encounter in Italy today, it can be said that development [...] passes through the problem of small groups. [...] This frontally invests large fields of human coexistence, such as politics, research, organization, and housing. [...] This is the logical center of group culture, that is, of the micro, of the parts, of existing social relations: never being able to be everything, but never being able to disappear [...] because groups mean different mindsets and different ways of living (Spaltro, 1985, pp. 9-10).

For Spaltro, therefore, the transition from couple culture to group culture requires a development of the ability to "do group work," which implies a transition to the social phase of a person's psychic development (Spaltro, 1982a; 1982b), and involves a shift of "eroticization" onto society and the community.

Let us briefly recall that, for Spaltro, couple culture has the Oedipal relationship as its basis and is characterized, among other things, by singleness of leadership, prevalence of competitiveness, dependence, and loyalty as central values. Group culture, on the other hand, has pluralism, leadership differentiation, group cohesion, and change as central values.

His contribution seems significant to us in pointing out an aspect often overlooked by those who work through work groups, namely that working in groups implies a shift to a new mindset, to a culture centered on alternative values.

3.1. Operational manuals for group work training

Completely different is the approach of some U.S. and French authors, who, in a manner more akin to today's community psychologists, pose the problem of selecting among the myriad knowledge about small groups, those that are most valuable and useful for those who must work in groups and, above all, attempt to explore through what ways this knowledge can not only be disseminated but become the concrete assets of those who must make their work groups function better.

Equally pragmatic is the work of U.S. social psychologists David and Frank Johnson, who, as early as 1975, developed one of the best manuals on learning theories and practices of work groups. Issues of leadership, decision-making, goal setting, communication, opinion disputes and conflicts of interest, use of power, norms, and problem-solving techniques are analyzed within groups.

The authors' ideology is humanistic in inspiration; they conceive of group work in a similar way to Spaltro-as a way to live more democratically, to improve the quality of life, group settings, and one's well-being. The knowledge on which they are based is derived from sociology and small-group psychology. The prevailing emphasis is on the conscious components of behavior.

4. Microsystem and macrosystem: location and role of the small group in the social context

The small group is a relational dimension whose potential is being discovered and rediscovered in the world of work, and not only a progressive assumption of the idea of the group as an organizational model is taking place. To better understand the motivations and potential of this phenomenon, it is necessary to dwell for a moment on some events that have characterized and characterized our social context.

The course of the 20th century witnessed the contrast between liberalist and capitalist ideologies, which favored the rights of the individual over the interests of public affairs, and totalitarian ideologies, which emphasized collective demands while sacrificing those of the individual. This juxtaposition has given rise to dualistic thinking that is currently predominant. Particularly in capitalist nations, there has been an excessive faith in individual initiative and a gradual decrease in solidaristic attitudes.

Circularly, it has been seen how cultures with individualistic values encourage a greater focus on the self (Twenge, Campbell, 2009) and depower the development of social values (Paris, 2014).

We believe that, with respect to this problem, small groups can play an extremely important role precisely because they represent intermediate structures between the individual and the community. Being made up of a limited number of people who know each other makes the small group a context capable of satisfying everyone's needs for the individuation of strengthening the identity and uniqueness of each participant. At the same time, it allows them to actively experience conditions of operational interdependence and, if the group is relationally sound, emotional solidarity, as well as to develop attitudes of respect for difference and cooperation for common ends. To work for their growth, both qualitative and quantitative, is to work for community development.

Group mentality, however, Spaltro (1999) reminds us, must be approached with caution because the group requires the ability to develop a "plural" mentality. Collective behaviors are possible only if individuals are capable of feelings of belonging, and belonging is a difficult choice because it requires the ability to self-limit and a process of growth and differentiation. The social development of individuals passes through the development of relational modes of increasing complexity: "First we learn the relationship of couple, then of small group (micro), then of large collective group, organization or institution (macro), then again of community, that is, of unbounded collective (mega)" (*ibid.*, p. 28). The transition from one level to another, Spaltro argues, occurs by leaps, that is, through a change of mindset, and involves more complex attitudes on the part of subjects but also the growth of their well-being.

The small group represents the ideal tool for facilitating the transition from one level to another precisely because it functions as a "transmission belt" between the individual and society. In this regard, the author uses a very effective expression and calls the group a "multiplier" since, through the first systems of belonging experienced by everyone (family, school), it has for millennia *multiplied* the pressures of the macro (society) on the individual.

However, he recognizes, in addition to the risks, the potential of the group tool and emphasizes how the group's multiplying capabilities can have an innovative, as well as conformist, direction and enable the individual to reverse the direction of pressure and exert power over the large group (macro) or community (mega).

5. Focus on small groups: bridging theory and practice

Scientific understanding of teams has grown significantly in recent decades (Salas, Stagl, Burke, 2004; Ilgen *et al.*, 2005). Numerous models and theories have been produced to explain the effectiveness of teams; they illustrate the relationships among input variables (e.g., individual and team characteristics), process variables (e.g., communication, coordination, and decision-making), and outcome variables (e.g., safety, member satisfaction, productivity) (see Salas, Rosen, King, 2007; Marks, Mathieu, Zaccaro, 2001). These *input-output process* models illustrate the dynamic and multidimensional nature of teamwork and the importance of process variables in achieving team effectiveness. A theoretical contribution by

Maynard and co-workers (2015) provided a summary of the main factors around which team success revolves:

- internal cohesion within the group;
- external support system (i.e., provided by the organizational structure);
- shared purpose and goals;
- members' knowledge, skills, and abilities;
- member and team motivation;
- collaboration;
- communication among members;
- member and team productivity;
- integration;
- member learning.

Research recognizes teamwork as an essential component of effectiveness in most work environments, especially in healthcare (Salas, Rosen, King, 2007), but the attention (in terms of training, care, process, and effectiveness evaluation) given to it in different organizations (healthcare, education, manufacturing, and third sector) cannot always be said to be equivalent.

One area where teamwork orientation and training are most urgently needed and, at the same time, also most neglected is healthcare. In healthcare, the continuing spread of chronic diseases in recent decades makes a multidisciplinary approach and the continuous interaction of many specialists increasingly necessary. The ability of the team to work effectively, sharing information and responsibilities and seeking solutions collaboratively, is considered a key element in achieving quality goals in healthcare sectors (Qaddumi *et al.*, 2021); teams that engage in teamwork processes are 2.8 times more likely to achieve high performance than teams that do not (Schmutz, Meier, Manser, 2019). Nevertheless, there is ample evidence that an extensive focus on clinical work and a relative disregard for teamwork is a latent error in health systems (see, e.g., Hamman, 2004; Leonard, Graham, Bonacum, 2004). Significant progress has been made toward teamwork approaches to healthcare, but there remains a large knowledge base that can be spread and especially training and evaluation of the mode of teamwork performed by different teams that can be developed.

The increased knowledge and methodologies produced by scholars from various social disciplines in recent decades now make it possible to learn more about the complex relationships between the individual and the small group, to use this intervention tool more fruitfully, to limit its risks, to develop its potential, and to identify ways to make traditional groups settings conducive to the growth of the people who attend them.

6. A proposed training model for group work

In the early 1980s, we developed and tested in a variety of work settings a model of group work training, which sought to make a synthesis between the more pragmatic and the more theoretical-ideological approaches.

Consistent with the initial Lewisian model, our model privileges the aspect of gaining knowledge about groups and one's behavior in task-centered groups rather than aiming for therapeutic change in participants through exposure to self-centered group experiences. From the literature on T-groups, we mainly borrow knowledge about the stages of group life, certain process phenomena (affective group life), and subjective experiences in groups.

The model includes an initial theoretical-informative phase in which the basic principles of community psychology and the main intervention techniques, particularly organizational analysis, are discussed to position work groups within the broader system to which they belong. The differences between therapy groups, group dynamics, work groups, and teamwork are clarified. It is emphasized that the community approach tends to see work as a tool for improving group work settings, which can be both positive and negative for the individuals involved. It is reiterated that learning to work in a group is a particular type of theoretical-experiential learning that enhances each member's ability to become a better participant and allows for improved functioning of the workgroup. A better-functioning work group, in turn, becomes a work setting that offers better opportunities for well-being to those who operate within it. Models of analysis and intervention in community psychology are aimed at understanding the complex multidimensional dynamics occurring in macro and micro-social systems, working to increase engagement and responsibility among participants, and promoting competent communities – capable of developing and utilizing existing resources while simultaneously creating new ones. Finally, it is highlighted that it is preferable to implement preventive interventions that promote greater functionality rather than intervening in already deteriorated group situations. The specific objectives of the training courses for facilitating work groups (as we have named them) are to help participants become better members of the work groups they are part of and better facilitators in the groups they coordinate. In particular, the first part of the course aims to provide awareness, a shared language, and, above all, tools that enable individuals with diverse professional backgrounds to work better together while respecting each other's expertise.

The training model includes several modules with different objectives. The first module aims to provide participants with theoretical frameworks and tools for observing and classifying various types of groups, analyzing their structural, task, process, and individual characteristics to capture their potential and problem areas. The different functions that the facilitator can perform to improve the detected problem areas with respect to the task or group maintenance or highlight critical issues related to the characteristics of individual participants are then identified. Participants are evaluated according to their ability to perform these different functions and to observe their behavior in working groups to improve their functioning.

Other training modules attempt to address problematic group situations by exposing participants to modes of decision-making, constructive communication, problem-solving, and individual and group creativity. The teaching methodology employed allows each participant to train in the different functions that can be performed during group work meetings: participating, observing, leading, and verbalizing. Each unit of work is divided into three moments:

1. brief theoretical introduction on the content, concisely summarizing various views and the results of the main research on the topic at hand;

2. practical exercise;

3. final discussion on the group experience made during the exercise, with training on the detection of indicators of behaviors and attitudes that denote the presence or absence of a certain group phenomenon and on the style of communication or participation experienced.

This training model for the facilitation of work groups has so far been tested in various work settings (schools, associations, production companies, social health, and educational-recreational services), and courses have been attended by business executives, middle managers, trade unionists, leisure workers, school operators, and managers to psychiatrists, gynecologists, pediatricians, midwives, psychologists, social workers, volunteer groups, etc.

7. Congruence of group work training techniques with the principles of community psychology

Participants using this technique learn to "see" the transactions that occur between individuals and group contexts and to look at group problems from multiple perspectives. Typically, this group work training, when done with real working groups, fosters cohesion and interdependence among members, helps them promote and implement joint empowerment projects, and identifies strengths to leverage to achieve desired changes.

When the history of the group and the types of power exercised are examined, there is some increase in sociopolitical empowerment. The methodology also integrates knowledge derived from the positivist and constructionist models. People with less power and different histories are also given a voice in the group, and the emergence of interpretations that break the tacit consensus by which members accept the systems of convention in which they are embedded is encouraged. This occurs, for example, when the group examines the explicit and implicit norms that govern its functioning.

However, rarely with this kind of technique can one see the historical origins of social problems and the link between individual empowerment and social struggles. Moreover, the small group can turn into a closed group isolated from the social context and unable to see the relationship between its functioning and that of the organizations and local communities, macro-communities, and virtual communities in which it is embedded. Therefore, the use of this technique reflects only some of the guiding principles for a theory of technique in community psychology.

9
Self-help groups

1. Introduction

The term *self-help* can create some misunderstanding, as it is used to refer to both *individual and group forms of support*. There are books that aim to offer the individual tools and knowledge to increase personal resources, such as cognitive, emotional, and behavioral skills. This conception of individual self-help is distant from that typical of groups, in which mutual and collective support is one of the main resources. In common between the two forms of help is the belief in people's potential for development and the rejection, when faced with problems or desires, of attitudes of delegation, especially when acting toward professionals.

Numerous definitions of self-help groups have been provided; overall, they are small groups interacting in face-to-face settings or online and composed of members who share common conditions, situations, legacies, hardships, or experiences (Borman, 1975; Katz, Bender, 1976; Noventa, Nava, Oliva, 1990). Some authors who consider self-help not as an activity but as a *nonprofessional health resource* consciously mobilized in a community to address health problems (Noventa, 1996) or as *artificial social networks* that are created spontaneously to produce social help/support (Folgheraiter, 1990).

The ethos of a self-help group is grounded in a peer-oriented, experiential approach, where members share similar distress or needs. These groups encourage activism and foster responsibility and active participation among their members, who simultaneously serve as both providers and recipients of care. Relationships within self-help groups are based on reciprocity, with members alternately giving and receiving support. They stand apart from other support systems due to four key characteristics: peer relationships, reciprocity, experiential knowledge, and autonomy (Ornelas, 2008).

Self-help groups are a phenomenon that emerged in the 1970s and developed over the next two decades in several countries. Sociocultural motivations, which we will examine during the chapter, have favored this evolution. It is certainly an extremely widespread phenomenon in the United States and increasingly in Europe and Italy. A survey by Katz (1981) estimated that by the late 1970s, there were at least 500,000 self-help groups in the United States, with a membership of about 23 million people. In Munich alone, there were more than 1,000 self-help groups in 1995 (Stark, 1995b), while even in Italy, several researchers have documented the existence of thousands of such realities (Prezza, Drahorad, Tomai, 1993; Fondazione Istituto Andrea Devoto, 1999).

Self-help groups have been an important cultural heritage in Italy for more than four decades. The first formal interest of national political bodies in the reality of self-help dates to 1999, when the Ministry of Social Affairs commissioned the Tuscan Regional Coordination of Self-Help Groups[9] and the Andrea Devoto Institute Foundation to carry out a detailed quantification and description of the phenomenon in the national territory. This first survey made it possible to track a total of 1,603 groups and identify the most innovative experiences. In 2006, the Volunteer Service Center of Tuscany promoted and published a second national monitoring (Focardi *et al.*, 2006). At present, this document contains the most up-to-date available data. The survey surveyed a total of 3,265 self-help groups, registering an increase since the first survey of 203 percent. Groups infrequent in the first survey (e.g., dedicated to bereavement and organic disorders) and groups that have sprung up around new addictions (to the internet, gambling, and compulsive shopping.)

With respect to the distribution on the Italian territory, the North gathers about 63 percent of self-help groups, the Center shows a presence of 24 percent, while in the South and islands, the percentage reaches 13 percent.

An interesting issue in this second monitoring is the considerable increase (from 38 to 62 percent) in the presence of the a professional facilitator within self-help groups. What is thought-provoking is the professionalization of this role. As many as 71 percent (up from 13 percent in 1999) of the facilitators of self-help groups are professionals from the-health field. A closer relationship with the healthcare world is an element that can be read in favor of greater recognition and use of the reality of self-help by the professional world. However, it is a phenomenon that should be viewed with caution. Self-help groups arise from below as spontaneous peer groups endowed with a strong emotional and ideological charge founded on direct experimentation

9. The Tuscan Regional Coordination of Self-Help Groups Association was founded in 1996 and is made up of associations, private social agencies, and individual citizens working to promote self-help.

with a problematic condition. The risk of ceding the role of facilitator to a health professional is to distort the phenomenon of self-help and deprive it of its distinctiveness: to strengthen social support, cultivate self-efficacy, and solicit processes of individual and collective liberation in settings connected to the real world (Humphreys, Rappaport, 1993; Humphreys, Moos, 2001; Orford, 2018).

2. Characteristics of self-help groups

2.1. Factors favoring the emergence of self-help groups At the root of self-help groups is the perception that official services, for the social-historical reasons previously discussed, do not sufficiently meet the needs of citizens (Stark, 1995a). The demotivation and/or mistrust of professionals, detachment and/or estrangement from users' problems, and the tendency to establish relationships of excessive dependence are especially emphasized. In Europe, self-help groups developed out of dissatisfaction with a depersonalized healthcare system. They successfully complement existing social and health services, but care must be taken that they are not instrumentalized to improve health outcomes to reduce health spending (Nayar, Kyobutungi, Razum, 2004).

Compared to formal systems of care, self-help groups are less bureaucratic, more affordable, and generally more accessible for users. However, they are still more organized and structured than informal social support networks. Defining their boundaries can be challenging, as they represent a complex and dynamic phenomenon, varying across organizational structures (e.g., autonomous groups or part of larger local or national organizations) and group characteristics (such as size, purpose, recruitment methods, boundary rigidity, and relationship to formal care systems). Additionally, distinguishing self-help groups from other alternative social support models can be difficult. For instance, in Italy, genuine self-help groups are less common than in the United States, with most resembling mixed-interest or service groups that incorporate volunteer participation and only partially share the traits of pure self-help groups.

Levy (1979) identifies the presence of five conditions for a group to be properly defined as self-help.

1. Purpose. The explicit and primary purpose of the group is to provide help and support to its members in dealing with their problems and improving their psychological skills and behavioral efficiency.

2. Origin and sanction. Its origin and the sanction of its existence reside in the group itself rather than depending on any external authority or institution.

3. Source of help. The primary source of help relies on the work, skills, knowledge, and interests of the members themselves. Group members establish a peer relationship so that they are all involved in requesting and offering care and support to each other.

4. Composition. The group is composed of members who share a common core of experiences and problems or a similar situation of distress.

5. Control. The structure and activities of the group are under the control of the members themselves, although they may occasionally seek advice or supervision from outside experts.

With respect to experiences similar to self-help, such as volunteer groups, we could thus distinguish the two categories in this way. Groups characterized by:

- sharing of a certain condition by all members;
- Absence of rigid ascribed technical roles.
- On the other hand, aggregations characterized by:
- simultaneous presence of members who directly experience a certain condition and members who participate to offer their support and help;
- coexistence of rigid technical roles and professional figures, involved as such with practices of function rotation.

From the perspective of preventive strategy, it is possible to consider self-help groups as secondary prevention tools, that is, addressing or limiting already ongoing distress through a process of mutual support among those experiencing such distress.

2.2.   Types of self-help groups The best-known typology is that proposed by Levy (1979), who, with the composition and purpose of self-help groups in mind, developed a classification into four types of groups.

1.   *Behavior control* or *behavior reorganization* groups are composed of members whose sole or overriding purpose is to eliminate or control certain problem behaviors.

2.   *Stress support and coping* groups are composed of members who share a stressful situation or condition. The goal of these groups is not so much to change this condition or *status* but to increase coping skills in managing stress through mutual support and the sharing of effective strategies. This category also includes all groups formed by people who share a chronic illness or disability or who are facing a period of crisis due to an unexpected event, whether negative (bereavement, separation, job loss, relocation, etc.) or positive (birth or adoption of a child, etc.).

3.   *Social action* groups that struggle against marginalization and for the defense of their rights and include categories discriminated against because of their values, lifestyles, or characteristics such as gender, race, or social

class. These groups, rather than an internal activity aimed at mutual support, mainly carry out an activity of information and promotion outside to stimulate changes in attitudes and/or institutional programs.

4. *Personal growth and self-actualization* groups are formed by members who do not share a common problem but have goals of increasing skills related to the emotional, sexual, and interpersonal spheres through mutual support. To this category belong the experiences of self-managed groups and are inspired by humanistic psychotherapies or group dynamics.

Overall, Levy's classification seems to have weaknesses because it counts within self-help social movement groups (category 3) and groups formed by people carrying not a problem they seek to remedy but a project for growth (category 4). Category 4 seems rather foreign to the experience of self-help groups because, although it shares some of their methodologies, it operates essentially from a primary prevention perspective (it is thus a type of empowerment akin to the experience of humanistic groups).

Francescato and Putton (1995) propose a typology that is more centered on the characteristics of the problems faced by the participants and the mechanisms of help and/or change that are activated and make the group action effective.

1. *Behavior control* groups consist of people who want to eliminate or control some of their problem behaviors; the task of the group is to help each participant leave this starting condition and not repeat the disturbing behavior. The group members are the direct carriers of the problem and are in the position of being able to choose to change but not being able to do so. The people who make up these groups generally believe themselves to be impotent and are then also such because they believe they cannot dominate events. Convinced that the control of their behavior does not depend on them but is external, they are incapable of reacting. The oldest and most established group of this type is Alcoholics Anonymous, founded in 1935 and with more than a million members worldwide. Similar aggregations are those of parents who abuse their children, drug addicts, people with eating disorders (*over-eaters anonymous*), violent behavior, excessive sexual needs, addiction to sex, nicotine, sports, etc.

2. Groups of people *with disabilities or chronic illnesses*: In these groups, individuals directly face the challenges of their condition. Unlike other groups, participants cannot change their situation, so the focus is on adapting to and managing their circumstances effectively. This often involves adhering to specific diets, undergoing uncomfortable or painful treatments, avoiding or controlling risky behaviors, and changing habits. Examples of such groups include those for individuals who have experienced heart attacks, schizophrenia, diabetes, cancer, AIDS, mastectomy, or permanent physical disabilities.

3. Groups of *relatives of people with serious problems*: this type of group aims to bring support and help to those who are affectively related to and/or live with people who have a problem (behavioral or health); this condition is a source of stress that is often very serious and heavy to bear. The first to organize groups for family members (partners and children) was Alcoholics Anonymous. Since then, these groups have multiplied, partly because public and private facilities focusing on the sick person often neglect the needs and wishes of relatives and friends.

4. Groups of *people going through a period of crisis*: a crisis refers to events involving a sudden change, whether negative (bereavement, separation, job loss, etc.) or positive (birth or adoption of a child, etc.), or even a change that is anticipated but affects one's identity (menopause, retirement, major operation). A crisis represents a temporary state of psychic upheaval and disorganization, often the consequence of particularly acute stresses or less intense but repeated wearisome events. Getting out of a crisis requires mobilizing internal and external resources, and it is important in these phases to be able to rely on sources of social support. Because these groups are formed to accompany people through a period of crisis, they are unlikely to be long-lasting groups or to give rise to more stable associations.

Finally, of note is the emergence of self-help groups via the internet. Many e-groups, virtual self-help groups that exchange e-mails, or chat discussion groups can be found on the Web. Analysis of the content of messages indicates that participants communicate in ways characteristic of face-to-face groups, favoring emotional support and confession. These groups are used more by men than women.

Such services are offered by groups for internet addicts, narcotics anonymous, co-dependencies, sex addicts, gambling addicts, transsexuals, homosexuals, and alcoholics anonymous; the latter has opened many groups on the internet in different languages.

Online peer support has evolved rapidly in recent years (Ilioudi *et al.*, 2012; Wright, 2016). Communication in online peer support groups is often conducted through e-mail, bulletin boards, or specific software for synchronous interaction with other group members that can be accessed 24/7. Several studies document that in these settings, people are more willing and comfortable sharing sensitive information or asking sensitive questions (Ilioudi *et al.*, 2012; Wright, 2016).

Online and offline peer support groups report similar positive aspects: connection, help in coping with daily challenges, reduction of stigma (Smith-Merry *et al.*, 2019), promotion of empowerment (Barak, Boniel-Nissim, Suler, 2008), and facilitation of recovery processes (Thomas *et al.*, 2016; Smith-Merry *et al.*, 2019). However, while social relationships that have evolved through online communities can migrate into the real world, online interactions can counteract exposure to real-life social exchanges and relationships (Wright, Bell, 2003). Excessive online relationship use appears to promote decreased offline interactions (Eysenbach *et al.*, 2004) and increased social avoidance (Chung, 2013).

2.3. Mechanisms of change: key factors in the action of self-help groups

Within self-help groups, there are, of course, specific and diverse ways of functioning and meeting management.[10] Beyond the variety of types and arrangements, there are general mechanisms and characteristics inherent in the structure and functioning of self-help groups that explain their undoubted (though difficult to measure) transformative effectiveness. Based on observations and some investigations made (Katz, 1981; Gartner, Riessman, 1984), we believe we can point out three primary and common key factors involved in the action of self-help groups.

1.   The function of informational exchange, support, reinforcement, and identification is inherent in the *socioemotional dynamics of the peer group*.

The presence of a peer group, without a priori-defined *status* or institutional roles or internal hierarchies and composed of individuals who experience or have experienced firsthand a certain condition, fosters some significant processes on the socio-emotional level. Members of the self-help group tend, in fact, to:

- lower habitual psychological defenses and resistance, no longer feeling negatively judged for one's own "diversity";

- communicate more directly based on common experience;

- identify with the evolutionary journey of people perceived to be more like themselves than the eventual experts, who are generally experienced as excessively foreign and detached;

- Mutually exchange information, knowledge, skills, emotional support, feedback, positive and negative reinforcement, structuring together rules of life and behavioral strategies for controlling and modifying behavior;

- Develop and facilitate opportunities for socialization both within and outside the group. Many associations are characterized by a range of recreational and social activities that go beyond being in the group; unlike

---

10. In this regard, it may be interesting to read the appendix of Norwood's (1985) text, in which she identifies some guidelines for the conduct of self-help group meetings. Among the most significant proposals are the call for confidentiality about the content that emerges in meetings, rotation of the role of facilitator, the use of a neutral venue – not a member's home – and a determined and limited time for meetings, and the sharing of personal experiences as a method of working, avoiding extraneous topics, value judgments about participants, and intellectualistic discussions.

therapeutic groups, in self-help groups there is an insistence that people meet outside the group, helping each other and organizing activities independently;

- The therapeutic value connected to the possibility of fulfilling the role of helper, that is, caregiver (Riessman, 1965; Pagano, Post, Johnson, 2010). Every member of self-help groups eventually plays the role of someone who helps, supports, and provides care to another person "with a problem." In this regard, since 1965, Riessman has developed the concept of *helper therapy*, or therapy of the caregiver. The hypothesis is that those who help also receive help themselves: an alcoholic who supports another alcoholic gains more self-esteem playing this active and selfless role. This implication concerns all caregivers, whether they are professionals or volunteers, but it is particularly evident among caregivers who have the same problem as those they assist. While all caregivers receive non-specific help from fulfilling their helping role, those who provide care to someone with a similar problem gain benefits in that area themselves. Taking on the role of helper enhances our sense of control, self-esteem, and competence. Skovholt (1974) argued that the following four factors motivate the effectiveness of the *helper therapy* principle.

- Feels a high level of interpersonal competence given the result of his action on another person;

- Generally, feels a satisfactory balance in his interpersonal relationships between giving and receiving;

- Learns and introjects learning strategies for change in working with their caregiver;

- Receives social recognition and approval for the role he or she plays, including being able to observe "from a distance" problems similar to their own. Being able to serve as a *helper* essentially allows one to move, as in a *role-play* experience, from the role of someone experiencing a certain problem to the role of someone who has acquired the experience and skills to help himself and another deal with it; and

- It has been estimated that participants in a self-help group exchange six types of resources: affection, *status*, information, money, goods, and services. Of these six classes of resources, affection, *status,* and information are usually exchanged in the context of each self-help group meeting. Based on this resource theory (Converse, Foa, 1993), Brown, Tang, and Hollman (2014) constructed a scale measuring social exchange in self-help groups.

2. The ideological drive that moves self-help groups is a significant driving force.

To varying degrees, all self-help groups establish a system of shared principles, teachings, and values that are highly persuasive within the group. The creation and adherence to these ideals allow group members to take on an active and transformative role, helping them eliminate prejudices and self-destructive attitudes while expanding the range of constructive alternatives, goals, and strategies. This fosters collective rather than purely individualistic or indifferent criticism. The group's ideology, combined with the socio-emotional pressure of peer relationships, significantly enhances the learning of skills and problem-solving methods typical of the group. Psychological literature is rich in descriptive observations and hypotheses explaining the transformative mechanisms at work in self-help groups. However, there is a notable lack of research on the evaluation of outcomes and the ability to document general effects. Various aspects of self-help groups have been highlighted by different authors as probable factors that promote change in participants.

From a review of the literature, Francescato and Putton (1995) identified eight mechanisms that make self-help groups effective:

1. shaping: the improvement of the living conditions of people who share a similar problematic condition; the example of positive behaviors enacted by other group members increases the hope that they can change (Folgheraiter, 1990);

2. affective and informational support given and received among members (Cecchi, 1993);

3. sharing: knowing people who share the same problem or situation reduces feelings of loneliness and isolation (Folgheraiter 1990);

4. sense of belonging: feeling accepted and part of a group;

5. *helper-therapy*: being helpful to another person in the group (Cecchi, 1993);

6. coping strategies (behavior control): having learned to control one's behaviors and be able to express one's negative emotions (Noventa, Nava, Oliva, 1990);

7. network restoration: befriending group members and receiving support outside the group from someone in the group (Pera, 1990);

8. responsibility: having gained more responsibility over one's condition and having contributed responsibly to the functioning of the group (Noventa, 1996).

Francescato, Tomai, and Foddis (2002) conducted a study of 560 individuals belonging to Italian self-help groups to identify which factors are considered most relevant by people who attend these groups. The research highlighted four: empowerment, identification, information/advice exchange, and friendship exchange. Empowerment and identification are the factors that seem to have the most impact. The factors listed have different weights and functions within the various types of groups identified earlier; we will try to highlight the specific and most pregnant ones for each type of group.

In *behavior control groups*, it is the mechanisms of identification and modeling that carry the most weight. Belonging to a group of individuals perceived to be similar but at different stages of problem-solving encourages the establishment of *processes of identification* with people who have succeeded in controlling behavior and processes of learning by modeling that increase confidence that they can change their behaviors. Other functional mechanisms in this type of group include learning to focus on the present and the ability to set small, realistic, gradual, and attainable goals with the help of others. These mechanisms enable the individual to become more aware of the power he or she has in changing their life situation and facilitate the transition from a situation of *learned helplessness* to one of *self-efficacy*

In *groups for the disabled or chronically ill*, the key factors that make the group effective are the presence of emotional support, informational exchange, identification with the peer group, the stimulus given by the ideological charge of the group to the modification of attitudes and prejudices, and the therapeutic value associated with the opportunity to help others. This last factor, which is present and active in all self-help groups, appears particularly beneficial for people who, because of their condition of illness or disability, are often dependent on others or forced to be cared for and helped. For these individuals, experiencing that they, in turn, can be of help or support to someone is often a source of pride and satisfaction because it is an opportunity to make the relationships experienced reciprocal and more equal. Being bearers or even promoters of helping initiatives fosters a significant evolution of self-perception, an improvement in depressive or self-injurious experiences.

In the third type of group (*groups of relatives of people with serious problems*), the most effective mechanisms are those of affective support but also instrumental and informational support. Fundamental to the members of the groups are exchanges of information on where to find resources, material help in caring for the family member, organizing and obtaining services (day care centers, housing, summer stays, etc.) that allow families to "step back and temporarily delegate" the problem to avoid the phenomena of depression, exhaustion, *burn-out*. The social support given and received has a buffering effect with respect to the consequences of prolonged stress, reduces its negativity, alleviates the emotional impact of distressing situations, and encourages active and adaptive responses. In the last category described (*groups of people going through a period of crisis*), for the participants, the self-help group comes to represent a source of social support; therefore, mechanisms of identification and mutual help are in action above all, which enable people to strengthen themselves, to become more competent in coping with the difficulties of their situation, and to redefine the crisis as a positive and growth moment.

3.   The evaluation of results and estimation of effectiveness

The intervention of self-help groups is one of those topics on which there is now, in the Anglo-Saxon world, a rich literature with interesting descriptive observations and hypotheses explaining the results of transformative mechanisms. What remains lacking, however, is research on the evaluation of outcomes and the possibility of drawing valid directions for different types of groups.

One reason for this shortcoming probably lies in the very unprofessional and, in some ways, spontaneous nature of self-help groups. The goal of objectively assessing the effects is thus seen as an academic or purely statistical requirement: the subjective and personal impression of the participants, the involvement and satisfaction felt by those who join and attend such groups would suffice - according to this view - to testify to the effectiveness of the groups. The interesting question to ask is whether self-help groups are effective in their goals and whether they are more so than other forms of support. The answers deduced from the literature are currently leaning mostly in the positive direction.

The present literature on the subject is not particularly extensive, yet the various research to date seems to show the effectiveness of all the various types of self-help groups mentioned. We will bring a few examples from each category.

1.   *Groups for the control of undesirable behavior.* The most popular and well-known groups in this category are Alcoholics Anonymous. They are very effective in maintaining long-term recovery and preventing the need for further treatment (Moos, Moos, 2006; Humphreys, Blodgett, and Wagner, 2014). Alcoholics in Treatment Clubs (now Territorial Alcohol Clubs) have a 70% sober lifestyle maintenance rate from the beginning of group participation. Self-help groups are considered an effective and cost-effective way to sustain long-term recovery (Zarkin *et al.*, 2005). Active participation and involvement in Alcoholics Anonymous groups allow recovery to be maintained beyond the end of structured support provided by the professional and/or formal system of care (Moos, Moos, 2006; Parkman, Lloyd, Splisbury, 2015). In addition, group participation provides access to a social environment within which to practice one's recovery (Parkman, Lloyd, Splisbury, 2015).

Finally, the use of self-help groups after addiction treatment has been associated with higher rates of abstinence. Participants acquire new abstinent friends and learn new coping strategies, achieve higher social *status*, and improve their self-esteem (Zemore, Kaskutas, Ammon, 2004)

2. *Groups for the disabled or chronically ill.* Compared with this category, self-help groups are increasingly being used in oncology. A landmark study by Fawzy and coworkers (1993) showed that participation in a self-help group for people in the early stages of melanoma can increase the chance of survival up to three times over five years. From the perspective of cancer survivors, the benefits of self-help groups include sharing information and common experiences, opportunities to help others, feeling encouraged and strengthened, and learning about complementary and alternative therapies (Gray *et al.*, 1997). For breast cancer, self-help groups play a very important role for women after surgery. Participants who enrich their social network report better levels of empowerment and quality of life (Yun, Song, 2013; Lee, Ryu, Hwang, 2014; Shin, Park, 2017), including more effective emotional control (Sowa *et al.*, 2018). The experience of self-help is also widely used in the management of chronic conditions. In people with diabetes, for example, participation in self-help groups enhances the sense of empowerment, ability to control, and sense of self-efficacy; this, consequently, promotes more effective management of the disease and better quality of life (Chaveepojnkamjorn *et al.*, 2009; Sukwatjanee *et al.*, 2009).

Finally, the formation of self-help groups has become an important component of mental health programs run by nongovernmental organizations (NGOs) in low-income countries (Eaton, Radtke, 2010). Randomized studies show that outcomes from self-help groups are equivalent to intervention in disorders such as anxiety, depression, social phobia, and panic disorder (Pistrang, Barker, Humphreys, 2008; Lewis, Pearce, Bisson, 2012). However, people with psychosis (Castelein *et al.*, 2008) or schizophrenia (*hearing voices*) can also benefit from participation in self-help groups by improving social, emotional, and expressive skills (Longden, Read, Dillon, 2018).

3. *Groups of relatives of people with serious problems.* Over the past four decades, there has also been a growing interest in self-help groups among caregivers of people with significant problems of both mental and organic nature. However, relatively little research has examined the benefits of self-help groups in this context. Through participation with others, caregivers gain a better understanding of the factors affecting their lives, extend their knowledge of available resources, and improve their problem-solving skills. Participation with others helps them (re)take control of their lives and the lives of their children in meaningful and sustainable ways (Cohen *et al.*, 2012). Benefits reported by caregivers of children with disabilities include increased social support, reduced severity of the child's disability, and

improved social connectedness and resource mobilization (Bunning *et al.*, 2020).

The formation of self-help groups for caregivers has become an important component of mental health programs run by NGOs in low-income countries.

4. *Groups of people going through a crisis.* Several studies have shown that self-help activities prevent or reduce the long-term effects related to traumatic events. The effectiveness of self-help groups in coping with crises for those suffering from the effects of grief is well documented. By providing a supportive communicative satiation in which to find each other and share their stories, self-help groups facilitate the process of processing and adjustment (McCreight, 2007).

Self-help experiences promote recovery of emotional stability and reduces symptoms of stress and physical pain in middle-aged widows (Stewart *et al.*, 2001; Kang, Yoo, 2007). Finally, a recent use of the self-help group concerns people who have suffered the suicide of a family member or acquaintance. Again, the self-help experience allows people to come out of isolation, freely discuss the possible reasons for suicide, and experience the universality of feelings, empathy, and mutual support (Testoni *et al.*, 2019; Tosini Fraccaro, 2022).

4.   Self-help in Europe

The literature on self-help groups offers a broad overview of their activities in the United States and a substantial amount of research conducted since the 1970s. As far as Europe is concerned, there is no similarly detailed literature. On our continent, self-help is understood as it was conceived in the United States, i.e., groups are conceived as a collection of people sharing the same problem, the conduct of which is not entrusted to an expert, a professional, but to a "peer." The figure of the professional has the marginal role of a consultant, an expert who merely provides specific information about the problem where he or she is asked to intervene. Many Italian self-help groups, on the contrary, seem to easily accept the presence of a professional who has the role of a facilitator and not of a counselor, indeed, sometimes, he is precisely sought after an experience initiated independently. These often integrate so well with the participants that they are no longer perceived as professional figures but as the participants, in no way affecting the benefits derived from *helper therapy*.

Few studies provide up-to-date statistical data on the development of self-help in a European context. Again, the most recent data are provided by the Tuscan Regional Coordination of Self-Help Groups in collaboration with the Andrea Devoto Institute Foundation. The European reality survey was published in November 2006 (Focardi *et al.*, 2006). Self-help groups have grown exponentially over the past three decades in Europe. This can be attributed to the decline of natural support systems, such as the church, the neighborhood, and the family, and the entry into crisis of healthcare systems due to marked professionalization, high fragmentation and specialization of services, bureaucratization, and the accentuation of patients' position of dependence.

It is not easy to give a completely exhaustive answer with respect to the number of groups in Europe because of the great variety of self-help experiences: there are associations composed of many members, large, well-structured, well-organized, and with divisions throughout the country, but there are also very small, informal, and locally sized groups and even temporary groups. Authors underline a difference in the development process of the self-help phenomenon between northern and southern Europe. In Northwestern Europe (UK, Netherlands, Germany, etc.), the reaction to individualism fostered a rapid growth of self-help groups in the late 1970s. In Southern Europe, on the contrary, a greater orientation toward collectivism and a sense of community, together with a greater attitude of dependence on the state, slowed the development of the self-help phenomenon. Finally, in Eastern Europe, self-help groups were able to grow only after the collapse of communism, as they were previously prohibited. At present, self-help is developing much more rapidly in those countries than in Western Europe. Similarly, in Southern Europe (Italy, Spain, and Greece), many self-help realities that had remained unknown for a long time have also been traced.

Some data. In Europe, the self-help movement is particularly developed in Germany, where the number of groups is estimated at 100,000, about 75 percent of them in the health sector (Matzat, 2013). In the United Kingdom, the number of self-help groups is 50,000. In Austria, according to the National Support Center, there are about 1,000 groups. In Poland, it seems that in about eight years, the number has gone from zero to 13,000. In Greece, there are 71 groups and a copious (but inaccurate) number of groups for the handicapped.

In most European countries, self-help groups for people with illness or physical problems are the most numerous. In second place come those dealing with particular social problems, such as situations of widowhood, divorce, single-parent families, loneliness, abuse, and bereavement.

5.   Development trends in self-help groups

5.1.   The relationship with formal systems of care Recognized as an intervention tool, albeit with validity established more at the experiential and application level than at the experimental level, self-help groups are

evolving toward integration with the formal system of care. This relationship presents itself as potentially fruitful though equally complex (Salzer, Rappaport, Segre, 2001). Suffice it to mention that the emergence of the self-help movement is motivated by a desire for differentiation from professional intervention: this allows us to understand the different assumptions and operational logics of the two systems of care and the tensions that can be expected when attempting to connect two such different realities.

The basic prerequisite for such an encounter is an attitude of mutual esteem, of acceptance of diversity rather than a claim to superiority. The ethic of self-help is proposed as activist and democratic, that is, peer-centered and the "power of sharing and reciprocity"; fear of losing one's role and/or the professional's lack of confidence in people's ability to be able to work effectively to solve their problems are attitudes that do not support self-help and undermine its possibility of collaboration (Croce, Oliva, 1995).

One of the risks in the relationship between professionals and self-help is that each of the two systems of care may consider its level of knowledge (professional and experiential) superior to that of the other party and may, based on that superiority, aspire to control the work of the other. It is important to limit the temptation, most conceivable in professionals, to co-opt and dominate self-help groups, perhaps in a good-faith desire to increase their scientific competence. Above all, the risk is to distort some of their fundamental characteristics, namely, spontaneity and motivation, based on a personal basis rather than on the coldest scientific interest. The high degree of emotional involvement of the members of a self-help group may appear counterproductive to the professional whose rule is objectivity and emotional distance; it represents the strength of this experience.

The relationship between professionals and self-help groups is facilitated by positive attitudes of appreciation and recognition of different skills and the control of negative attitudes that are not always conscious. The exchange and interaction between the two worlds must be conducted, in a situation of lucid awareness, as a relationship of equals, allowing for mutual control and criticism, thus limiting the risks of isolation or degeneration of one or the other avenue of intervention. How can the work of the professional coordinate with that of the self-help group? What resources can professionals and the formal system provide? First, the professional can support but also promote the birth and development of self-help groups in the area in which he works, learning to modulate and differentiate his intervention according to the different stages of group development. He will have to shift from the role of catalyst to that of occasional adviser to the group, maintaining an active but indirect role. In the *stage prior to the birth of* a group he will have to identify the need, raise awareness, and organize meetings. In the *start-up phase* he will have to support it instrumentally (find the venue, funding, technical supports, etc.) and process-wise (reinforce membership and rewarding experiences, help identify norms, encourage mutual identifications, safeguard participants' individualities) by taking a leadership and organizing role. With a *group initiated* and able to self-manage, the role of the practitioner will be marginal, limited in time, at the request of the group, and with a consultative relationship (Silverman, 1989; Cecchi, 1993).

Among the resources that professionals can provide to self-help groups, two suggestions offered by Maguire (1987) seem interesting to us. In addition to the already mentioned instrumental, training and emotional support supports, the author points to *credibility in the community* and *credibility among other social workers*. The support of professional social workers, in fact, makes the activity of self-help groups known to the whole community. Many people or associations are unaware of the existence or goals and modes of action of various groups. The practitioner is usually not lacking in opportunities to "advertise" the groups he or she works with. This awareness work can be done both among the public and among other social workers.

Finally, with respect to the relationship between professionals and self-help, it seems important to us to note the specificity of the approach of the mechanisms of action present in self-help groups and how this heterogeneity does not necessarily imply that the effects and participation in a group are antagonistic to psychotherapy or another type of social intervention. For example, Yalom (1978) notes that participation in an Alcoholics Anonymous group is complementary to traditional group psychotherapy: in research, in fact, the self-help group is functional for maintaining sobriety, while psychotherapy provides a context for interpersonal and interpsychic change. Within group methodologies, therapeutic and self-help groups are increasingly being defined as potentially complementary; in this regard, experiments involving forms of integration between traditional methodologies and self-help groups have been present for several years (Shepherd *et al.*, 1999)

Cooperation between healthcare staff and patient organizations can be seen as a concrete way to put into practice a modern concept of patient-centeredness, as defined by the Institute of Medicine (National Research Council, 2001) and give rise to numerous opportunities for shared participation and governance (Kofahl *et al.*, 2014). In this regard, some studies document how mental health services produce better outcomes in terms of long-term recovery and stabilization of mental pathology when they work in collaboration with self-help groups in the community (Segal, Silverman, Temkin, 2010). For example, an interesting longitudinal study conducted in Italy (Burti *et al.*, 2005) compared the clinical and social outcome, service utilization and direct costs of two groups of patients in the Verona Sud community psychiatric service. Only one of the two groups attended a self-help group. Two years later, members of the self-help group reduced their hospital stay (in terms of number of hospitalizations and days in the hospital) and contained the costs of their care; they were, in addition, more satisfied with their work activities. In contrast, the members of the group who had not experienced self-help presented an increase in unmet needs. As the authors also comment, self-help should not be considered as an additional form of psychiatric treatment or rehabilitation, but rather as a form of intervention to be integrated with traditional treatments to promote users' resources and drives for recovery.

This "hidden healthcare system," as the phenomenon of self-help was called more than 30 years ago (Levin, Idler, 1981), is highly developed and involved in healthcare governance in Germany – a leading European nation in this area, especially in advocacy, evaluation, and occasionally service planning and delivery (Schulz-Nieswandt, 2011; Trojan, Nickel, 2011).

The German experience shows that self-help and user groups can play an important role in developing partnerships between "patients" and "professionals" for improving the quality of healthcare services (Kofahl *et al.*, 2014).

5.2.   Clearing houses to support self-help groups   The loosely structured and organized nature of self-help groups and their continuing growth in numbers have given rise to the need for liaison structures having the same territorial reference. Such liaison structures are called *clearing houses* or, in the Italian translation, *self-help group support centers*, and they provide self-help groups belonging to the reference territory with organized support systems of various kinds, including health information centers (Caselli, 1993).

The idea of establishing clearing houses initially originated in countries where the self-help phenomenon is oldest and most widespread. We refer, of course, to the United States, where in 1980 alone there were about ten *clearing houses*, which by 1985 already appeared to have tripled in size. In Europe, where self-help groups spread later, *clearing houses* also have a more recent evolution, developing mainly in Great Britain and Germany.

At the same time, the magnitude of the phenomenon and its gradual positive evaluation have prompted the oms, and more precisely the Regional Office for Europe, to promote initiatives to support self-help groups and thus also encourage the appearance of support centers. Among the many initiatives promoted we highlight the creation of an international self-help and health information center, which deals with the links between organizations and self-help support centers.

From cognitive research conducted by the International Center (Wollert, 1987), it was possible to outline the main characteristics of existing *clearing houses* worldwide with reference to European *clearing houses*. Common organizational principles of *clearing houses* include *flexibility* and *autonomy*. Indeed, the clearing house organizational model, to adapt to the needs of existing groups, varies according to local and national characteristics, the level of welfare, and the characteristics of the service system in the relevant territory. It also emerges clearly that the main purpose of these support centers is to *facilitate the various self-help groups* to achieve the goals they independently set for themselves, seeking to pursue them in the ways they identify. The main functions of the coordination centers can be summarized in the following areas:

- *information and documentation*: the *clearing house* collects a wealth of documentation on self-help; develops study projects and research activities; organizes seminars and meetings between different self-help groups or between groups and their interlocutors; and provides information to the potential user on the characteristics, location, and procedures for accessing the different groups in the area in which it operates;

- *Facilitation in acquiring material resources* takes action to find the primary resources needed by a group, such as places to meet, secretarial services (e.g., telephone, photocopying, printing), economic resources;

- *Consulting and training* offer administrative, legal, tax, psychological and marketing consulting services and organizes workshops for animating and leading groups at different stages of development;

- *Promotion and networking work* maintains relations with government bodies and ministries, follows legislative and implementation activities at national and territorial levels. With respect to territorial organization, we can say that the first three functions are mainly carried out at the local level, while at the national and international level *clearing houses* mainly perform functions of research, study, promotion of development policies, and liaison between the various existing *clearing houses*.

The acquisition of a territorial dimension represents a significant step in the evolution of self-help groups in any country or nation, since it implies the recognition by territorial communities of the importance of the social function they perform and thus marks the achievement of a stage of maturity in the development process. Territorialization initiates inter-group coordination initiatives and collaborative research with local communities that strengthen self-help groups and enhance their function within the public health system.

At present, the implementation of *clearing houses* in Italy is still fragmented and isolated, but it is to be hoped that appropriate policies to support self-help groups, which are now very numerous, can also be developed in our country.

5.3. How to create a self-help group   Due to its structural and functional characteristics, the self-help group can be considered a valuable intervention tool for community psychology, which hopes for an increasing knowledge and ability to use it. In addition to studying, supporting and utilizing the self-help group, a psychologist or community worker should be able to create new self-help groups according to the needs and requirements of their local area. With respect to this goal, useful and certainly still valid are the suggestions provided by Silverman (1989):

Forming new self-help groups

- *identification of at-risk groups* requires local knowledge obtainable through a community study, use of data already collected by services, or interviewing significant people;

- *definition of need*: possession of this information will enable the practitioner to identify the presence of categories or groups of people with an ongoing or potential problem and for whom there are no already defined forms of support or there are inadequate or insufficient forms of intervention;

- *involvement of primary caregivers*: this is the stage of contacting people and propagating the group instrument. It is necessary to make sure that the people contacted are clear about the goals of the group and the difference between it and the therapeutic group, that they do not develop expectations of delegation and passivity toward the management of the problem, and that they have an interest in dealing with people in the same condition as themselves.

Once a potential area of need has been identified in the territory, it is necessary to proceed with the establishment of the group reality. For this reason, especially for the start-up of groups belonging to the second and third typology, Francescato and Putton (1995) proposed a model of intervention that includes moments of support and training. Indeed, it is believed that it is necessary for members of a new group to master some basic rules that guide and facilitate group participation and conduct. At the same time, participants must learn to perceive the group as a source of informational, affective and emotional support. In this case, the role of the promoter is of fundamental importance, who, through a process of *modeling* and rejection of any delegation processes, must educate the group to "rely on itself," to identify and use the resources in it. For this reason and to avoid the dangers of "over-attachment" of the group to its promoter and vice versa, the training and accompaniment phase envisaged in the proposed model is very short (an average of four meetings) and the operator must be able to "plan for his death."

6.   Trends in the development of self-help groups

Over the past two decades, the self-help group concept has garnered considerable interest for its potential, worldwide, in a few areas in particular: in the health sector, to foster user participation in the planning/management of health services; in processes of women's development and empowerment; and in psychosocial recovery and rehabilitation pathways, especially in gender-based violence.

6.1.   Health sector The importance of empowering communities to take care of their own health problems has long been widely recognized and recommended even by international bodies (who, 2012). Nonetheless, at the level of individual healthcare institutions, there are rarely general provisions for the systematic participation and involvement of patients in services for them. The horizontal and peer-oriented structure of collaboration and support of self-help groups has led many scholars and oms itself to consider them as bodies capable of fostering change in health services and the development of participatory processes (Baggott, Forster, 2008; Brown, Tang, Hollman, 2014).

Consistent with this process, for example, in Germany, the *Self-Help Friendliness (SHF) and Patient-Centeredness in Health Care* initiative was launched in 2004[11] to establish and implement quality criteria related to collaboration with patient groups (Trojan, Nickel, Kofahl, 2014). Between 2004 and 2013, many self-help groups and numerous healthcare organizations, both hospital and community-based, were involved in developing and agreeing on criteria for good collaboration between healthcare facilities and groups and guidelines to facilitate their implementation. The intensive and prolonged involvement of self-help representatives allowed SHF to develop as a systematic approach and initiate closer collaboration between professionals and groups (Nickel, Trojan, Kofahl, 2017)

6.2.   Women's development and empowerment Especially in the last decade, self-help groups have come to play a prominent role in women's empowerment processes, with a focus on improving health conditions (Chakravarty, Jha, 2012). Currently, both governmental and nongovernmental institutions deploy considerable resources to facilitate self-help groups in low- and middle-income and developing countries. Indeed, numerous studies have documented how women's self-help groups have positive effects on economic and political empowerment, women's mobility and women's control over family planning. Qualitative research also suggests positive effects of self-

11. *Self-Help Friendliness* (SHF) describes a strategy to institutionalize the cooperation of healthcare institutions with self-help groups of chronic patients.

help groups on empowerment, fostered by increased familiarity in money management, independence in financial decision-making, and acquired respect from family and other community members (Brody *et al.*, 2017; Setia, Singh Tandon, Brijpal, 2017).

For these reasons, the potential of self-help groups to be an effective tool for alleviating poverty and promoting women's empowerment and empowerment has garnered considerable interest worldwide. Self-help groups have come to dominate the development landscape, particularly in South Asia. The Indian government, along with NGOs, uses these groups as platforms for development programs (Gugerty, Biscaye, Anderson, 2019; Nichols, 2021). They are an attractive phenomenon for governments for multiple reasons. First, they ensure more effective transmission/dissemination of information (e.g., water sanitation or hygiene practices in infant care). In addition, conveying information through self-help groups leads to greater cost-effectiveness and more rapid change in the behavior being promoted (whether it be health and nutrition practices, agricultural improvement, or demands for government entitlements) because groups with more social capital can better regulate norms or sanction "deviant" behavior (Nichols, 2021). Finally, the regular interactions that the group setting promotes generate important social capital resources (in terms of trust and reciprocity; Kumar *et al.*, 2019). Indeed, it has been seen that self-help groups with high levels of social capital are able to take collective action to demand public interest interventions (Sanyal, 2009).

6.3. Pathways to recovery in domestic violence The *continuum of* domestic violence and its gradual escalation over time is also reflected in the increasing number of victims of femicide by a partner, spouse or ex-partner (United Nations General Assembly, 2012). In this regard, self-help is proposed as an interesting methodology that is still relatively unexplored in its implications in relation to situations of violence in an intimate relationship (*Intimate Partner Violence*, IPV). Overcoming the experience of IPV in this case is conceived not as a corrective process but as a process of recovery, in which self-responsibility, participation, and taking an active role are a key factor. In this understanding, recovery is understood as a process of regaining hope, personal empowerment, and control over important decisions in one's life. Associations and professionals involved in the recovery processes of women with IPV experiences have used both individual and group interventions. The individual-centered intervention strategy is crucial in the early stages of the process of escaping violence, when the woman begins to break the cycle of violence and process feelings of pain, loss and helplessness (Leitão,

2014). However, in the advanced stages of the recovery process from violence, when the woman has already left her violent partner, a group intervention seems more appropriate to meet other emerging needs of the victim (Esposito, 2012).

Overcoming the condition of isolation, promoting self-esteem and self-efficacy (Tutty, Babins-Wagner, Rothery, 2016), building a network of positive relationships for the exchange of social support and coping strategies (McWhirter, 2011), reclaiming meaningful social roles, and using the experience of abuse to help other victims (Flasch, Murray, Crowe, 2017) are the needs of women survivors of IPV that the self-help experience can take on.

By promoting self-help experiences among women victims of IPV, it happens that the group begins to be a functional relational context that becomes a source of different forms of support (informational, emotional, material). This allows these women to rebuild a social network and reduce the sense of loneliness and guilt common in IPV situations.

Through the process of comparing, identifying and reflecting with other women who share or have shared the same experience, participants can feel less guilty, less inadequate and less alone. Discussing among peers and feeling that violence is also a common experience for others helps these women regain their self-respect. The experience of a self-help group can lead participants to carry out advocacy actions in defense of women's rights and gender equity (Esposito, 2012). In this regard, advocacy has been indicated as a particularly relevant factor in the process of escaping violence (Murray *et al.*, 2015).

A recent study (Esposito *et al.*, 2018; see also Box 1) analyzed the impact of a self-help experience on the quality of life of women survivors of IPV. The analysis showed significant group impact due to the affiliation and membership in a new social network, the interdependence and mutual support developed among participants, and a renewed ability to control the environment that fostered significant engagement in advocacy activities.

With respect to our stated theoretical principles, self-help groups create bonds between people who share a problem and provide opportunities to expand their social networks. They may also sometimes promote initiatives and actions to obtain better services or more innovative responses to their needs. They are environmental contexts that originate around a problem (use of *stressors*) but give much room for sharing positive experiences (use of *meliors*). In self-help groups, in fact, individual narratives are compared, and group narratives often emerge that make new life scripts and behaviors possible.

The limitation we find in self-help groups is their lack of ability to foster the sociopolitical empowerment of their members by exploring the historical roots and social aspects of individual problems.

BOX 1 Regaining quality of life after violence: a self-help experience in the Roman territory (Francesca Esposito)

Framing

Gender-based violence against women is a complex phenomenon that takes various forms and connotations and includes domestic violence or violence within intimate relationships. Looking at the intervention models used, mostly individual-type models are described in the literature (Mancoske, Standifer, Cauley, 1994; Johnson, Zlotnick, 2009). In general, an individual-type intervention appears to be crucial, especially in the early stages of the process of getting out of violence (Hirigoyen, 2000; Reale, 2000; Campbell, 2002; Leitão, 2014). Once this first phase is overcome, needs of a different nature emerge in the journey of reconstructing oneself and one's life. The building of positive social relationships, which allow the exchange of social support and the use of one's experience for advocacy actions and to help other survivors, is a crucial interpersonal process in the path out of violence (Flasch *et al.*, 2017). Participation in self-help groups has been proposed as one way to facilitate this process (Wilson, *Baglioni, Downing*, 1989).

Nevertheless, most of the studies in the literature on group interventions targeting women survivors of domestic violence refer to *face-to-face* support groups (Costantino, Kim, Crane, 2005; Martinez, Wong, 2009; Molina *et al.*, 2009; Morales-Campos, Casillas, McCurdy, 2009), or even online (Hurley, Sullivan, McCarthy, 2007), mostly characterized by a psychoeducational approach (Abel, 2000; Sullivan, 2012). Although self-help groups have long been studied in various fields, the literature on self-help among women survivors of intimate partner violence is rather scarce.

Self-help and gender-based violence in intimate relationships.

One of the few contributions in the literature on self-help and intimate partner violence is that of Hartman (1987). Although the author refers to the self-mutual-help model, the group is described as characterized by the joint leadership of a professional (*sponsor*) and a peer (*chairperson*). In this sense, this connotes itself as a "guided" self-help group (Ghaderi, Scott, 2003), distancing itself from the definition of "pure" self-help (i.e., without leadership by a professional), proposed in community psychology (Francescato, Tomai, Ghirelli, 2002; Ornelas, 2008). Despite this significant difference element, Hartman's analysis provides interesting insights into the potential inherent in using the self-help model in this area.

In general, the activation of such groups among women survivors of gender-based violence in intimate relationships allows for relational communities that, as highlighted by Ornelas (2008), respond to participants' needs for affiliation and sharing, promoting supportive contexts that facilitate adaptation and generate a sense of community. Comparing oneself with other women in the group who have experienced similar situations and feeling acknowledged and believed while sharing one's own story, has a profoundly transformative effect. It helps participants regain self-esteem and process the deep emotions of guilt, and to give new meaning to their experiences. Sharing with other women also provides an opportunity to learn new and/or different strategies for coping with the often-common difficulties that characterize the journey out of violence (e.g., with respect to the judicial process). Finally, witnessing the experience of female partners who have succeeded in rebuilding their lives after the violence they suffered provides group members, including newcomers, with a direct experience that change is possible, allowing them to regain confidence in their own abilities and to promote a process of individual and collective empowerment. In this way, through the group, women gradually move from the identity/narrative of "victim" to that of "survivor," an active protagonist of their own lives (Costa, Duarte, 2000).

The context

The context in which this self-help experience was born is SOSDonna h24, an anti-violence service of the City of Rome run by BeFree – *Cooperativa sociale contro tratta, violenza e discriminazio*ni, a feminist women's cooperative. Like most anti-violence services, SOSDonna h24 aims to offer support to women survivors of violence and their sons/daughters through the provision of integrated interventions (legal, psychological and social support). In January 2011, within the service a self-mutual-help experience was initiated that gave rise to the group The Phoenixes Flying to Ithaca, composed of Italian and foreign women with experience of intimate partner violence.

The group The Phoenicians flying to Ithaca

When I was approached by the members of the BeFree cooperative as to whether I would be willing to start a self-help group within the SOSDonna h24 service, I was immediately keen to establish that the model of self-help that would guide the experience I conducted was the one proposed by community psychology, that is, the model of a peer group without professional leadership. The professional (in this case, me), in agreement with the model proposed by Francescato and Putton (2022), would have facilitated the establishment of the group and accompanied it in the start-up phase, promoting the autonomy of the participants from the beginning and scheduling her own "death" (i.e., the end of the necessity of the exercise of her function as facilitator). The presence of an expert or expert in the group, Francescato and Putton argued, should be reduced to a few meetings, mainly concentrated in the initial phase; these meetings should be aimed at promoting the psychological identity of the group, the establishment among the participants of explicit and implicit norms, the identification of goals, roles and functions, and the development of skills to participate in and lead group discussions. It was clear to me the importance of following this model and resisting the easy trap of falling into dynamics, often rewarding for the practitioner(s), of dependency: in this sense I asked Professor Manuela Tomai to supervise my work to help me develop "meta thinking" about the process.

There were initially five women selected by the workers to participate in the group,all Italian except for one woman of Colombian origin. All were being followed individually by SOSDonna h24 on the path out of violence and all had expressed a desire to participate in a self-help group and get to know other women with the same experience. Over time, however, the participants almost all changed, except for one who remained until the end of the experience (in 2017) as a *helper*.

The process of starting the group proved neither easy nor straightforward and involved reformulating the guidelines for the establishment and start-up of self-help groups proposed by Francescato and Putton. The inexperience of the participants, who were involved in a self-help experience for the first time, was one of the critical elements, making it more complex to relinquish the reassuring and containing function I performed as a psychologist/professional. Some aspects I had to work on more were the group's difficulty in managing time, identifying discussion topics, and generally managing the group process. In addition to this, an additional complexity was connected - given the very nature of the group, formed based on sharing a hopefully transitory "crisis experience" (such as that of violence) - with participation from the outset being characterized, in most cases, as transitory. This has meant that cyclically the group has had to deal with the processing of grief resulting from the exit of some participants. This processing, however, is physiological, if not evolutionary, since it marks the exiting participants' attainment of a greater degree of autonomy.

The meetings, held biweekly and lasting two hours, were held at the SOSDonna h24 service, in a room equipped with comfortable sofas and chairs. The choice of the place to meet, as also noted in the guidance document *Starting a Self-Help/ Advocacy Group* (National Mental Health Consumers' Self-Help Clearinghouse, 2011), is a very important step. Indeed, such a place must be a "safe space" for participants to feel comfortable sharing often very painful experiences and narratives. It must also be an informal and welcoming place, as well as easily accessible by car or public transportation (a particularly important aspect in cities such as Rome). The group, by definition and choice of the participants themselves, was defined as open, that is, always open to new entries (a key aspect of successful self-help).

The first meetings, as is often the case in the start-up phase of a self-help experience, were devoted to getting to know and tell their stories. The participants were incredibly excited at the idea of meeting other women with similar experiences, and they felt the urge to tell their stories, somewhat like "rivers in flood." As the facilitator of this start-up phase of the group, I remember the difficulty of trying to accommodate this need to share individual stories, but at the same time foster a circular process of collective discussion and confrontation. It took several meetings for the group to start communicating in a more circular way and, more importantly, for them to start discussing their own modes of communication/discussion and shared rules for managing that space so that it would be a safe space for all. Initially, moreover, women often referred to me as being the "expert," the anti-violence worker who is knowledgeable about the issues. So, a big job was, from the beginning, to make it clear that my presence in the group was temporary and aimed at helping them move from being a "collection of individuals" to thinking of themselves and functioning as a group, and specifically a self-help group. In other words, my goal was to work with them on the "how," the process of the group, and not on the "what," the content of the discussion, an aspect of their exclusive choice. To do this, it was also necessary to discuss with the participants what it meant to create a self-help group and what the distinguishing features of such a model were (reciprocity, autonomy, the role of *helpers*, etc.). In other words, it was necessary to co-define the organizational dimensions of the setting, as understood by Carli and Paniccia (2003).

With the participants, we therefore gradually began to reflect on the different tasks that characterize the function of facilitating a group process (e.g., keeping time, setting meeting dates, establishing a small agenda with a list of topics to be addressed, making sure that all participants, if they want, have a chance to speak), as well as on shared values, goals, and rules that define it. A crucial step in this process of building a group identity was the choice of a name. Naming themselves offered women the opportunity to confront the meaning of their group. The Phoenixes was one of the first names proposed because, as expressed by one participant, "we feel that we will be reborn, albeit from our ashes, and that together we will succeed." Another participant, however, proposed the poem *Ithaca*, from 1911 (Kavafis, 1992), a text that symbolizes origin, reason and at the same time the destination of the long journey, similar to that of the legendary Ulysses, that each person makes through life. After much discussion, the group chose to merge the two proposals into a single name: The Phoenicians Flying to Ithaca. One component painted a picture depicting *The Phoenicians Flying to Ithaca*. The image was taken as a symbol of the group and was later used in a brochure created by the women to publicize an happening they organized for the March 8, 2016 holiday (see the figure on the next page).

In the summer of 2011, the women began holding their first self-directed meetings. In the beginning resistance was common, especially the participants did not believe themselves capable of running the group on their own. They were afraid of failing or activating dysfunctional dynamics among themselves. The same reluctance, to some extent, was also expressed by the service workers, with whom it was necessary to work on the fact that the exit from the group of the facilitator, that is, me, did not mean "abandoning" the women, but rather promoting their autonomy and confidence in their own abilities. I believe that the crucial factor in breaking the *impasse* and overcoming resistance was my deep-rooted belief that the participants could self-manage the group, in complete autonomy. The first *helper* was spontaneously and unanimously identified: she was among the longest-serving participants in the group and the only one with prior experience of leading groups (though not self-help). It was she who facilitated the first self-managed meetings, always interspersed with meetings with me, until it was decided that other participants should join her in the role of *helper*, to assume a more collective responsibility for the organization and functioning of the group. For the first few months, at the end of each self-managed meeting, the participants would draw up a small report that would be used in meetings with me to discuss the progress of the group and any difficulties encountered. Gradually my function became completely external, taking on the role of a consultant who provides moments of supervision and intervenes directly in the group only in case of crisis situations.

PHOENIXES FLYING TO ITACA
This self-help group was
born in January 2011.
In saying group, we refer to all
Phoenixes who are reborn from
their own ashes and have
chosen to no longer break down,
to no longer succumb to
any kind of violence.
We believe that sharing
life experiences is an
irreplaceable resource for
our group and an important tool
for change for other women.
Contact person:
Stefania 3396326949
betefae.segreteria@gmail.com

Phoenixes flying to Itaca

BE
free

Social Cooperative against trafficking,
violence, discrimination

www.bete.cooperativa.org
email: betefae.segreteria@gmail.com

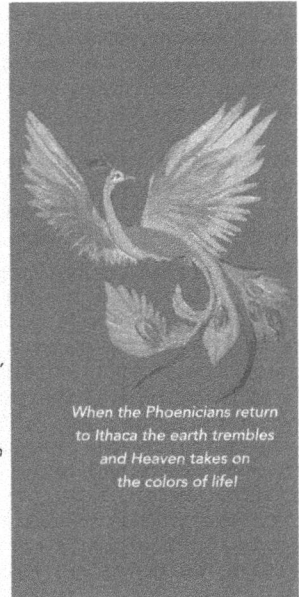

When the Phoenicians return
to Ithaca the earth trembles
and Heaven takes on
the colors of life!

And then, one day
we met and were
born...slowly we
rose into the air and
flew to Ithaca.

I am a Phoenix and
as such I am reborn
and fly high again. I
do not allow anyone
to put me in a
corner, to judge my
choices.

To be a Phoenix is to
have shared one's
story all together.
And to have found
the strength to
resume flying toward
a new future.

And no voice is
hidden I like the
group, and I am
thankful that it exists
because I can
express myself.

There is a happy island
that does not isolate you
and can make you happy
It listens to you not
judge you It welcomes
you and helps you pick
up the pieces into which
your life may have
turned. Confront, listen
and be listened. It will
quietly bring you back
toward self-respect
allowing you to look at
others but like a
reflection on a glass also
see yourself to find the
strength to return to the
world and from the
world always come back
ALIVE

One day straw dolls
bent by the wind
crossed their different
stories and
rediscovered
themselves as women
and mothers Today
none deeply hides the
other story Yet theirs is
a pink army with great
wings open to the
world with arms
outstretched to
welcome

We Phoenicians who in
the group find respect
dignity and sensitivity
for each of us.

I participate in the
group because I like to
listen and be heard

The truth that liberates
the soul and brings
order back

An example of such situations occurred in the summer of 2014 when the SOSDonna h24 service was closed and subsequently reopened in another area of Rome. On that occasion, my intervention, urged by the women themselves, was aimed at identifying another location to ensure the continuity of the group's meetings; a space that was close to the original one of the service and that provided an appropriate setting for the meetings. I hoped at the time that the "crisis," produced by the closure of SOSDonna h24, could turn into a moment of further growth for the group, so that it would finally become a self-help group autonomous from formal services and rooted in the territory. Unfortunately, things did not turn out that way. Although we were able to find a new, welcoming venue for the bimonthly meetings, the fact that we moved away from the service where the group had been born and raised triggered a phase of disorientation and disintegration, which almost ended in the demise of the group itself. It was only in the fall of 2014, when SOSDonna h24 reopened its doors and was able to start hosting Phoenix meetings again, that the group was able to effectively -- and efficiently -- restart. In the meantime, however, a tension had arisen between two of the *helpers*, a dynamic that had contributed to a shift in focus from the original goal (reaching out to as many women as possible and establishing a node in a larger network of support) to a power struggle for internal leadership. This problem, too, not without several meetings with the participants and supervision by me with Professor Tomai, was finally resolved and the group returned to meeting and functioning normally.

A significant event to remember is March 8, 2016, when women, invited to various meetings and broadcasts on the topic of gender-based violence, they decided to create a brochure of the group to be disseminated in the territories to communicate its existence and invite other women to join. In the summer of 2016, the SOSDonna h24 service was decommissioned, the funding period arranged by the City of Rome having ended. A few months later, the group, forced again to change locations, was discontinued. The difficulty of regrouping in this new space and within a new organizational framework (they were no longer receiving referrals of new women from SOSDonna h24) unfortunately contributed to the end of the experience in early 2017.

Nevertheless, this experience, which lasted for a full six years, was extremely meaningful for the women who were, at various times, part of it. In this regard we report an excerpt of testimony from one of the participants:

Yes, yes, I think so, of course it has impacted very positively, it's like living in a dark night and seeing the day appear. Thank to their experience, through the fact that they made it and so knowing that if they made it, you could make it too and so it reopens your hope towards life (A., 60).

In the group, and through the group, some women were able to solve their problems at the legal or work level, thanks to the support of others. Frequently, participants exchanged advice and information on how to deal with their respective situations (e.g., court paths), as well as coping strategies to protect themselves from violence. How to structure one's testimony, how to record calls and file messages, how to use the internet to find information for reporting, and how to behave on the street if the aggressor approaches are just a few of the many topics addressed in the meetings. Also crucial was the exchange of forms of material support, which ranged from hosting each other at home in times of difficulty and danger to helping each other with children, e.g., picking them up from school, or even accompanying each other to court-on-court days, offering support in case of health problems (e.g., doing overnights in the hospital). This aspect, which is specific to self-help groups, is very important in the case of women who, often alone, are facing the severe consequences of the violence they have suffered. In some cases, the group has also played a key role in protecting women from violence by their ex-partners where the state's response seemed to be delayed. As reported by one of the women:

and so, they followed me [...] even outside Rome, [...] they almost acted as a barrier of protection because he kept being violent ... and so he couldn't there because my friends were there [...]. Yes, they really did escort me, escort me (I., 49).

Through the sharing of the experiences of injustice and oppression suffered, contextualized within the heteropatriarchal matrix of the society in which we live, the group also helped stimulate in participants a desire to fight for women's rights and against all forms of gender-based violence. This has been expressed through a willingness to help other women survivors of violence, keeping the group constantly open to new entries, and in the engagement of many of the participants in activism and advocacy actions.

Fostering Brighter Futures

I would like to conclude by reporting the testimony of one of the first participants in the group:

What always happens, when we see each other, no matter how many of us there are, is a total exchange of trust in our future and that of the children who are undergoing the heavy situations we have. Sometimes, being there, I completely forget what happened to me and feel like I want to start living "normally" again. When I think back to when I first arrived, I see myself as a lonely, hopeless woman who didn't know where to bang her head to find some confidence in the future. Today I feel designed and eager to live. The exchange of our experiences and the fortitude that the others conveyed gave me back the drive for life that I lacked. During our meetings there is no judgment, no disbelief in listening, because we all know, on our own skin, that unfortunately what the others say is possible to happen, and we do not always have to try to convince our listeners that we are telling the truth and not lying. [...] We wanted the group to be open to other women who need it because it is only fair that others can meet like us, to begin a path of rising again. An abused woman can be recognized by the way she looks at you, her posture and the fear she has in expressing a thought; if others like us come, we will help them to look up again and look forward and not at their own feet with their heads down, and I am sure they will have a lot to say and pass on to us through their experiences (E., 49).

10
Networking as the practice of the community psychologist

1.   Network intervention models

The term "network" is a metaphor widely used in various disciplinary fields, which stands for a set of subjects (individuals, organizations, communities, financial markets, computers, etc.) connected by contributions of a different nature. In the psychological and social fields, the term is used to describe and represent the set of relationships between people, organizations, and social systems; in this case, the focus is more on the social ties existing between the subjects in the relationship than on the characteristics of the subjects themselves. In the psychosocial field, the term denotes methodological interventions at different levels; we can identify a level centered on the individual and the characteristics of their social network and a level centered on the interaction between organizations.

1.1.   Individual-centered network intervention The *first level* concerns the theoretical insights and operational practices of social work that recognize the preventive and rehabilitative potential of the network of belonging, its supportive functions, and the use of the resources present in the community. These methodologies of intervention share the Lewisian assumption that an individual's behavior and social mobility are determined by the forces present in their field (understood not in spatial but in psychological terms) at a given time; the position a subject occupies in the network and the characteristics of the network may have, therefore, more influence than individual variables or cultural attitude (Amerio, Croce, 2000).

An individual network intervention is based on a preliminary analysis of the structure of the social connections of the individual on whom attention is focused (ego). The first phase of the intervention involves mapping the network. In this phase, the practitioner uses tools such as interviews and semi structured questionnaires. The information collected can be supplemented with observational data. Such analysis may reveal conditions of marginality and a notable lack of ties and social support. The need arises in such cases to enlarge or even create from scratch a sufficiently rich and extensive personal network. A typical example is the deinstitutionalization of a prisoner or psychiatric patient. An appropriate community intervention will have to offer the person opportunities for socialization, especially in the early stages of reintegration: from the possibility of living in residential housing to that of attending a day hospital, from facilitating the search for employment to the opportunity to participate in initiatives and attend social centers in their free time. Indeed, community integration is based on opportunities to access resources, develop social networks, contribute to society, and engage in activities that connect people to their community (Ornelas, 2008; Aubry *et al.*, 2013).

Community integration is also a particularly relevant issue for people who have experienced either chronic or long-term mental illness or housing distress. They often experience feelings of loneliness, rejection, and isolation and do not have the same opportunities to engage in community activities or develop social networks (Siegel *et al.*, 2006; Townley, Kloos, Wright, 2009; Tsai, Mares, Rosenheck, 2012; Tsai, Rosenheck, 2012). The participation of people with mental illness in meaningful activities fosters a greater sense of community belonging, which, in turn, has a positive effect on the quality of life, satisfaction, and psychological well-being (Prince, Gerber, 2005; Townley, Kloos, Wright, 2009). As evidenced by some studies (Gulcur *et al.*, 2007; Yanos, Stefanic, Tsemberis, 2012), clinical diagnosis histories of hospitalization or psychiatric hospitalization do not necessarily determine the quantity and quality of a patient's social support network. A significant and prominent example is the *Casas Primeiro* (*Housing First*) program implemented by Ornelas and collaborators (2014). The project provides individual, distributed apartments within residential neighborhoods to people who are homeless and/or have severe mental illness. The program seeks to connect participants to community resources, facilitate relationships with the neighborhood, and support participants' projects and activities in the community. Several comparative studies (O'Shaughnessy *et al.*, 2021; Greenwood *et al.*, 2022) documented how participants engaged with *Housing First* programs perceived the services offered as more likely to develop skills and offer opportunities for improvement than participants engaged in traditional treatment.

In less radical cases, there may be a need to restructure a personal network that is perhaps quite dense but offers an inadequate or ambivalent type of support. In situations such as this, formal support systems (practitioners and social and health services) need to sense which "potential supporters" within the informal systems need to be stimulated, trained, or supported so that they more effectively or correctly help the target person or group. As much as possible, informal, supportive figures are chosen who have significant and privileged ties, starting with parental figures; it is important to emphasize the positivity of their role, noting the supportive effects and interdependence existing among members of a system.

Individual-centered network intervention refers to network therapy, *community care*, ego-centered network intervention (Speck, Attneave, 1976), social network work (Sgarro, 1988; Croce, Merlo, 1989, 1991; Ferrario, 1993; Maguire, 1994). In the 1990s, in Italy, Merlo and Croce, inspired by the work of Spek and Attneave (1976), developed a clinically derived network intervention scheme that effectively summarizes procedures commonly used in social work. So-called network therapy or network therapy (Attneave, 1980; Oliva, Croce, Merlo, 1995) is a useful technique for enhancing the supportive resources of the subject's closest network of ties, whose labeling as a "patient" is also intended to be avoided. Practitioners structure a series of sessions or group meetings, inviting significant members of their current network together with the person in crisis. These meetings, in addition to elaborating a social view of the person's crisis, are primarily aimed at identifying positive conduct and supportive strategies that can be practically implemented by the network surrounding the person. Interventions of this kind have been tried out in the United States on families in which incidents of mistreatment and violence had occurred, in which individual or family-only psychological intervention is seldom sufficient, but rather it is necessary to broaden and increase the core's ties with external systems (both formal and informal).

Network therapy

Network therapy tends to operate not only within informal systems, such as families, schools, or friendship groups but to enhance the action of formal systems (e.g., hospitals and community centers). Network-oriented strategies have been widely used in public health for several types of interventions, including but not limited to smoking cessation, cervical cancer screening, diet, and weight management (Shelton *et al.*, 2019). They have also been used with other vulnerable populations, such as substance abusers (Latkin, Sherman, Knowlton, 2003), in HIV risk reduction (Ghosh *et al.*, 2017), and to provide support for individuals with chronic conditions such as diabetes (Vissenberg *et al.*, 2016). A more recent and innovative use involves the use of *network therapy* with vulnerable groups such as asylum seekers and refugees (Weine *et al.*, 2008; Tappis *et al.*, 2016); in these cases, network strategies have improved social support and factors that reduce repeat victimization (Ogbe *et al.*, 2021).

1.2. Social network-centered network intervention The second level of network intervention relates to the interaction between organizations and refers to operational ways of working on the ground that configure a *networked organizational design*. This strand of studies places the focus not on the individual or the organization and the relationships these two entities may have with the external environment but on the set of relationships that exist between different organizations in the area.

Before presenting the network intervention model, we would like to discuss the historical and cultural factors that led to the development of this indispensable psychosocial intervention model.

2.   The how and why of networking

The development and spread of networking have been facilitated basically by three orders of events: the process of democratization of society, the development of social policies, and significant social changes.

1.   The development of "modern society" has gone through a progressive modification of the way of conceiving existence, which appears to be increasingly characterized by demands for psychological emancipation and the pursuit of individuation processes. The accelerated environmental changes to which modern Western man is exposed legitimize an increased expectation of discretion in the life of everyone. We seem to be able to hypothesize that in the paleo-industrial period, personal **identity** is **undergone**, that is, determined almost entirely by strong social and clan affiliations, with severely limited needs for individuation. Subsequently, the spread of industrial consumer products facilitates the search for an identity expressed through the purchase and possession of goods and services that constitute *status symbols*, an identity that we might call **bought**. These trends still appear to be dominant at the mass level today, but a tendency to seek a quality of life independent of the objects one buys is beginning to develop. This process leads toward the establishment of a **created** identity: an attempt to forge one's personality through a personal itinerary of formation and experience. External changes seem to have legitimized an expectation of personal and inner change and definition.

An individualistic culture seems to be asserting itself on whose consequences (in terms of positive or negative evolution) modern sociologists would show discordant opinions (Giddens, 1991; Lasch, 1991, 2019, 2021) but which nevertheless seems to be accompanied by events of democratization and collective emancipation (Amerio, 2000).

Such a situation also makes its influence felt in the public sphere, where there is an increasingly clear shift from the dominance of a single interpretation to the co-presence and interdependence of various ways of conceiving life and social welfare, and where there is a progressive shift of power, increasingly centered on the territory and voluntarist bodies and ties. These conditions facilitate and enable the development of networking, which, by its very nature, requires breaking rigid "chains of command," overcoming too one-sided definitions of problems and interventions, and exercising confrontation and collaboration.

2.   Looking at the historical course of social policies, it is possible to see the succession of stages of growth and critical rethinking of the changes brought about by the logic introduced as time goes on.

Social policy reforms, initiated since the 1970s, have had among their most significant guiding principles desegregation and deinstitutionalization, which put under indictment the one-size-fits-all, nonspecific, and dehumanized response that characterized social interventions before this period.

The responses provided to social distress were rigid, predefined, and organized, all embedded within a welfare-type circuit. We refer to social problems of various kinds, from child welfare to interventions in mental illness to the management of marginality and deviance.

Overcoming this situation has been implemented mainly by proceeding along two lines: the specialization and differentiation of services and their humanization and personalization. The consequence of such a development was the gradual creation of a formal network of community support. In this wake, since the late 1960s, there has been both quantitative and qualitative growth of increasingly differentiated services. This process of differentiation historically has been favored by the effort to adapt to the needs of different territorial realities and the evolution of the professionalism of the operators, which has also developed in the sense of specialization of skills.

In the present day, we are currently in the presence of a functional specialization of services that adds to the professional specialization of practitioners. The outcome of this evolutionary process has been to diversify and enrich the capacity to grasp and respond to the needs of the population, but at the same time to create conditions of overlapping competencies and/or compartmentalization of user demand. The loss of a comprehensive and systemic view of the situations under consideration has set the stage for a crisis in this "Tayloristic model" of distress management.

At present, services are mostly organized according to rather rigid *field* and *competence* criteria that incentivize the development of parallel paths of intervention and accentuate communication problems. While it is becoming increasingly evident how the new social and health needs and the new way of conceiving them can rarely be answered by a single professional figure, the need for integration and coordination of specific professional contributions is increasingly emerging.

In almost all cases, the new social and health needs require the construction of a personalized intervention project on which the contributions of distinct professional skills converge. Cases presenting a multi-problem configuration are very frequent in the social and health field. Similarly, there is a growing awareness of how even interventions for the development of the territory or inherent to the labor area (orientation, promotion of business work, prevention of school dropout, etc.) require intensive collaboration between different agencies and professionals for their success. The implementation of public policies (in the fields of health, education, and employment) requires joint working and convergence of several organizations with very different objectives, legal nature, and competencies. The implementation of *networked organizational designs* seems to meet this need. Considering changes in social policies that increasingly require complex interventions that can only be implemented within an interinstitutional network, joint planning becomes an indispensable working methodology as a point of intersection between different institutions and services that fosters partnerships between local authorities and the private social sector. A methodology to be implemented with circular and interchange modalities to allow the activation of comparisons and co-design and the possibility of recognizing resources and limitations of the territory in which interventions are activated (Leone, Prezza, 2005).

3. Finally, significant social events seem to have necessitated the development of an increasingly integrated relationship between the public and private sectors. We mention among the most relevant, the evolution of the role of the family, the gender division of labor, the demographic expansion of the elderly population, and the development of the phenomenon of urbanization, all of which have altered the modes of relationship within small communities, often causing the impoverishment of the network and its informal support functions. The presence of new social emergencies, such as, for example, the phenomenon of immigration and cultural integration, combined with the growth of the population's welfare expectations, has initiated a progressive difficulty in the current welfare system. The enormous deployment of professional resources in the mental health field of social services is predominantly a phenomenon of the 20th century. In previous centuries, people referred almost entirely to family members, friends, or neighbors for forms of social, economic, and emotional support. The impoverishment of the informal support network incentivized the population to "use" the formal network, and public institutions found themselves less and less able to respond to progressively more complex needs addressed to them by the population. This situation has highlighted the presence of empty spaces in service offerings, and this has fostered the emergence of new social initiatives and the development of the many private social organizations in which various protagonists operate social cooperatives, associations, and volunteerism.

The current state of services makes it increasingly clear that there is a need to identify ways of integrating public services and public and private agencies and to develop, among the nodes of these potential networks, non-competitive but complementary relationships that can offer diversified responses to community needs.

Today's society needs networked organizations to move toward ways of integrating formal and informal, societal and community, political, civic, and social, both in daily life and at the level of socializing agencies. Institutions are asked to move beyond the provision of services designed in a self-referential logic in favor of flexible responses that consider the interdependence of service systems. It is easy to see why the capacity for integration and communication a strategic element in the effectiveness of the work is currently: the more complex or richer the intervention program, the more weight is given to collaboration between different professionals and territorial services or agencies.

3.   What networking involves

We have seen how socio-historical factors have made the demands placed on formal systems of care more complex and have led to a perceived need for collaborative relationships between different agencies in the area. The cultural background, knowledge, and work practices of practitioners have not always proved, and are proving, adequate to the change useful for new social policies. We can say that the shift from a summative to an integrative model of intervention is not yet complete.

The need for better communication and more efficient exchange of resources between services is an increasingly recognized need among social workers, who often form informal or "natural" networks in the sense of unrecognized and/or legitimized. They frequently practice networking in an "empirical" way, creating networks made up of relationships linked to the personal and informal knowledge of colleagues whose expertise they recognize. However, the potential of such networks is marred precisely by the occasionality and informality of the relationships, by the lack of what Butera (1990) calls a "center of development and governance."

In network work, a key element of effectiveness is the identification of a center of responsibility that performs collectively recognized coordinating functions and is responsible for the execution, monitoring, and evaluation of the intervention.

The outcomes of a network intervention program depend on the policy constraints and resources available but also on how the organizations involved implement it and the type of relationships (reciprocal, intense, collaborative, or conflictual) between them.

Interactions within the network can be marked by both cooperation and intense hostilities. Conflict, much like in work groups, is a natural dimension that only becomes dysfunctional when it is not anticipated and effectively managed.

Collaboration between different services and/or agencies is not an immediately realizable reality; it is most frequently activated based on the utility found in professional practices, from the recognition that there are goals and methodologies of work that require specific organizational arrangements.

Networking constitutes a process that requires changes concerning all the different levels of an organization:

- *individual* level: change of professional identity. Networking is not only a way of working, but also above all a "mindset," a point of view from which practice can be better understood;

- *systemic* level: change in the representation of the service and its location in the territory and within individual intervention projects. Developing a culture of integration requires each service delivery center to learn to perceive itself as a "necessary but not sufficient" element in achieving the goal, as a "node" within a network of exchanges;

- *functional* level, of rethinking the functions and roles of individual departments, of the work process understood in less top-down terms;

- *structural* level, of facilitating the integration process through legislative innovations that allow for more flexible regulation of work activities and that recognize and provide a "place" for coordination functions;

- *psychosocial* level: change in roles, relationships between operators (in the same service and between different services), leadership styles, power distribution, and communication patterns.

The networked organization appears to be a current trend line in social and health services (Folgheraiter, 1997, 2004; Messina, 2011), but it represents a change, in some cases a radical one, in work procedures and professional identity, and therefore requires time and development of new skills.

Several authors (e.g., Casale, Piva, 2005; Taverna, 2014) point out that an effective and comprehensive response to the population's new social and health needs requires the contribution of distinct professional figures and the use of different methodologies. This way of working helps professionals formulate comprehensive and consistent responses to the multi-problem issues that the user usually presents.

3.1. Problem areas in the execution of network work
The concept of a network can evoke a dual image: one positive-protective and another negative-contingent.

In the former, network indicates a web whose connections configure relationships, interdependencies, and reciprocal ties between individuals and social groups. They generate a web of protective relationships that can activate experiences and exchanges and increase knowledge and knowledge, information, and communication. In the second meaning, the network is defined as a trap, a tool that harnesses and removes power. In this case, it is presented as a web of ties and constraints that condition, weaken, and hinder the freedom of the individual and the community. This negative perception probably accompanies the practice of professionals and practitioners who often fear "being bridled" in network work, losing autonomy and management space. More often, the two dimensions seem to coexist in that the importance of the network model is declared, and self-referential work practices continue.

The relationship problems that most often recur in the initiation of service network work relate mainly to the following areas (Maguire, 1994):

Frequent problems in starting work

- Competence;
- *Status*;
- Power;
- Mutual trust.

Reasons for interdepartmental conflict may emerge in determining "who does what" or who has "precedence" (by seniority, prestige, ability, etc.) in responding to certain user needs; in the distribution of tasks, certain areas of work may be deemed by practitioners to have greater prestige or to be able to garner greater public recognition and visibility. Another frequent problem is, in fact, that of *status* and interprofessional competition, often related to recognition needs and *power* dynamics.

Power is an important dimension and is more functional when it is distributed without excessive inequality and supported by the ability to activate processes of delegation. In the work of promoting a network, it is essential to question the underlying power dynamics and the changes that more or less directly the introduction of a network organization might prompt; if participation in the network is perceived by an organization only as a loss of autonomy and/or a decrease in its credibility, the functionality of that node will be greatly affected (Fosti, 2013).

Network work is placed in relationships of collaboration and confrontation, tending to be equal; however, the larger the network, the more actions are needed, to coordinate the relationships between the nodes of the network decision-making, the flow of communication, and the execution of the work process, but maintaining a democratic approach. The relationships between different nodes in the network may differ in terms of breadth (number of nodes making up the network), directionality (unidirectional or reciprocal relationships), frequency (number of contacts with a point in the network), intensity (degree of cohesion between nodes), strength (amount of time and emotional involvement), etc. It is important to emphasize, however, that there is not, and cannot be, a type of "a priori adequate ties" since the adequacy of the characteristics of the different ties existing within a network must be evaluated according to whether it is functional to the goal that the different nodes aim to pursue.

Some networks, for example, work best when governed by weak ties, while others require frequent and intense relationships. An example of the first type of network is provided by the many projects concerned with preventing adolescent distress. These organizations, especially if they work in the same area, have a shared user base but do not require consistent and lengthy intake. The network between these organizations will be more functional if it is characterized by weak links that still allow for the avoidance of waste of resources (organizations offering similar services, or services that do not meet the needs of the target population, etc.) but ensure a wide organizational autonomy; in these cases, the failure or failure of a node in the network does not affect the survival of the network itself.

Projects always aimed at youth and adolescents may need, on the other hand, networks characterized by *ongoing, intense, and factual collaborative relationships*. An example in this regard may be secondary and tertiary prevention projects for deviant teens that set the goal of social or labor insertion; in this case, the organizations called upon to collaborate may be many (police, juvenile justice, prison, families, local social services, voluntary associations, educational centers, local businesses, etc.) and will need constant and frequent communication flows, sharing of intentions and operating methods.

It goes without saying that within the same network, there may be a need to develop different types of relationships with various nodes, depending on the role each node must play in achieving the common goal. For example, an educational institution, in pursuing its educational objectives, may need to establish continuous relationships with some nodes in its network (e.g., with the students' families), *collaborative but not necessarily frequent relationships* with other organizations (e.g., with schools that precede and follow it during the planning of educational continuity activities at the beginning and end of the school year), *frequent and collaborative relationships* with higher bureaucratic bodies (such as the school board), and *collaborative but occasional relationships* with local services (e.g., when promoting specific projects or requesting intervention for students with particular issues). The effectiveness of the network is linked to the ability to create functional connections aligned with the goal and the ability to adapt them as needs change.

The links between services, however, are mediated by people-to-people relationships, availability, perceptions and shared attitudes within the organization. Indeed, among the prevailing problem areas, we find the lack of an attitude of mutual trust and esteem among the services that need to interact, which allows for an assessment of each other's priorities and competencies and fosters processes of delegation and the creation of a collaborative logic.

One mode of overcoming the difficulties associated with service integration that has recently been increasingly experimented with is project work. This tool facilitates the overcoming of power and competence issues because it allows for a non-rigid and definitive allocation of roles and spaces assigned to network nodes: they change as goals change. Project work opposes the logic of "treatment ownership" and instead supports the idea that each service is deputed to achieve more than one function while retaining sector specificity. Centrality in the work process and network coordination will be changeable and linked from time to time to the contribution each service can make to the specific project.

In these cases, the integration process takes place through the *project team*, which becomes the prevailing center of reference for the operator as opposed to membership in their service.

3.2. Key factors for effective network functioning

The development of a culture of integration and the emergence of collaborative logic can be facilitated by the presence of certain factors, both internal and external, in the various organizations involved.

The first element that facilitates the integration of a network is the presence it of a shared *culture*; this term refers to values, codes, communication processes, and the type of language used. Particularly relevant elements of culture include the value given to *diversity*, whether understood as a potential enrichment or as a factor of exclusion or disruption. Along with culture, *operational practices* should be mentioned, such as how effectiveness is *planned* and monitored, the *reward systems* used, and the ability of practitioners to perceive benefits and opportunities and not just limitations in participating in in-network group projects.

Other essential components of network organization are proposed by Benson (1988), who identifies four:

1. *Consensus on scope*. It refers to the degree of agreement among practitioners of different services on identifying the roles and functions of their own and others' organizations within the network. The definition of the scope of an intervention does not imply the identification of operational goals and areas

of responsibility, over which services and operators have a wide margin of discretion. This space is a terrain for bargaining and agreement, on which many factors influence; among them the "power" of a service, from the skills of operators to agreement on the effectiveness of the methodologies to be used;

2. *ideological consensus*. Here we mean the sharing of values, implicit and explicit, that characterize the purposes, logic, and modes of action of network organizations;

3. *positive evaluation*. Reference is made to the image one has of other organizations, the evaluations and esteem that those involved in networking have for each other, which, if positive, reinforce collaborative processes, as is easy to guess;

4. *operational coordination*. This refers to the coordination methods that each organization implements to initiate effective collaboration processes.

Leone (1993) finally identifies indispensable contextual elements for the promotion of networking, which we can summarize in the following points:

- the presence of a program-promoting organization that manages indispensable resources for all other organizations;

- the inability for them to take advantage of the resources except by joining the program;

- the undisputed credibility of the sponsoring organization and its sponsoring members, who should be regarded as leaders in the field;

- The overall sharing of the program;

- the ability to recognize and use existing natural networks and to strengthen ties deemed more functional only at a later stage;

- the development of a *design logic* (Leone, Prezza, 2005) that enables practitioners, organizations, and consultants to more effectively manage processes of internal change and adaptation to external changes (territories, users, environmental constraints).

4. Organizational influence styles

While the collaborative attitude is a decisive facilitator, the democratic and collaborative strategy is not the only one that can be identified. Indeed, among the different agencies that find themselves working in the network, it is possible to find the use of different ways of influencing other organizations and directing their behavior.

We present, by way of example, the typology developed by Benson (1988), which includes four different change strategies. As is the case whenever a classification effort is made, the identified behavioral categories are unlikely to occur "in a pure state," while they are more likely to merge, overlap, develop over time, or succeed one another through sequences.

- *Cooperative strategies.* These strategies are based on the premise that there is no power imbalance between the parties and that everyone has value to exchange. The exchange process, i.e., the benefits gained, and values surrendered, is mostly explicit and shared, as is the definition of common goals and projects. Exchange values mean information, professional skills, means and tools, and the number or type of users.

- *Breakup strategies.* These are strategies most easily implemented in the presence of strong power disparities among networked organizations. They consist of behaviors that intentionally tend to weaken the position of an organization or its possibility of intervention. Interference can be acted upon in various ways (detour of funds, execution of activities not attributed to that organization, detour of communication flows, etc.), mostly explicit.

- *Manipulation strategies.* In this case, disruptive actions are acted premeditatedly and not explicitly; usually, interference is caused in the flow of resources or the channel of access to them (e.g., changing the procedures for accessing necessary funding without equally informing all affected organizations).

- *Authoritarian strategy.* This strategy occurs only when power is highly centralized in a network node and used by this organization in an authoritarian way in defining projects, resource flows, activities of operators, and relationships among participants.

5. Community psychology and networking

Community psychology conceives of reality as a set of relationships between elements of increasing complexity: individuals, small groups, organizations, and communities; it has the task of studying transactions between networks of social systems, populations, and individuals (Murrell, 1973).

Community psychology has a complex view of the human-environment relationship, a universalistic and systemic conception of social problems, and a mode of reading and intervention that is always multi-dimensional. The community psychologist, because of his training, habitually perceives himself and his activity within a larger context with which he is in constant interaction and in which he is interested in grasping dynamics, potentials, and constraints. He is aware that each problem is multidimensional and, therefore, carries out extensive and complex processes of analysis.

In this theoretical and applied framework, the network concept fosters an understanding of the circular interactions between individual and social levels and facilitates intervention at the interface between the psychic and the social.

It is easy to see how current trends in the organization of services, the orientation of relationships between public and private agencies in the area, and the basic logic of network intervention are deeply consistent and congruent with the theoretical and methodological principles of this discipline.

The contribution of community psychology to the facilitation and implementation of networking can, therefore, be significant. One task that falls to the community worker is to integrate, through networking strategies, formal support systems with informal support systems and those belonging to the nonprofit area to create multiple ties and social networks.

His intervention can be aimed, therefore, not only at identifying resources in the area but also at building networks of relationships among them. In the implementation of networking, he collaborates:

- To the development of the project;
- To eliminate resistance and opposition between professionals belonging to different territorial services or agencies;
- To use pluralistic logic not only within but also outside their organization;
- To build a common culture related to working by projects and not by tasks;
- To the production of shared rules on intervention methodologies and evaluation criteria.

6. Concluding remarks

Network work applies some of the guiding principles that underlie a theory of technique in community psychology. Indeed, network work encourages pluralistic interpretations of a social problem and integrates different types of knowledge and professional skills. It activates forms of collaboration and participation among different organizations in an area and can bring out locally produced knowledge from the people involved in the social problem. Networking also lends itself to giving voice to minority narratives and promoting the production of new metaphors and new narratives that make additional scripts and roles thinkable for individuals and social groups.

However, it is not necessarily the case that networking per se forces or helps to critically reflect on the way dominant narratives describe a certain social problem, nor especially that it automatically leads to examining the historical origins of the social problem and the unequal distribution of power to access resources in the social context. It does, however, make it more likely that at least one of the actors involved performs these important functions, which should be, as argued in this book, the primary task of the community psychologist. Networking lends itself particularly well to promoting and implementing empowerment projects that create links between people who share a problem and between the services that deal with it; therefore, it increases the social capital of a community when effective.

11
Empowering training: a multifaceted intervention strategy

1.   Theoretical and methodological roots of empowering training

In a world characterized by continuous and rapid technological, environmental, and sociopolitical changes, individuals, groups, and organizations must develop the skills to adapt to these changing contexts and possibly transform them into new opportunities. Italian community psychologists have developed some tools to facilitate these transformations. In chapter 7, we described how using a tool such as the PMOA can help organizations of all kinds (micro, small, and medium enterprises, public administration, associations in the nonprofit world, social cooperatives, and voluntary associations) to become more creative, empowered and empowering (Francescato, Tomai, Solimeno, 2008). In this chapter, we will describe another method of intervention that aims to foster individual and group empowerment: empowering training.

The empowering training model integrates knowledge and certain tools developed by a range of disciplines, from neuroscience to community psychology, from the sociology of new media to adult education, to foster the overall growth of the individual and their ability to influence their future and the contexts in which he or she operates. Empowering training includes theoretical contributions from community psychology (Rappaport, 1995; Zimmerman, 1995) and occupational psychology, drawing on Colletti's (1991) service sector training models and Bruscaglioni's (2005; 2007) innovative training proposal for the development of personal empowerment. Also traced are references to the practices of feminist self-awareness groups[12]

12. See https://efferivistafemminista.it/.

and the formative autobiographical experiences of Knowles (1996) and Rogers (1971). The methodology aims to increase the psychosocial empowerment of participants, and that is, to stimulate a process of generative empowerment, carried out through the enhancement of the ability to actively control one's own life and influence decisions by implementing actions useful for achieving desired results.

A variety of collaborative training strategies centered on both face-to-face and asynchronous online training are experimented in empowering training, possibly diluted over time, to provide opportunities to consolidate individual learning and promote mutual help among participants. Empowering trainings usually explore both **family microcultures** and new ways to analyze **personal and generational media assets** to uncover desires and better manage negative emotions that sometimes cause us to make suboptimal decisions. The aim is to improve the ability to network and read the changes that have occurred in local, national, and international contexts, using the various existing social networks in creative ways.

Empowering training fosters change by addition rather than rupture; people are encouraged to imagine a broad spectrum of possibilities for their future, including the set of previous ones but adding significant new ones. Empowering orientation and training aim to bring out desires and address problematic nodes, exploring the positive emotions that promote empowerment and the negative emotions that can hinder the achievement of desirable changes. They help distinguish between needs related to experience, which evoke discomfort and lack, and desires that project into the future and tend to create "new possibilities" (Bruscaglioni, 2005). In this age of advertising that tries to make products and services become "needs" that are only "induced desires" created for maximum profit, it is vital to reflect on this difference so that we do not waste our limited energies pursuing false desires and neglecting deep needs.

Empowering training is a path of change in stages. First, it brings out desires and explores multiple desirable futures. Then, it helps to explore one's past, and that of the group one belongs to identify negative emotions that can hinder change and positive emotions that can strengthen capacities that create new possibilities, as Bruscaglioni (2007) postulates.

2.   The stages of empowering training

The empowering training developed by Francescato, Tomai, and Solimeno (2008) (pioneered by Francescato over the past two decades and conceived and reconceived by her as a work in progress) always takes place in a small group. It includes four phases, which can be of different time lengths, depending on the type of groups, their needs, and desires. For example, with groups of girls in prison, with multiple traumas behind them, one course lasted three months; with people who had lost their jobs, on the other hand, six days in-person and three weeks online; with disadvantaged school children one week, in secondary schools as little as two days. The important thing is that these phases are always explored sequentially:

- Step 1: Explore desires and imagine different possibilities for desired changes;

- Step 2: Reflection on acquired skills and experiences to identify one's strengths and weaknesses;

- Step 3: Examination of the environmental and virtual contexts to grasp and enhance the resources of the various contexts, while also reflecting on their critical issues;

- Step 4: Identification of congruencies between individual desires, acquired skills and demands, limitations, and opportunities of environmental and virtual contexts. Empowering training can be done individually, but in groups, its effectiveness is enhanced.

2.1.   Step 1. Objectives of empowering training and exploration of desires and future possibilities The overall goal is to increase participants' empowerment. Empirical surveys (Francescato, 2010) have indeed shown that empowered people can let hope predominate over fear, set and achieve challenging goals, have better relationships with partners and children, and use humor to cope with problems. They also have higher self-esteem and are more satisfied with their lives. There may then be various sub-goals, to be pursued with different exercises or activities:

- Exploring desires and futures possibilities: rediscovering multiple desires and posing proactively and positively in various areas (work, relationships, leisure, political-social engagement, health, and mental and physical well-being);

- Explore the desires for change within the family microculture. Through techniques such as the 'family novel,' 'work novel,' and 'school novel,' the aim is to enhance the understanding of family microcultures, reevaluating their positive aspects and enabling energies while neutralizing negative family myths;

- Examine generational and personal media novels ("tell me what you watch, and I'll tell you what you want");

- Reframe particularly frustrating and negative work experiences (multiple layoffs, years in layoffs, oppressive work environments, etc.);

- Broaden horizons: bring out desires and multiple possibilities for work or engagement in specific areas (leisure, hobbies, and interests; health and mental and physical well-being; civic engagement, volunteerism, political-social associationism) after listening to other trainees;

- After the course, choose priority desires to begin implementing throughout the course.

Fictional novels: The exercises are called "novels" to emphasize that memories can also be not completely true but "fictionalized stories or narratives" Therefore participants feel freer to tell what emerges without censoring themselves, more freely identifying needs (unresolved remnants of the past), desires, and various possible futures.

To this end, after each exercise, in a special envelope called the **"chest of wishes and various possible futures,"** each participant inserts slips of paper with the wishes and needs that emerged during the exercise. Depending on the type of group you are working with, you can also invite participants to work in pairs and, in turn, ask the questions relevant to each exercise. At the end of each exercise group, you can briefly share the results achieved and specify (if you want to make them public) what needs and desires were included in the "treasure chest." Some examples of exercises on family microcultures are given in Box 4.

BOX 1 Sample exercises on family microcultures.

Each participant is invited to individually re-explore their family roots through analysis of the family novel, with the help of prompting questions and suggestions.

1. *Ideas, interests and characteristics of grandparents and parents*

What are/were your grandparents' interests, hobbies, and passions? What were grandparents' ideas and values about money, work, politics, male and female roles, and goals to pursue in life? What personality traits and positive aspects of grandparents do you have or would like to develop? What about the negative ones, which you do not want? What family myths emerge from family stories or proverbs mentioned in grandparents' homes?
What were your parents' ideas and values about money, work, politics, male and female roles, and goals to pursue in life? What personality traits and positive aspects of your parents do you have or wish you had? What family myths emerge from family stories or proverbs often quoted by your parents?

2. *Tales from parents or grandparents about your birth*

What are the circumstances of your birth? Were your parents trying to have a child or did you arrive unexpectedly? Did they want a boy or a girl? Who chose your name? Do you like it?

3. *Very first memories of childhood*

What are your childhood memories? Photos, home movies, etc.?

4. *Memories of how you were talked about in the family as a student.*
Were you considered studious and intelligent? Or not too interested in studying? How did your family react to bad grades and possible flunking? Do you have memories of secondary school? Better or worse grades? What were the main fears and sources of joy? What dreams and interests did you have then? If you still want to fulfill any that you have not explored, put them in the wish envelope.

5. *"Inherited" affective and emotional heritage*

What were the prevailing attitudes, observed by you or heard about, of your maternal and paternal grandparents about male-female and parent-child relationships? What about those of your parents?

Have grandparents and parents influenced your current relationships and behaviors as partners in a couple or as parents, siblings?

6. *"Killer phrases" heard in childhood and adolescence (e.g., "you're worthless," "you're denied," "you'll never make it," etc.).*

How to get rid of negative memories if they still affect you today? Write down the killer phrases that diminish your empowerment and write in large print the opposite of what they state (e.g., "you are a jerk" becomes "I am not a jerk, instead I am very smart in various areas"). The slips of paper on which the sentences were written should be placed in the wish chest.

7. *Family emotional attitudes toward negative and positive events and management of negative and positive emotions*

If events such as accidents, job loss, relationship breakdown, etc., happened, to what were they attributed, fatality, responsibility of others, or personal behavior? Did reproach and guilt or anger and rage toward others prevail? How were negative emotions (envy, anger, wrath, sadness, sorrow) and positive emotions (joy, enthusiasm, laughter) handled by both parents? Who do you resemble today in managing emotions?

8. *Your grandparents*

Were they impulsive or reflective? Did they tend to see the future in positive or negative terms? Did they tend to worry about events worth worrying about or even minimal events? Were there differences between grandparents and grandmothers?

How much did irrational factors weigh in their behaviors, such as assuming the role of helpless victim, enjoying complaining or, conversely, avoiding all concern believing that if a problem was put aside, it would solve itself? Were they superstitious?

Did they often remember and repeat negative events from the past? Or did they more often recall positive events?

Current affective relations area

What desires do you have now about improving your emotional relationships with friends, relatives, partners and children? Write down the ones you could begin to realize during this empowering course.

2.1.1. Media narratives In our empowering training, we use several strategies to examine how the media have influenced the macro-culture in which we are all immersed and, therefore, our generational and personal media novels. Media novels are effective tools for increasing our empowerment; examining them can help us discover desires and aspirations of which we are often not fully aware. By comparing ourselves to the media models of others, we can, in addition, also be stimulated to change the media we see. By exposing ourselves to different stimuli, it becomes possible to change the patterns that influence us as well. We first try to *identify generational media novels* by comparing what types of media were accessible in the different historical periods when our participants were preadolescents (11-15 years old). Participants are divided by age (so that they were teenagers in the same years) into pairs or small groups who must remember specifically what movies, songs, TV programs, blogs, internet sites, or social networking sites nurtured their microcultures. If desires emerge, they are placed in the casket.

Participants who were early teens in the years (1954–1962) remembered the first children's programs put on the air), such as *children's TV, Carousel, Lascia o Raddoppia, and Il Musichiere*. The early scripted shows were also beloved.

Television in those years had not yet entered every home. It was watched by friends, but even more often in bars and cafes. The commercials of *Carosello*, the contestants of *Musichiere*, or *Lascia o Raddoppia*, the speeches of politicians on *Tribuna Politica*, and the events recounted by the news stimulated animated conversations among people of different opinions. People would compare each other and argue, in a choral enjoyment of the same programs, which were the only ones aired by one channel and then later by the three RAI channels. In addition to recalling with nostalgia this choral fruition that encouraged the comparison of opinions, many trainees recalled with pleasure how women, with the excuse of going to see Mike Bongiorno or Mario Riva or the Sanremo Festival, came out of their homes en masse and invaded bars and cafes, previously frequented mainly by males.

Here is how Maria (67), a resident of a town in northern Italy, describes this choral atmosphere:

We would eat supper early to go to the bar well in advance to grab the best seats, and listening to the broadcasts was punctuated by constant comments and jokes and great collective expressions of emotion. We would share laughs, and then we would all go back together. The streets of the village, which were always silent before, were traversed late in the evening by groups of people of all ages, children, adults and old people, men, and women, who kept discussing the broadcasts we had just seen. I still remember the good feelings I had as a young girl listening to the different opinions; I also met many people that my parents usually did not frequent. Television in those days opened me up to the world, both the magical world of the screen and the convivial world of the people in my village.

The desire that emerged from this testimony was to find situations outside the family where they could experience conviviality among different generations.

People who are now 35 to 45 years old as early teenagers mostly consumed many cartoons, which they remember with undiminished affection. Almost all these teenagers also watched, very frequently, television programs such as the Sanremo Festival and *Fantastico*, but they remember, among the programs that affected them most, those that talked about romantic relationships, such as *The Couples Game* or *He loves me, he loves me not*. They also included many movies in their generational media novels: *Flash Dance* and *The Time of the Apples* were among the most popular. These teenagers spent a lot of time listening to music: De André, Guccini, Zucchero, Vasco Rossi, Duran Duran, and Renato Zero.

Discussing as a group about themselves and how these media acquaintances of theirs had influenced them, they came to some interesting conclusions. They all agreed that the way they dressed, combed their hair, and groomed their physical appearance was greatly determined by the protagonists of their favorite shows. Cartoons, which often featured orphans among the main characters, reinforced in our trainees the value of a close-knit family. In general, watching movies or shows about couples has encouraged some to talk more freely about sexuality with their peer friends.

Among the TV programs they remembered, not surprisingly, *Non è la RAI* and *Amici*, which they watched because they also talked about sex and showed many beautiful girls for boys to admire and for girls to model themselves after. The image and looks of girls seen on TV heavily influenced perceptions of their adolescent bodies, so much so that some girls went on fierce diets to achieve the standards of beauty and thinness admired on TV.

These adolescents also made heavy use of video games and watched and rewatched cult movies such as *Blue Lagoon*, *Top Gun*, *9 1/2 Weeks*, and *Trainspotting*. Such films also showed scenes of transgressive sex, violence, and drugs, which provoked perturbations and strong emotions, but also reflections on the sex roles of women and men. For some girls, films such as *Blue Lagoon* fostered an idealization of romance, which later did not match reality and fostered their first major romantic disappointments.

Many kids of this generation have religiously watched every episode of their favorite TV shows, such as *Beverly Hills*, *The Robinsons,* and other familiar sagas like *Beautiful*, comparing their parents and homes to those seen on TV.

Mara, 30, from Rome, living in Magliana, a working-class neighborhood in Rome, recalls the following:

I was a *Beverly Hills* fan, and as a young girl, I had an agitated relationship with my parents, especially my mother, who had dropped out of school in junior high school. I used to compare her to the mothers of the *Beverly Hills* protagonists and wished I had their mothers and their homes. Mine looked and felt shabby in comparison! My mother seemed ignorant and domineering, I didn't feel understood, I didn't want to be like her, I was unhappy in my own home, and I took refuge in the world of TV.

In general, members of this generation admitted that they have watched too much TV, which has "dumbed them down," and suffered from loneliness in front of their screens.

Those who were early teenagers only a few years ago and who are now 18 to 25 years old have grown up with smartphones and the internet, as well as often with satellite TV. They are part of that generation that is exposed to many "media temptations" and must resist the strong attraction that media exerts on them. Media they used in the privacy of their bedrooms, often in virtual company with friends contacted online. Members of this group spent about seven hours a day as teenagers chatting on internet sites such as YouTube, socials such as Facebook and TikTok, or video games, texting, and listening to radio and TV. They watched less television than the other groups, especially soaps, movies, and satirical programs. The latter have conveyed to them a view of politicians as corrupt, interested only in their careers, far from the problems of young people.

Programs such as *Big Brother* and *Temptation Island* and internet sites fostered the development in their adolescence of a paradoxical view of sexuality: strong temptations to be sexually very free to experiment with promiscuous relationships on the one hand, but also early desire to get into a steady partnership at 14-15 years of age (relationships that often lasted only a short time, due to infidelity or sexual needs that were not appreciated or shared). The perceived influence of media on clothing also continues for this generation; for many, the scruffy look of hip-hop dancers is fashionable, but several cited teenage frustrations at not being able to buy "amphibians" or "spikes" (shoes and jackets) worn by their idols and particularly coveted.

One fact distinguishes this generation: they are not only users but often producers of movies, videos, and images that they spread on YouTube, TikTok, and other sites. Some pointed out that the strongest emotions were derived from their active participation in games or in producing movies and pictures with cell phones

Here is how Luca, now a student enrolled in a discipline of arts, music, and entertainment (DAMS) program in Bologna, talks about his media experience, "I had a lot of fun filming with my cell phone what was happening in the classroom or around. As early as 12 or 13 years old, my friends and I were already sending each other pictures and home movies and compiling rankings of the coolest, which we then put in a video to show to everyone."Despite the differences in age, gender, profession, schooling, nationality (immigrants from Africa, Asia, and South America, as well as citizens of various European countries), and social *status* usually present in our groups, members identified the same areas as those most influenced by the media: the use of leisure time, values that give meaning to life, violence, health, politics, gender and ethnic stereotypes, erotic and love relationships, and between parents and children, attitudes toward school and work.

In comparing different media generations, it became increasingly apparent that while older people remembered the face of their high school classmate, 21st-century adolescents remembered more of the physical appearance and behaviors of well-known cartoon characters, video games, TV shows, or internet sites, or influencers, but had "forgotten" the face of the classmates with whom they had spent three years of their lives in middle school.

Virtual emotional experiences, that is, the emotions we feel seeing someone else doing a certain action in a television program or video game, might have a similar impact to when we do this same action in our own lives, suggest neuroscience scholars. Many of our trainees supported this hypothesis: they recalled as their first significant sexual thrills not only, and not so much, childhood games with peers, episodes of collective masturbation, first kisses, or falling in love that they had experienced, but episodes they had seen, images of protagonists in videos, movies, YouTube clips in which others, and not them, had strong erotic experiences. The same mechanism might act in encouraging the imitation of other complex interpersonal behaviors, from positive ones like hugging another to negative ones like hitting them. We are neurologically predisposed to imitate what we see others do. Imitation plays a key role in learning not only in childhood but throughout life. Moreover, after viewing certain behaviors in others, we will tend to imitate them both intentionally and automatically and unconsciously. In fact, from our experiences in socio-affective and sexual education, it emerged that many people had taken as role models, often unconsciously, characters seen in the virtual world to an increasing extent as technology made access to others' sexual and romantic experiences more available. In the group, participants also had much discussion about whether the increase in violence seen in various media had influenced them.

2.1.2.   The work narrative In addition to family and media narratives, for those who have lost their jobs, the school narrative is also examined to understand how the educational path was chosen (and by whom). Positive and negative experiences, as well as key positive and negative figures among teachers and peers, are identified, along with 'killer phrases' about the individual's future potential. Significant time is also devoted to reviewing past work experiences, analyzing positive and negative organizational variables using the PMOA framework, highlighting areas where the greatest incongruences existed between company demands and the individual's desires, and exploring possible alternative developments. Finally, at least two or three options for desired work (and/or new skills to acquire to pursue the desired jobs) are formulated, which are then placed in the 'treasure' chest of desires for change.

You can conclude Phase 1 of the empowering course by choosing priority desires to begin fulfilling throughout the course: "Carefully reread all the desires listed during the meetings. Choose at least three or four priority desires that you can begin to fulfill in this course. With the help of a colleague, identify who and what can support you and how to overcome any environmental and internal obstacles."

In longer courses, we also used various platforms such as Yahoo! Groups, where participants continued to practice followed by a tutor in the intervals between in-person meetings. In Box 2, we reproduce an example of an online course.

BOX 2 Online pathway: first stimulus interventions of the tutor to the online group.

1. Read some material on empowering training (Francescato, Tomai, Solimeno, 2008), send comments and any clarifying questions.

2. Conduct a group exercise on "current emotional assets"; importance of emotions as a driving or hindering factor in achieving a goal. Stimulus questions:

- Can you remember the last time you felt a strong positive emotion (joy, satisfaction, excitement, cheerfulness, a feeling of well-being)? What triggered it? How did you express it? Did you share it with anyone?

- Can you remember the last time you felt a strong negative emotion (jealousy, anger, envy, anger, sadness, anxiety, fear, feeling unwell)? What triggered it? How did you express it? Have you shared it with anyone?

- Are you similar to someone in your family in managing emotions? Do you want to change something about your current emotional capital?

3. Send encouraging group comments and even individual e-mails about activities performed and optimal management of negative and positive emotions. Recommend using and/or watching specific programs, movies, scripts, reading books and poems that can help to have more positive emotions and better manage negative ones.

4. Finally, each participant should send a description of the activities performed to achieve the identified desires and, if he or she found obstacles in doing so, describe them. Other participants and mentors can offer suggestions on how to overcome the obstacles and celebrate the progress made.

2.2. Step 2. Skills assessment

Skills assessment is an orientation technique that aims to encourage reflection on acquired competencies, interests, skills, values, and experiences to identify one's strengths or, conversely, weaknesses. It originates as an action aimed at professional development and is related to continuing education and active labor policy pathways. It explores the various aspects that influence a person's work: motivations, skills, interests, values, and, in this growing knowledge society, increasingly also personal qualities and flaws, behaviors favorable or detrimental to physical and mental health, relational modes, and awareness of one's main strengths and resources.

Taking a skills assessment in a broad sense means valuing personal experiences, not only professional, formal, and informal, what you know and can do, but also understanding where you can transfer your skills and abilities and where and how you can use and develop your potential.

Through the skills assessment, the person is then able to work out the various stages of different desirable professional projects, including through reflection, discussion, and comparison with the group of trainees.

Various tools are used: questionnaires on interests and skills, tests on work behavior, etc. (Francescato, Tomai, Solimeno, 2008). In some cases, depending on the objectives and duration of the course, in the interval between one day of in-person training and the second, which may take place after one or two weeks, the trainees are tutored by an online tutor. The tutor elaborates on the topics covered in person and suggests new exercises.

Following Gardner's theory of multiple intelligences (logical-mathematical, linguistic, spatial, musical, kinesthetic, interpersonal, and intrapersonal; Gardner, 1994a; 1994b), the tutor, for example, proposes games and exercises aimed at allowing the group to experience the various types of intelligence. Participants can verify their different modes of knowledge and approaches to reality, confirming, refuting, or enriching their self-assessment. Alternatively, if the tutor is facilitating a group of young people who have never worked, they will help them explore the various skills acquired in organizing a party, a trip, an event, a sports tournament, helping a peer with homework, or calming a fight among friends. The intent is to identify their strengths and problem areas to enhance more desirable and feasible professional projects and to identify educational projects to undertake to fill gaps or improve identified problem areas.

Emotional Assessment: At the end of the two phases just outlined, the participants, first individually and then in groups, try to answer the following question, "What is your current emotional asset after the first phases of the course?"

After going through the various exercises, participants are asked to recall how they have handled negative emotions such as anger, wrath, sadness, jealousy, envy, fear, and positive emotions such as joy, enthusiasm, and hope over the past few days, and are asked to identify what strengths and problem areas have emerged and how they could improve their emotional assets They also explore how parents, siblings, and grandparents express negative and positive emotions and what they want to keep of their emotional heritage and what they want to eliminate.

This exercise on current emotional assets aroused much interest and involvement among a group of young job seekers and produced specific additional desires for change and positive emotions, such as confidence that they would be able to find at least one of their desired career paths. However, a large minority wanted instead to develop better management of some strong negative emotions, very much present in their current lives, dominated by the fear that none of their desired jobs would be found, making their personal and professional futures increasingly bleak. These desires for change should also be placed in the caskets.

BOX 3 Assessment of psychosocial attitudes, skills and emotional assets.

Area of behavior
- Risk protection and entrepreneurship
- Personal empowerment
- Interpersonal empowerment and positive attitude
- Taste for achievement
- Willingness to change
- Personal initiative
- Positive health habits

Cognitive skills area
- Logical intelligence:
— linguistics
— space
— bodily
— musical
— interpersonal
— intrapersonal
- Creativity
- Problem solving
- Use of cognitive and differentiated strategies

Language skills area
- Oral communication (informal and formal)
- Written communication
- Group communication
- Public speaking

Group relationship area
- Willingness to work in teams
- Participation in the work of the group
- Leading working groups
- Collaboration and integration of goals
- Motivation, listening, and development of colleagues' contributions
- Relations with external stakeholders, conflict management
- Political sensitivity, shrewdness

Affective-emotional skills area
- Recognition of one's positive and negative emotions
- Appropriate expression of positive and negative emotions
- Controlling destructive emotions
- Ability to give and receive support

Contextual skills
- Can find information
- He knows people and entities
- Knows the institutions in the area

2.3.   Step 3. Understanding the contexts: discovering the world outside We use several strategies at this third stage.

1. Our empowering journey continues by providing participants with effective tools developed by community psychologists to increase the sociopolitical empowerment of individuals, small groups, organizations, and local communities in this book (see chapters 5, 7, 8,9. Also, in chapters 16 and 17, several empowering training experiences carried out by community psychologists involved in training in public administration, schools, social health facilities, the nonprofit sector, and small and medium-sized enterprises are illustrated.

2. During the course, participants can also search the web or social media for pages that describe contexts of work in which they may be interested. For example, after viewing community profiles, they will understand how in many different areas they can find organizations to contact to expand their social capital and networks. Depending on their interests, participants can locate one or more associations or other structures in their area that deal with the issue they want to explore and visit their location, reporting as a group what they found and sharing information in person or online. In addition, each participant can send a description of the activities they have done to achieve the identified desires and, if they found obstacles in achieving them, describe them. Other participants and mentors offer suggestions on how to overcome them, both when meeting in present and when working online.

3. Also, at this stage, some participants who were uncertain about which type of work interested them used generational and personal media novels in their work aspects. The questions asked were: what jobs interested you in the movies or TV shows you watched as a kid? What kinds of jobs do today's favorite movies, most-watched shows, and websites visited offer? Possible life projects that may have emerged should then be put into the casket.

4. Using the familiar and personal work novel, each participant reflects alone on past work experiences and desired future work experiences. Then, in pairs, they use the PMOA grid (see chapter 7, para. 2) to examine strengths and problem areas of already experienced and desired empowering work contexts. Again, possible life projects that may have emerged are placed in the casket.

5. Each participant, keeping in mind the place he or she wants to give to work in relation to other desires in other areas, begins to elaborate on the "daisy chain of multiple possibilities" (Bruscaglioni, 2007), briefly listing at least two or three desirable life projects that will be reworked and included in the casket.

2.4.   Step 4. Identifying congruencies between individual and/or organizational desires and the demands of their contexts This phase is organized into the following successive steps.

1.   *Group presentation of one's possible life plans.* Identification of how to overcome possible obstacles and how and where to find support in family, in our networks, and the real and virtual outside world.

2.   *How to increase your social capital.* On the internet, you can find free courses on a variety of topics and meet new people. In addition, new platforms allow collaborative learning modes in small groups (*Computer Supported Collaborative Learning* [CSCL]). In various investigations (Francescato, Tomai, Mebane, 2004; Francescato *et al.*, 2012), it has been documented how online one can acquire professional knowledge and skills and increase social capital to an equal or greater extent than small group face-to-face training.

3.   *Use of creativity techniques.* A variety of creativity techniques can be used to overcome perceived obstacles, from problem-solving to synectics (Francescato, Putton, 2022).

4.   *Modifying our media use.* By modifying their media use--by choosing, for example, to stop watching certain movies that increase the propensity for violent behavior in the family but to watch programs that illustrate more constructive ways of handling conflict between partners and parents and children--participants achieved the desired changes and improved relationships within and across generations.

2.5.   First steps to take after the end of the course The empowering training course ends with each participant describing the steps to be taken to achieve specific goals. To monitor progress made and any possible obstacles, one may choose to conduct pair or group mentoring. In the former, two participants agree to stay in periodic contact to support each other on the outlined paths by scheduling exchange and/or supervision meetings. For example: after two months, a collective online meeting with the lecturer and tutor can be implemented to monitor progress made and obstacles encountered. Alternatively, one can meet as a group six months after the last meeting for an overall evaluation of the training experience.

These follow-up appointments, in our experience, have been conducted both in-person and remotely with satisfactory results for participants.

12

Exploring emotions: social-affective education in action from school to cohousing

1.   Introduction

The method of intervention we developed in the 1980s (Francescato, Putton, Cudini, 1986) and reworked in the last decade (Francescato, Putton, 2022) combines theoretical insights from humanistic psychology and community psychology to promote well-being in school by enhancing that part of the educational process that deals with students' attitudes, feelings, beliefs and emotions. Recognizing the interdependent relationship between the affective and cognitive components (Piaget, 1967), the integrated method aims to improve each student's self-knowledge and to facilitate communication among members of the class group. At the individual level it aims to develop feelings of acceptance, security and trust in oneself and others, interpersonal problem-solving skills and the ability to cope with situations of emotional stress. On the group level, on the other hand, it wants to promote behaviors and attitudes of cooperation, solidarity, mutual respect, and acceptance of diversity. When we created it, we had two categories of recipients in mind: on the one hand, teachers who wanted to acquire a valid operational technique to implement socio-affective education in their classrooms, prevent maladjustment of some pupils and promote the psycho-physical well-being of all children; on the other hand, psychologists working in the school who were interested in training.

The book Star bene insieme a scuola (Enjoying wellbeing in school) has been reprinted twenty times and, as we document in the new 2022 edition, social-affective education has spread to all school levels, from nursery to secondary, and groups of parents, as well as trainers, sports coaches, and members of associations of all kinds, have benefited. Moreover, the initiatives we describe in *Part Two* of the new edition show that passionate leaders and teachers are creating a new school of relationship and resilience. Finally, in various social, health, sports and cultural services, social-affective education techniques have been used to develop relationship skills (effective communication, problem solving, training and small group management); to carry out interventions aimed at the prevention of distress (preventing cyberbullying, violent behavior, substance use, dropping out of school, repositioning of unschooled and unemployed, spreading empowerment in disadvantaged contexts). Mayors, managers and operators of territorial services have fostered the development of **proximity welfare**, supporting territorial educational pacts, networking in territories, mutual aid groups, courses on environmental education including online and the development of democratic participation, and funded courses against violence and feminicide and for the promotion of a sexual education that respects feelings and discusses the dangers of pornography (now available also online) that portrays that violence against women and minors.

Social-affective education strategies have been used to support the **well-being of "intentional communities"** such as cohousing experiments, which originated in Denmark about sixty years ago and have spread to many countries on various continents, developing new ways of intergenerational coexistence, but also new types of conflicts. For these reasons, in paragraphs 2-3 of this chapter we will outline common theoretical foundations and different experiences implemented in schools and territorial and virtual communities, and in paragraph 4 we will describe the historical evolution of intentional cohousing communities.

2. Educating on emotions and effective relationships to promote individual, group, and collective well-being

2.1.   Theoretical foundations With the *humanistic psychology* of authors such as Rogers, Maslow, May, and Perls, community psychology shares an emphasis on potential, positive resources to be enhanced and developed rather than dysfunctions and disorders to be treated. Both disciplines aim to promote people's psychophysical well-being rather than to treat their illnesses; they conceive of the individual as a bearer of potential to be developed; they place great value on experience in interpersonal relationships and interaction with the environment; and they recognize the importance of the group as a learning context. Both approaches originated in the United States in the 1960s with an innovative and emancipatory intentionality with respect to U.S. culture and spread to Europe, finding considerable conceptual contributions from European colleagues (Orford, 1992; Caprara, Accursio, 1994; Sánchez Vidal, Musito, 1996; Zani, Palmonari, 1996; Amerio, 2000, Francescato, Tomai, Ghirelli, 2002; Ornelas, 2008).

To elaborate our *integrated method*, we drew inspiration from several sources: from some circle time experiences, observed by Francescato during her doctoral work, used in kindergartens in the United States to improve the teacher-student relationship; from the work of Gordon (1974; 1979) and from community psychology for theories on working and mutual-help groups. We also used models of multidimensional participatory organizational analysis and network work developed by community psychologists and discussed in this text (see chapters 7 and 10; Francescato, Tomai, Solimeno, 2008).

In general, these different theoretical approaches suggest strategies and methodologies that promote the coping (active coping) skills of individuals, strengthen the skills of key persons and nonprofessional practitioners, and support spontaneous and self-help groups in the community.

In recent years, "positive psychology," which aims to develop positive individual qualities, such as self-determination, self-efficacy, optimism and individual and social happiness, has also become widespread. Seligman (1990) identified five factors through which it is possible to achieve optimal functioning, increase one's personal well-being and pursue one's happiness: positive emotions that increase satisfaction with one's life; involvement in life events related to the awareness of exercising an active role in the pursuit of one's well-being; positive interpersonal relationships; the meaning individuals give to their lives, that is, the perception and awareness that life has meaning; and personal fulfillment, that is, the awareness of having achieved a goal in one's life.

For a detailed discussion of the different theoretical approaches inspired by humanistic and community psychology we suggest reading the new edition of *Enjoying wellbeing in scho*ol (Francescato, Putton, 2022), while in this chapter we present the salient aspects of the integrated method.

2.2.   The integrated method The integrated method proposes three different modes of social-affective education, which can be used all together or separately, focusing on a particular aspect:

1. To **improve communication between adults** (teacher, parent, educator, etc.) **and child-child**, the methodology applied is that proposed by Gordon (1974; 1979), the methodology applied is that proposed by Gordon (1974; 1979), which involves the use of three techniques: message-me, active listening, and the "nobody-loses" method;

2. To **foster the relationship between members of the class group**,

especially to increase mutual acquaintance and to provide a different relational space, the technique of circle time, or "circle time," is applied;

3. to **improve children's understanding** of experiences, sensations, attitudes, feelings, and fantasies, a set of psychomotor exercises is used (see chapter 13) that aim to develop the ability to get in touch with and recognize the emotions felt by focusing on self, the external environment and relationships with others (Francescato, Putton, Cudini, 1986). In recent years we have also used generational and personal media novels (see chapter 11).

2.2.1. The Gordon Method Gordon's techniques aim to foster the development of an effective relationship between teacher and pupil. The scholar stresses the importance of acceptance, authenticity and empathy in any human relationship. He also believes that parents and teachers often fail to help children resolve the inevitable conflicts that arise in peer and intergenerational interactions. To communicate effectively, it is necessary to understand to whom the problem belongs: if the problem is perceived by the teacher, he or she may use the *message-me*; if it is more perceived by the student, one intervenes using *active listening*; if it is perceived as belonging to both, one can use the "nobody-loses" method.

The message-me allows for constructive criticism in the face of inappropriate behavior, as opposed to the message-you, which tends to blame, scold and humiliate. For example, a teacher, confronting a boy who has pushed another on the floor, is afraid that the boy has hurt himself, gets angry and says, "You're always the usual wild one, you don't know how to play quietly, recess is over." Instead of expressing anger (secondary feeling of fear) he may express fear (primary feeling) with the message-me, saying, "When you push yourself running, you may fall down, and I am afraid you will get seriously hurt." A boy, confronted with a teacher who openly says what he feels, does not feel unfairly attacked, reflects on the consequences of his actions and will try to act more consciously in the future. Issuing the message-me always involves expressing one's feelings and taking responsibility for one's experiences, and it changes the way one relates to others.

Active listening is used when the adult perceives that a child has a problem that does not disturb another, but reports discomfort (isolates himself, does not participate in games, does not listen to the teacher's explanations, etc.) and agrees to talk about it with the teacher or parent. Active listening is divided into four moments:

1. Silent listening allows the person to expound without being interrupted;

2. Messages of acceptance: these can be nonverbal (a nod of the head, a smile, etc.) or verbal ("I'm listening," "I'm trying to understand"), and they send confirmations of attention and interest;

3. Warm invitations that encourage the child to speak and elaborate on what they are saying;

4. Active listening (feedback): the adult "reflects" the child's message without issuing messages of their own but trying to return (as in a mirror) what difficulties, emotions, and states of mind he seems to have picked up from the story. It is not important whether the boy will recognize himself with the description presented to him by the adult; in either case, he will have one more piece of information on how to reflect more clearly on his problem and independently seek his solution.

The "nobody-loses" method is used in conflict situations that may arise between students and teachers and between children, to seek a mutually satisfactory solution. Gordon proposes using the creativity technique of problem-solving, which involves six successive steps that can be carried out in one or more group meetings:

1. Exposition of the problem;

2. Formulation of possible solutions;

3. Evaluation of the negative and positive aspects of each proposed solution;

4. Choosing the most suitable solution;

5. Implementation of the solution;

6. Verification of the results obtained.

With problem solving, one can also construct classroom regulations that reinforce behaviors acceptable to all, as Gordon suggests. For example, encourage *relational time*, i.e., set aside 10 minutes in which children can confide in the teacher or a classmate, decide what rules of behavior are useful in the *widespread time* (in which the teacher lectures collectively), but also how each *person* can use *personal time* when, having finished the common work (such as a drill), he or she can devote to different activities of their choice (e.g., reading a book, making a drawing, etc.). In addition, it is useful for the regulations to specify when one can talk quietly with classmates (*rest times*). Working out in groups in a participatory way requires an understanding of the rules that are to regulate coexistence in the classroom, writing them down on a poster to hang on the wall so that they are visible can serve as a daily guide of one's own and others' behavior, committing oneself to remembering and respecting them means practically experiencing democracy and learning to become an active citizen.

2.2.2. Circle time Circle time originated as a group intervention technique aimed at fostering the relationship between members of the class group and their knowledge of each other. During circle time, class members come together to discuss a topic or problem proposed by one or more pupils. This activity aims to develop acceptance, cooperation, and solidarity. Circle time is a special moment in school life when teachers and pupils step out of the cognitive realm and their role as "culture processors" into the affective one, becoming facilitators and participants respectively in a hetero-centered discussion group. In the classroom, the setting is changed: **chairs are placed in a circle**, and communication also changes, from descending to circular, characterized by listening and abstaining from judgment. The group is led by a specially trained teacher or psychologist. The children, to whom the objective and specificity of the activity have been explained, choose a topic of their interest and share lived experiences, feelings, experiences. "The criterion of choice is to accept all proposals and put the priority of discussion on the ballot; this is to prevent anyone from feeling excluded" (Francescato, Putton, 2022, pp. 70-1).

In the first circle time, the norms to be followed to carry out the activity in the best possible way are defined; the facilitator usually explains that "everyone has the right to speak, but no one is obliged to do so. Participants are to be listened to, not judged, not interrupted, not denigrated or mocked; you follow your turn to speak by a show of hands or by waiting until your neighbor has finished expressing their ideas, depending on the rules you have chosen" (*ibid.*, p. 71). Other rules are proposed by the children. Circle time meetings are held once a week or every two weeks, and last from 30 minutes, for younger children, to 60-70 minutes in secondary school classes; preferably they should be held for at least one or more school years to achieve the best results.

2.2.3. Objectives of circle time One of the major limitations of the educational models used in Italian schools is that they do not value the group enough as a formative and growth tool, contrary to what modern pedagogy advocates (Poli, 1998). Students are grouped together in the classroom for organizational and economic reasons, not because educational value is placed on the group; consequently, the class is conceived by teachers as a collection of learners with whom to relate independently and among whom to discourage informal interaction. Introducing an activity such as circle time into annual or multi-year educational programming is a priority *goal* for us as community psychologists because it is a transformative intervention that restores value to relational variables and importance to the development of all those prosocial skills defined as "emotional intelligence."

Improving the knowledge students have of each other becomes a formally recognized goal, to be pursued within a physical and temporal space. Learning to be an **effective participant in a working group**, learning to discuss together, listening without interrupting, accepting to hear all opinions, feeling free to express one's own, to promote mutual confrontation and enrichment; learning to resolve conflicts, analyzing the problem and finding possible solutions together using creative tools: these prosocial skills need to be promoted throughout life, in order to work together to improve individual and collective well-being and address multiple social problems. This is why "circle time meetings should not be only occasional but have a venue and a timetable, as other educational activities carried out within school hours" (Francescato, Tomai, Mebane, 2004, p. 113).

A second important goal concerns the transformation of the class into a primary mutual-help group so that each participant increases their social capital, friendship network, and personal empowerment and perceives the class group as a source of informal support. Circle time creates certain conditions that facilitate the achievement of this goal: it fosters mutual acquaintance, communication and cooperation among all members of the class group; it creates a climate of respect in which everyone fulfills their need for belonging and individuality; it teaches how to ask for and give help. This goal is achievable only when circle time is **used for at least several months**, once a week. Introducing a practice of collaboration and sharing can prepare the ground for dealing with sensitive and intimate topics such as sexuality, but also activate a supportive context in which boys offer concrete support to classmates in difficulty, to avoid dropping out of school or flunking out.

This development of the ability to help each other we have also found with groups of teachers, parents, social and health workers and third sector workers, with whom we have conducted empowering trainings, giving them the opportunity to explore their problems but also their resources, also using circle time in group discussions (Francescato, Putton, 1995; Francescato, Putton, Toraldo, 1997; Putton, 1999; Francescato *et al.*, 2000; Francescato, Putton, 2022).

Another priority objective concerns the promotion of certain knowledge with respect to group phenomena and specific skills in facilitating and observing discussion groups.

3.   Facilitating groups by knowing their life stages

The circle time facilitator can be a teacher, but also a community psychologist who uses this method in projects for parents, teachers, social and health workers, sports coaches, etc. (see chapters 17-18).

Its functions are to: help choose the topic and briefly introduce it; invite input; prompt, without coercion, those who are shy; restrain overly talkative participants; encourage everyone's participation; give feedback at the end of each meeting on how the group worked.

An important function is also to read the dynamics that can occur in a task-centered group so that various situations can be addressed in the most appropriate way (Mucchielli, 1980, 1986; Muti, 1986; Quaglino, Casagrande, Castellano, 1992; Francescato, Putton, 1995; Di Iullo, 1999;

Putton, 1999). A task-centered group, from establishment to completion, goes through four stages according to Jones and Gerard (1967):

1. **Dependence on facilitator:** the group starts a new experience and relies heavily on the facilitator for elucidation, stimulation, reassurance, each participant trying to figure out what to do;

2. **Conflict**: the timing of the onset of this phase depends on various factors, for example, children at school see each other every day and might bring disagreements or quarrels that have occurred previously back into the circle time; a training group for parents or social-health workers, on the other hand, might develop alliances or dislikes among group members, or struggles to impose their ideas. The facilitator can identify and clarify various situations by helping the group see diversity as an asset, develop mutual respect, and pursue the intent to find a solution to disputes. Conflicts should be treated as opportunities **for growth**, for a chance to see other sides of the task the group is undertaking;

3. **Cohesion**: the group recognizes the value of interaction and mutual acceptance. New initiatives can be proposed, subgroups can be made to achieve goals;

4. **Interdependence**: participants feel they belong to the group, which is perceived as a unit in which each member is related to the others; a sense of community emerges in which there is mutual respect and trust, security in dealing with new situations and, in groups with a set deadline such as training groups, peace of mind that they can profitably manage the end of the group, having acquired new social capital.

3.1. Learning to observe to become better participants and facilitators

We believe that knowing how to be good participants and facilitators of small groups in presence and online increases individual and collective well-being and empowerment and is a necessary skill not only in work but in many areas of our lives. Therefore, Francescato has modified the circle time tool by adding the possibility-both for children participating in school and for adults in empowering trainings-to observe group work in action.

To this end, she envisioned that at school, after the first few meetings, one or more children would take turns stepping out of the circle and getting used to observing group interactions, and then reporting the data at the end of the circle time and reflecting on what behaviors (not people) fostered or hindered the group's good performance. Teachers can provide observation grids with which to note, for example, which behaviors are most functional and which, on the other hand, are disturbing. The pattern will vary according to school level, starting from simple formulations to more complex ones

To exemplify, we present a complex scheme inspired by Bales (a well-known Harvard professor whose course Francescato took in 1970)-and modified by

De Grada (1969)-which lists twelve behaviors: the first four are functional for affective maintenance of the group, the second four promote the cooperative performance of the task, and the last four hinder group work. The twelve items are:

1. Shows himself friendly and encourages teammates;

2. Expresses esteem;

3. Mediates between multiple ideas;

4. Listens/expresses agreement/does not judge;

5. Talks about personal experiences;

6. Listen to the experiences of others;

7. Intervenes only when another has finished speaking;

8. Does not interrupt;

9. Demonstrates tension/laughter/disturbance;

10. Interrupts the others;

11. Criticizes continually; and

12. Shows hostility/does not intervene.

In secondary schools and empowering training, peer groups, voluntary and private social work, learning to observe and return one's observations to group members in a constructive way without evaluating has proven to be very fruitful (Francescato, Tomai, Mebane, 2004; Francescato, Tomai, Solimeno, 2008). Introducing the role of the observer helps to develop the skills of effective group participation; we have been able to verify that starting to experiment with this role as early as secondary school allows some children to become competent facilitators at secondary school.

4. Affective education strategies in present-day intentional communities: ecovillages and urban cohousing

4.1. Historical background The contemporary cohousing movement, as well as the family communes of the 1960s or the community movement of the 19th century in the United States, constitute an attempt, in historical eras marked by an overemphasis on the "individual," to swing the pendulum back toward the "social." In all three periods, these movements can be seen as a reaction to the exacerbation of the individualist ideology sustained by the socioeconomic system. In the nineteenth century it was based on *laissez faire* capitalism that favored private enrichment and frontier expansionist strategy, in the 1960s it rested on corporate capitalism, and in the 2000s on financial turbo-capitalism.

The U.S. Constitution proclaims that all men are born equal, and the majority American culture mythologizes the self-made man (the famous *self-made man*) competing with all his peers to achieve success. The dominant ideology sees the family as the basic core of the private sphere, where values different from those practiced in the public sphere prevail. Competitiveness, aggressiveness, and cunning are seen as valid values in the field of work and politics and in the public sphere in general, while values of solidarity, cooperation, and generosity are practiced in the private sphere, but are seen as decidedly out of place in the public sphere. However, in different historical periods there have been minority groups that have challenged this prevailing view of social organization peculiar to the *American dream* and advocated models of coexistence that allowed for different ways of practicing the pursuit of happiness. The kingdom of heaven was believed to be attainable on earth, usable daily through communal living, with rules to guarantee all members equal rights of access to material and spiritual goods. Horace Greeley, an editor of the "New York Tribune" whose ideas had considerable influence on the nineteenth-century movement, believed that there should be no more poor or unemployed.

The advocates of communities in the nineteenth century had lofty ambitions; they wanted to create model communities that overcame the dichotomy of values between the private and public spheres and could show how to gradually change society. In fact, Robert Owen, one of the pioneers of the commune movement and founder of New Harmony, who arrived in Washington in 1825, declared, "I have come to this country to introduce an entirely new social system, to change an ignorant and selfish system into an enlightened system which will unite all individuals and remove the motives of strife among them" (Tyler, 1962, p. 217). Unfortunately, the eleven Oweni-inspired communes failed within a few years (Francescato, Francescato, 1974). In the 1960s, in the corporate era, motives for struggle among individuals remained high as, with the rise of the black and minority civil rights movement and the women's liberation movement, many groups became aware that they did not have equal opportunities to get to the top of the social pyramid. In addition, several members of the "counterculture" were criticizing both the need to participate in the *rat race*, this frantic race to get to the top of individual wealth and fame, and the traditional division of family roles, with the man plugged into the productive world and the woman seen primarily as housewife, wife and mother. Feminist theorist and activist Betty Friedan (1963) describes the isolation in which middle-class women raised their children alone in the suburbs and their dependence not only economically but also affectively and emotionally.

The communes of the 1960s also express a moment of that trend toward "reclaiming the collective": the nineteenth-century communes focused on the creation of a Christian or socialist *commonwealth*; the communes re-propose the same ideals of cooperation and brotherhood but have more limited aims.

While the former are micro-communities that aim to restructure society, the latter tend for the most part to recreate a particular social structure: the family. They have less ambitious goals; they have *wiped* the *slate clean* of the grandiose visions, shot through with a deep faith in the future, that animated the communities of the past. The traditional concept of "salvation" has been traded for the more modest one of "personal growth," religious or political rhetoric has given way to psychological rhetoric. The new and more limited size of the community movement of the 1960s compared to that of the nineteenth century can also be seen in the physical structure of the communes; those of the nineteenth century were quite large, with hundreds of members of all ages, those of the 1960s, on the other hand, often have only a dozen or so members, the majority of whom are young or very young.

4.2. The cohousing movement The cohousing movement more closely resembles utopian communities in the sense that it focuses on larger dimensions than family communes. Here is how cohousing is described on the association's website: "cohousing communities usually include single-family houses or apartments along a common street or built around a central courtyard".

The first cohousing experiment in Europe took place in Denmark in 1972. In 1980 two U.S. architects went to study with Danish researchers and introduced cohousing to the United States (Durrett, 2022). During these years, Cohousing Associations were formed in several countries, including the most prominent in the United States, to spread a new dream of being surrounded by caring and cooperative neighbors and leading a lifestyle that respects the natural environment.

They range from a minimum of 7 up to 67 cohousing settings, with the majority including 20 to 40 housing units. In each type of cohousing there are many opportunities for casual gatherings among residents and planned meetings for parties and to make management decisions. A community house is the social center of the community and usually has a kitchen and large dining room, and recreational spaces for adults and children, often also a guest room, and rooms with common laundry facilities. Common meals are usually provided at least three times a week. Members take care of the common spaces, and this working together builds trust and support.

In addition, cohousing originated at the initiative of architects who want to attempt, through targeted urban planning projects, to foster social relations typical of rural pre-industrial societies in today's post-industrial cities. These often have historic centers overrun by tourists and increasingly unattractive suburbs, frequently just dormitory neighborhoods where residents do not know each other, fear going to the streets because of increased petty crime or are divided into rival ethnic groups. In such urban settings, it is difficult to form attachment to places and to develop solidarity and mutual-help relationships with neighbors whom one does not know and with whom one often has only adversarial relationships over condo or parking disputes. Cohousing aims to address these problems, to propose a **model of living coexistence** and to **foster a sense of community**. By frequenting each other habitually and maturing bonds of interdependence and mutual trust, people tend to become more engaged and empowered and develop higher perceptions of social security, as many community psychology scholars have found (Amerio, 2000; Francescato, Tomai, Ghirelli, 2002; Francescato, Tomai, Mebane, 2004). Both cohousing and family communes attempt to create alternative family types and develop affective and mutual help relationships but presented different strengths and problem areas. Therefore, in the following paragraphs we will discuss family communes and cohousing together.

The failure of the family communes of the 1960s had shown that the most frequent problems were caused by ideological disagreements, economic disputes over how much each person should pay for food and shelter, complaints about lack of privacy, quarrels over how to raise children, food diets to follow, criteria for selecting members, and decisions on controversial issues. Architects promoting cohousing aim to bring back this sense of place and attachment to a specific local community, while preserving each person's needs for autonomy and independence and ideological preferences, by designing both rural ecovillages and urban cohousing.

Residents of ecovillages love the countryside, are actively environmentalists and share with nineteenth-century communities the conviction that it is the whole of society that needs to change; they pose as models for implementing sustainable development, with a particular ecological-social vision of living. Many are convinced that in an era when climate change is causing more and more damage (see chapter 15) it has become exceedingly necessary to build ways of living that promote a "greener" lifestyle. For example, Sunward Cohousing in Ann Arbor, Michigan, aims to create a place where the earth is respected, diversity is welcome, children can play together without danger and live in community with their neighbors.

Some characteristics of cohousing that differentiate them from ordinary family households are:

1. Participation: residents organize and participate in the planning and design processes of the real estate operation and are responsible for the final decisions;

2. Intentional design: cohousing is designed in a way that encourages a strong sense of community;

3. Houses and apartments are purchased by individuals, and everyone can enjoy spaces of privacy;

4. Common spaces are an integral part of cohousing and are designed for daily use to complement private spaces;

5. Residents are directly involved in management.

The cohousing movement, like the family communes of the 1960s and 1970s, seeks to provide an efficient response to the management of a variety of practical living issues (childcare, care of environments, convivial living spaces, even for elderly people). Indeed, major demographic changes have occurred in recent decades: the crisis of the nuclear family has widened, and the number of elderly and single-parent households has increased. In Italy, for example, single-parent households and one-person households are about 34 percent, traditional families are about 33 percent, and the remainder, are extended families, childless cohabitants, cohabitants with other relatives, etc. (ISTAT, 2022).

The cohousing movement seeks to provide services and social occasions previously entrusted to families of origin; thus, in some respects, it aims to create, just like family communes, new ways of living and helping each other together. For example, after the 2008 financial crisis, one of the themes proposed by the international cohousing conference held in Seattle in June 2009 was how to make cohousing even more supportive of neighborly relations even in response to the global financial and employment crisis. It differs from the communes of the 1960s that sought to experiment with alternative forms of (*a*) socioeconomic organization based on cooperation and egalitarianism, (*b*) collective participation in decision-making, (*c*) interpersonal relationships based on overcoming traditional male and female roles, and (*d*) cooperative child-rearing.

Some communes were interested in developing new, freer love relationships. Cooper (1972, p. 47) defines the commune thus:

A microsocial structure that achieves a vital dialectic between solitude and being with others or implies a communal residence of work and experience. Love relationships spread among community members much more than in the family system; this naturally implies that sexual relationships are not restricted to the societally approved male-female couple and, most importantly, that children have free access to other adults besides the couple who are biologically their parents.

Heller and Vajda (1970) indicate three characteristics as essential:

1. **Mandatory employment** for every adult member (the commune, for example, will not be able to allow a man with a good income to support a woman who "belongs" to him);

2. The place occupied in the social division of labor does not exempt anyone from **performing the collective tasks** set by the commune;

3. Everyone is obliged to **take care of the** commune's **collective children in some way**, whether they have children of their own.

On the other hand, the commune will not be able to determine, as a collective, the behavior of the members about both their outside work and sexual relations (the commune should allow both a **stable couple relationship** and **complete sexual promiscuity**). According to the two Hungarian scholars, the important thing is that the end of a couple's relationship in the commune does not necessarily constitute a traumatic change in the children's lives and allows the two divorcees to remain in the original commune after the termination of their relationship.

In our research on 63 communes in Boston and 10 in Italian municipalities in the years 1970-72 (Francescato, Francescato, 1974), we examined whether the new formulas of community life succeeded in practicing economic cooperation among members with different incomes, the extent to which the division of male and female roles in paid and domestic work was diminished or eliminated, and how children were raised.

In terms of *economic cooperation,* none of the communes we examined practiced sharing all income. In 10 percent of cases, members contributed to group expenses in proportion to their respective incomes; in all other cases, overhead expenses were divided equally among adult members. Good cooperation, on the other hand, was evident in the common use of furniture, tools, cars, records, and books.

In overcoming male and female roles, the communes we examined had been much more successful. No woman was maintained by a man; each adult member provided his maintenance. Each adult member was entitled to a room, and everyone took turns cleaning and keeping order in the communal spaces, usually a kitchen and dining room, but also in different communes, music rooms, workshops, etc. In addition, again, taking turns, men and women purchased food from consumer cooperatives that bought groceries in bulk or from rural communes that supplied ecologically grown vegetables and fruits. Domestic work was considered an important part-time occupation for both men and women because the counterculture valued all activities aimed at building and improving the living environment. In the communes we surveyed, food preparation had become an act charged with special spiritual significance after the organic food boom; grocery shopping had also taken on political value because people turned to consumer cooperatives, and taking care of waste was considered important for the preservation of the Earth. Moreover, since in each commune, usually one member cooked once a week (only evening meals were eaten together), on average, the dishes were excellent and varied because each person tried to serve something special to the others. We found that when one commune had problems, the level of meals and cleanliness of shared spaces deteriorated.

We also ascertained that supportive, kind, and affectionate behaviors considered "female characteristics"-were attributed to both male and female members, confirming the results of another survey of communes (Kanter, Halter, 1973). These scholars found that, among the top ten characteristics attributed to men, six were found to be reported among the top ten attributed to women: supportive, warm, affectionate, kind, cooperative, and interested in culture and art. Overcoming traditional stereotypes also appears in the answers given to questions designed to ascertain what behaviors and characteristics members thought they had acquired during their community experience. Men had learned to cook, take an interest in furniture, and be more supportive, affectionate, cooperative, intuitive, and trusting. Women had learned to become more aggressive, independent, and assertive and to build and repair tools.

Although most cohousing projects to date do not set out at all to change the family or the gender roles within it, the issue is raised in some papers prepared for the 2009 international conference. On the contrary, cohabitants generally emphasize that each household maintains its independence, both economically and in terms of its vision of life. However, some cohousing experiences of the first phase (1970-80) questioned the nuclear family, and most cohabitants are still very much interested in discussing how best to raise children in a situation where each parent can observe the behavior of other parents in a variety of situations. I would say that cohabitants **maintain two of the great goals of family communes**: finding more cooperative forms of child-rearing and using cooperative decision-making processes. It is in these areas that social-affective education strategies have been most applied.

4.3.   Raising children more communally in cohousing: what the surveys reveal
The few surveys of children in communes in the 1960s had shown that in those where there were children with different parents, more disagreements occurred. In some, there emerged a keenly felt ambivalence among parents, divided between the desire to share educational responsibility with other adults and the fear of abdicating their authority, exposing their children to unwelcome influences (Francescato, Francescato, 1974).

In research that is still ongoing, three variables were identified in the cohousing settings visited in Europe that were particularly important in determining the degree of cooperation achieved: having participated in affective education courses, the age of the children, and the co-presence of multiple single-parent households.

Children under three naturally require a greater degree of care and attention than older children. While possessiveness issues emerged in some cases, we found that children under three were **also cared for a great deal by other adults**. Almost all the members we interviewed said they changed diapers, fed, and supervised the sleep of other parents' children as well.

For older children, the degree of involvement seemed to depend more on the attitude of the biological parents and the presence or absence of other parents. In some cohousing, parents made it clear that they wanted to maintain their exclusive authority over the children, and others could be friends and nothing more. These parents tended to use others as **occasional babysitters** but rejected group discussions about educational methodologies to be collectively preferred. In other cohousing settings, however, more adults were able to **collaborate and discuss the "educational philosophy"** to be followed and came to establish common norms. These cohabitants had been inspired by some of Gordon's social-affective education strategies and, above all, had created a climate of mutual help and mutual trust by using the circle time methodology in their meetings.

Both parents and other adults said that many **changes** have occurred **in the attitude of biological parents**, who have become more open and less authoritarian with their children. Here is how one father summarizes his experience:

I have become a more relaxed parent, less tense because I share responsibility with others. Living in cohousing with children means learning to continually compare how others would behave in your place and living with many adults reliving their childhood memories. It, therefore, means getting back in touch with your childhood. I repressed a big part of my childhood, not realizing that I was imitating my parents with my children. Having to discuss my parenting behavior with others, I became aware of my childhood again, and this helped me understand children better. For example, when one of them used to play with mud, I used to scold him for getting soiled; now, remembering the joy many of us had as children, I let him do it, I think it is a nice sensual game (interview done in a Berlin cohousing in 2020, research not yet published).

This same sense of relief is particularly present in separated parents, who were very happy that their children could have the friendship and companionship of adults of the opposite sex.

Despite the difficulties in establishing effective collaboration in child rearing, all the parents interviewed, both in urban cohousing and particularly in ecovillages, believe intentional communities **are the best environment to raise their children.** This opinion was also shared by the 18 children aged 3 to 13 whom Francescato personally interviewed. What they liked most was precisely "having many people around." For the younger ones, it meant having the opportunity to play with different people or to talk to "lots of people at the table." The older ones, on the other hand, appreciated that there was always someone who was doing something interesting, and this allowed them to learn and have more fun. Then, in times of arguments with parents or siblings, they could take refuge in the common areas and find something to do or someone to talk to. When asked whether they would have preferred to live in a cohousing or small family when they grew up, all the children of the separated, that is, children who had experienced family tensions, said they preferred cohousing. Almost all the children (15 out of 18) reported that they had found at least one other adult in cohousing to whom they became attached.

4.4.   Making better decisions together is possible but complicated How and to what extent are the ideals of cooperation and egalitarianism put into practice in today's cohousing decision-making processes?

All communes made use of the same decision-making mechanism: the weekly meeting of members. This meeting had a twofold purpose: that of making decisions regarding the management of the house and that of constituting the place and time when the group shared and tried to resolve various conflicts and relational problems. In short, the weekly meeting was a special type of group work that included elements of both a business group and a T-group (interpersonal dynamics group) so that it could function both as an operational coordination center and as the affective and ideological focus of the commune. Several members described their attitudes toward these meetings as ambivalent. On the one hand, there is the threat of the overly strong presence of the group, the node on which individual energies are focused, and, on the other, the attraction brought about by this convergence of individuals into a single organism. The ambivalence of individuals was highlighted by the fact that weekly meetings often took place late and that there was a tendency to deal with trivial, run-of-the-mill things first to give the group time to warm up and then face the important problems. One girl described these meetings in the commune in her diary as follows, "They are a time of coming together. Sometimes, that means breaking down chasms of non-communication. Sometimes, the bridges are strong, and we celebrate them. The magic fact is that different people have different needs to give and receive, so when we gather to share, there is usually enough for everyone" (Francescato, Francescato, 1974, p. 194)

In many communes, meetings began with a ritual called "preprocess": each member, in turn, would tell the others how he or she was feeling at that moment physically and psychologically so that the others would know what their present reality was (whether one is still suffering from an ended love affair, whether another has had trouble at work, whether the third is satisfied with some positive event, etc.). Then, the actual meeting would begin. Usually, decisions were made by consensus. There were two types of consensuses: first order, when all members agreed; second order when some member did not fully agree but was not against the group's decision to the point of exercising their veto power. Consensus decisions were almost always reached after lengthy discussions. Indeed, participatory democracy requires, as one Cambridge commune member commented, "a difficult process because each 'self' takes time... It is important to understand how the person speaking feels, to understand that the same silence can mean participation in one and rejection in another" (*ibid.*).

In all the communes analyzed, there was a widespread belief that thanks to these decision-making mechanisms, power was equally divided. Even in communes where some people enjoyed greater influence, members perceived that this kind of power varied from person to person: depending on circumstances, different individuals took on dominant roles, but power was never institutionalized or concentrated in the hands of a single person.

When meetings were attended by all members, commitment to the collective was very strong, and relationships were good; in contrast, when meetings were deserted by more than a third of the members, this was a strong indicator of unresolved difficulties. In the communes that had existed for several years, we found that there were cycles in what concerned meetings as a decision-making tool: at first when the need to create a sense of group membership and mutual solidarity was most intense, weekly meetings were a norm to which everyone had to submit. When members felt more confident about their and others' commitment to the commune, meetings were made non-mandatory and could be called upon request. In times of crisis, pressure was again found to make meetings regular and attended.

Thus, in each commune, there was a need to combine spontaneity (adherence to the needs of the moment) with structure (a now-codified history of past needs): an attempt was made to develop a way of dealing with problems that ensured continuity and effectiveness on the one hand (ensuring that decisions were made and executed to ensure the survival of the commune) and respect for individual needs on the other (ensuring that rules did not become impersonal and coercive).

That decision-making processes are also a challenge for current cohousing is also evident from the high number of seminars and discussion groups scheduled on the topic for international conferences. Entire panel discussions are devoted to analyzing group processes, the importance of using circle time to create a favorable affective atmosphere, how to best use consensus as a decision-making mode, how to conduct effective meetings, what to decide in plenary sessions, and what to delegate to committees or a manager, how to persuade each cohabitant to do their part, and how to hold people accountable in keeping agreements made. The presence of conflict is addressed in different ways in different cohousing settings with **problem-solving techniques** and Gordon's **nobody-loses method**, but also by holding innovative workshops on the "**art of apologizing**," a skill in high demand among cohabitants.

4.5.   Membership selection In the nineteenth century, the majority of the

members of utopian communities were intellectuals critical of the myths of private property and expansion to the West, fundamentalist Christians who were repulsed by the increasing secularization and materialistic emphasis of American society, and various groups of "deviants," that is, people who were unwilling or unable to fit successfully into the system (unemployed, underemployed, new immigrants, etc.). The failure to select participants is pointed out by the son of Owen as one of the reasons for the failure of the New Harmony community (Tyler, 1962).

Members of the communes in the 1960s were also predominantly self-selected, but more homogeneous in age and social class: predominantly young people in their 20s and 30s (70 percent) with peaks in their 50s, middle-class and mostly white. The average number of people in family communes was about 10 (ranging from a minimum of 6 to a maximum of 25). In our research in Boston (Francescato, Francescato, 1974), 85% of the members had completed high school and 55% even a few years of college. Most came from Protestant families; about one-fifth were of Jewish origin. The jobs held by members of these communes ranged from typical middle-class occupations (social workers, teachers, lawyers, students) to jobs chosen based on rejection of elitism and careerism (taxi drivers, waiters, alternative service center operators). A small minority lived on unemployment benefits. The **most common reasons given for** membership were in order of frequency:

- Political-economic (rejection of the system, anti-capitalism, socialism, women's liberation, refusal to perform military service, etc.);

- Financial (to save money, to experiment with economic cooperation);

- Social and personal (for companionship and support, escaping the pressures of the nuclear family, fear of isolation, and desire to realize one's potential and improve one's relationships with others);

- Religious (to live together a life set on religious or mystical ideals);

- Experimental (curiosity, desire to try a different lifestyle);

- Work (desire to work together with other members - for a newspaper, alternative center, etc.).

When one considers that as many as 45 out of 63 subjects stated that one of the greatest satisfactions of communal living was the companionship, safety, and warmth given by the presence of other people, it becomes clear that for the members of urban communes, these aspects constitute above all an attempt to form families in which the positive aspects of traditional households (affection, warmth, protection, safety) are retained, and some negative aspects (authoritarianism and sexism) are eliminated instead.

In our Boston research, we found that members also differed in their assessment of whether the choice to live in the commune was permanent or temporary. The presence in most communes of people with different expectations of temporal stability often created conflict. The needs and expectations of those who wanted to make the commune a permanent home sometimes contrasted sharply with those of those seeking temporary refuge after having difficult life paths or those with a strong ideological motivation to create a model for changing society. Moreover, even in communes whose first members were driven by coincidental motivations, difficulties arose in the failure to select those who joined later, a problem that cohousing members also face.

In communes, proper selection was made difficult, on the one hand, for economic reasons: when they lost a member, they were forced to replace him or her quickly because they had to pay rent and thus did not pause much to consider whether or not he or she adhered to communal principles; on the other hand, members who adhered in theory to communal purposes could change their views once the communal experience began and try to set up life on a different basis. Moreover, many communes came into existence without preparatory work and with hasty methods of recruitment. Often, members were identified with ads like these: "Lee House seeks two new members, male or female, starting June 4" or "A house in Lexington with three women, two men, and three children seeks a man for early June." In our research, we found two major problems. Some would-be members, during the "courtship" phase of the commune they wished to join, had difficulty understanding or openly declaring their deepest motivations for fear of being excluded. They, therefore, tended to conform to perceived majority expectations among other commune members, only to try to change things once they were admitted. Others aspired to enter to resolve unresolved relationship issues in their previous romances. The presence of one or more "disturbed and disturbing" psychologically fragile members caused major conflicts and led to the failure of several communes (Francescato, Francescato, 1974).

Today, however, the cohousing movement seems to pay more attention to the processes of group preparation and formation. Perhaps the most innovative aspect of the movement, which bodes well for its growth, is to place great emphasis on **comparing experiences** at international conferences and on virtual platforms. Other strengths seem to us to be **continual reflection on daily experiences**, considering the difficulties and strengths that emerged in the communities of the nineteenth century and the family communes of the 1960s, using many consultants, frequently promoting meetings and conferences with a variety of workshop opportunities; and fostering **continuing education for members**, paying particular attention to relational, affective and emotional aspects.

From studies to date, it has been found that living in cohousing has many positive effects on members' social capital and their physical and mental health.

A great deal of social *bonding* activity (i.e., relationships are cultivated with people one already knows) and less frequent social *bridging* as well has been observed in various international research studies. Developing meaningful connections with others as parts of a network encourages adherence to salutogenic norms and behaviors and offers resources that moderate *stressors* (Warner *et al.*, 2000, p. 1). Cohousing communities, then, also often develop *bridging* capital by inviting neighbors to sports, cultural, and social activities. They also expand their network of acquaintances by opening common green spaces to others to use as gardens or small vegetable gardens and common rooms for meetings, becoming a resource for the neighborhood (Droste, 2015; Ruiu, 2015). A research study that examined ten cohousing settings found in eight of them a **positive impact on** mental and physical health or **quality of life and well-being**. Further investigation needs to be done to find out what factors did not promote positive effects in the other two cohousing settings (Carrere *et al.*, 2020).

There are already 165 intergenerational cohousing settings in the United States, and 140 are under construction; the vast majority are using **social-affective methodologies** such as circle time, message-me, no-lose method, problem-solving to **create a positive affective climate**, and various other strategies such as the gratitude technique. The increase in the number of elderly people over the age of 65 in all developed countries (in Italy, they are already 23 percent and are expected to reach 34.9 percent in 2050) may increase the desire to live in cohousing or ecovillages. In some countries such as Germany, local governments are donating land where they can build intergenerational cohousing settings, which also allow for a reduction in energy consumption and medical expenses because cohabitants stay healthy and, above all, enjoy a satisfying relational life.

13
Research methodology
in community psychology

1.   Research approaches in the community

The basic theoretical assumptions of community psychology conceive of knowledge production as an activity aimed at fostering and sustaining processes of social change. This highly emancipatory purpose of the discipline makes research inextricably linked to intervention.

What methodologies are congruent with these goals? Which research methods are functional and most used? Some research modes, for example, are more conducive to conducting a descriptive analysis of phenomena rather than guiding transformative action. Each approach varies along two key dimensions. The first of these dimensions reflects the degree of collaboration between researchers and participants. In some approaches, researchers have minimal contact with users: in epidemiological research, for example, data available from official sources and archives are sometimes sufficient. Equally, in the social indicators approach, citizens are involved only as they are subjected to questionnaires or interviews. In contrast, in change-oriented research (or participatory research-intervention), community members play a significant role in defining the problem, planning the intervention, and choosing the information to be collected. The second dimension (vert refers to the degree to which researchers have control over the variables or phenomena being studied. There are approaches, such as ethnographic approaches, in which researchers exercise no control over community members but merely observe, influencing events as little as possible. In field experiments, on the other hand, researchers exercise considerable control, for example, by randomly assigning participants to different treatments.

2. The complexity of research in community psychology: the object of study and methodology

The ecological perspective taken by community psychology imposes a complex vision on the research process. Indeed, community psychology has always strongly emphasized the need to examine people in their natural contexts. Being focused on the "person-in-context" rather than on individual variables or environmental characteristics makes the research immediately more complex; the focus must be simultaneously and synergistically on individuals, groups, and systems and the interrelationships between these three levels. The complexity of the community psychology research approach has been compared to Pandora's box, which, once opened, can no longer be closed (Hobfoll, 1990).

Tricket (2009a; 2009b) defines the epistemology behind the ecological approach as "contextualist"; he believes that the knowledge produced by research reflects the specific culture of the context in which it was generated and, for that very reason, is hardly generalizable.

Because it is situated and contextualized, community science must provide space for and enhance the subjective perspective of participants. Finally, the emphasis on the contextual dimension poses relevant problems of generalizability of results.

We can say that the raison d'être of community psychology is in the search for solutions to problems that connect communities and individuals. The community psychologist must, therefore, ask how the discipline supports the growth of community health and the search for existing factors of malaise and possibilities for change.

The rise of social problems (crime, drug use, AIDS, poverty) that occurred in the 1970s and 1980s accentuated attention to these issues and, in community psychology, encouraged research based on collaboration between scientists and citizens (Speer *et al.*, 1992; Rosen, Painter, 2019).

2.1. How do we measure these variables? In addition to the object of study, methodology is also a critical point of research in community psychology. Community phenomena are not simple and do not respond to linear causality, so methods of inquiry must reflect this complexity. As early as the 1980s, Novaco and Monahan (1980) pointed out the lack of research methods capable of capturing the specificity of the object of study and/or measuring it in a scientifically adequate way (validity problems).

Traditional quantitative methods are not the "essence" of community psychology (Rapkin, Mulvey, 1990). This is not because they are not used by community psychologists, who indeed often make use of many differentiated techniques generally used by social scholars, but rather because they are "borrowed" from quantitative approaches developed by other disciplines or areas of psychology. These approaches derived from quantitative methods have sneaky effects on the designs of much community research. Although the researcher wants to be interested in the complex relationships among behaviors, attitudes, and community structures, he or she is often in the situation of limiting the analysis of results to evaluate them with standardized methodological designs. In the name of empirical rigor, the use of quantitative techniques often restricts the kinds of questions the researcher would be interested in asking.

Quantitative methods require large, representative samples that place distances between the researcher and the phenomenon being investigated both because of their numerosity and because there is sometimes a tendency to overgeneralize explanations of events based only on the prevalence of correlations. Information about the variety of patterns and nonlinearly correlated variables is often lost. The concept of population, as used in inferential statistics, takes on limited meaning from an ecological perspective. An ecological approach requires research centered on contextual dependencies since traditional quantitative methods ignore them.

The above also challenges the traditional way of assessing the validity and reliability of a research model because an important aspect of these concepts is that they introduce elements of stability. Ecologically oriented inquiry requires events placed in their specific physical, temporal, and relational contexts. Community psychologists are called upon to consider whether research that accords with the dominant models of psychology is the kind of research they want to foster and develop. The choice is not toward less rigor but toward a better articulation of what rigor means for the community.

These considerations highlight the distance with traditional research models, which are usually more oriented toward controlling, balancing, and treating as "disturbance" the variables that connote context. As an alternative to quantitative methods, some authors (Stein, Mankowski, 2004; Orford, 2008) have argued for the validity and appropriateness of research in community psychology of qualitative methods.

One possible solution, increasingly proposed and tested, is to integrate quantitative methods with qualitative ones, considering them complementary and not exclusive, with hypotheses formulated using one methodology and answers provided using the other (Barker, Pistrang, 2005). The qualitative methodology can contribute to several research objectives in community psychology, such as:

- *Description of events*. A naturalistic description of a phenomenon of interest constitutes a fundamental work for any discipline. A fortiori, accurately describing the natural history of a community or social problem, could contribute to the effectiveness of social intervention. An early advantage of qualitative techniques is their ability to sketch participants' experiences exhaustively, to identify complex behaviors or social patterns, and to delineate the complex nature of the situational, organizational, or community context in which the phenomenon occurs;

- *Hypothesis formulation*. Hypothesis formulation represents a very creative task for the community researcher. Especially the use of interviews and observations can generate new alternative perspectives;

- *Hypothesis evaluation*. Hypothesis evaluation is traditionally based on statistical analysis of quantitative data. Qualitative analysis of qualitative data alone is usually not among the preferred procedures for evaluating a hypothesis. The only exception might be when qualitative observation or interviewing can reveal behavioral patterns or cultural norms whose existence is sufficient to disconfirm the given research hypotheses (Maton, 1990).

3.   Some thoughts on qualitative research and multi-method research designs

The last three decades have witnessed the growth of a progressive interest in qualitative inquiry, which, from being used in "weak" modes (describing phenomena rather than verifying their causes, facilitating the formulation of hypotheses rather than testing theories), has gained credibility, especially in the study of opinions and social representations. Factors that have facilitated this process include:

- *Sociocultural factors*: the rise in post-industrial society of the need for subjectivity, the focus on everyday experience and inter-individual differences;

- *Theoretical factors*: the need for a new, less artificial, and more ecological research, accompanied by criticism for the claim of objectivity and contextualization of quantitative methods and the development of theoretical options, often interdisciplinary, in support of qualitative paradigms (constructivism, symbolic interactionism, social representation theory, discourse analysis, narrative analysis, etc.);

- *Technological factors*: the development of recent technologies (computer, audiovisual, telematics) that have facilitated and made the qualitative investigation more reliable and accurate (Shah, Corley, 2006).

Positive regard is now well established toward this type of research because of the advantages it offers because of the wide range of methods developed, which include, in addition to the more traditional tools (individual and group interviews, observation techniques), content analysis methodologies, examinations of documents, cultural artifacts, life histories, and verbal accounts. Each of these tools produces a considerable amount of data, the analysis of which constitutes the crucial problem of qualitative research since the numerosity and complexity of the information that the researcher is faced with processing usually make it difficult to extrapolate data compatible with the requirements of the scientific method.

The orientation that currently seems to be asserting itself is that of a pluralistic and tolerant epistemology of qualitative research, an orientation that values a "weak approach" to research, flexible toward theories and methods to ensure the best coverage of the object of inquiry that is itself particularly complex. For this reason, the use of a single tool often gives the feeling of distance between research objectives and results, and hence the widely felt need to use "multi-method research designs" (Tuominen, 2018; Roulston, 2019). Compared to the experimental model, the qualitative paradigm, both in the collection and return of data, proposes a less aseptic and scientific role of the researcher. In the qualitative framework, the construction of a research report resembles more a work of data restructuring than a simple transfer of information. Moreover, the choice of a "narrative" or "objectivist" style presupposes different paradigmatic positions, either positivist or constructivist, of which the researcher should be aware.

The "participatory" modes used by the "qualitative researcher" in both returning and collecting results have value in rethinking and attributing meaning to the reality being investigated.

From these premises, it is easy to grasp the close relationship between knowledge and action inherent in qualitative research; frequently, it is linked to the research-action model either because it takes on the value of social practice or because it is implemented to gather information that then guides social practice. Applied qualitative research requires not only expertise in data production but also in managing the entire process. On such issues, community psychology has made and will continue to make a relevant contribution.

Studies have highlighted the relationship between pathology and environment, broadening the focus of observation and intervention to the context of the origin of the pathology and not only to individual characteristics. We refer to all those studies whose role is to highlight, for example, the relationship between socioeconomic *status* and psychiatric disorder (Dohrenwend *et al.*, 1992; Stansfeld *et al.*, 2011; Manstead, 2018), between social class and organic health (Townsend, Whitehead, Davidson, 1992; Lutfey, Freese, 2005; Mackenbach *et al.*, 2008).

4. Methods of diagnostic research, experimental and quasi experimental research

Diagnostic research includes different methods like epidemiologic and social indicators. The epidemiologic approach was born inside social medicine to study the frequency and causes of physical and mental problems, and to identify "risk groups" who were more likely develop certain diseases. Diagnostic research explores the relation between social context and illness, for instance exploring how socioeconomic status influence psychiatric problems (Dohrenwend *et al.*1992: Stansfeld *et al.* 2011; Manstead, 2018), and how social class affects organic health (Townsend, Whitehead, Davidson, 1992; Lutfey, Freese, 2005; Mackenbach er al.2008). The social indicator method explores how wellbeing and malaise are present in different social groups, using objective data (such as rates of crimes, suicides, unemployment; and presence of community services for small children) community psychologists use also subjective data asking community residents for perceive strong and weak points) (Privitera,1987; Arloti, Barberis, Kazepor, 2008).

Experimental research attempts to follow a rigorous statistical scheme to verify an experimental hypothesis using data from the field (Cambell and Stanley (1966), Fairweather (1967) and Cambell (1969). This enables one to infer causal hypotheses between variables, the data and the situation permitting:

- Extraction from the population a random sample sufficiently representative of the population;
- Dividing the sample into sufficiently random control and experimental groups;
- Submitting the experimental group to a specific treatment not administered to the control group;
- Eliminating as much as possible interference from disturbing factors.

The confrontation of the results of the two groups permits the verification (or non-verification) of the hypotheses made concerning the effects of the experimental treatment.

This is the valid scientific method but most difficult to apply because of the necessary rigor. Unfortunately, it is often difficult to make the rigorous extraction, the assignment of the two groups, and in the complex social reality, to control the disturbing variables.

Nonetheless, the experimental method is used relatively frequently to evaluate the efficacy of experimental programs.

Quasi-experimental research: In social research, very often, one encounters situations that do not allow for control and/or manipulation of all relevant variables. In such cases, an attempt can be made to approximate the experimental setting as closely as possible so that the research is sufficiently valid. This approach, termed "quasi-experimental," was particularly developed by Campbell and Stanley (1963; 1966) and Cook and Campbell (1979). As mentioned in the previous section, one of the most difficult experimental requirements to meet is random assignment to the experimental or control group: indeed, participation or non-participation in each program often cannot be freely decided or managed by researchers. The latter can, however, select control groups that, although "not equivalent" because they are not randomly divided, are as similar and comparable as possible to the experimental groups in terms of the most relevant variables (e.g., biographical data, setting, etc.).

In performing the research, the researcher introduces an independent variable that he or she hypothesizes will produce changes in the people under observation. If, indeed, the intervention changes these behaviors, the pre-treatment measurements will differ from the subsequent time series. This is the reason why many scholars (e.g., Linney, Reppucci, 1982) have advocated methodological synthesis, that is, the combined use of time series and nonequivalent control groups, to increase the validity of quasi-experimental research intervention.

14
Participatory action–research

1.   From Lewin's *action–research* to *participatory action research*

Action research has two theoretical matrices of reference: the tradition of philosophical pragmatism (particularly Dewey), which laid the first theoretical foundations, and later the work of Gestalt psychologist Lewin and radical psychoanalyst Moreno.

Lewin introduced the methodology of *action–research* after World War II, calling it research that leads to social action (Lewin, 1948). It can be considered a logical outgrowth of his experiences (from industrial psychology to group dynamics, from authority styles to ethnic minority problems) and his scientific conceptions (see Lewin, 1946, 1947, 1951; Marrow, 1977). According to Lewinian's "field theory," in fact, understanding social and psychological phenomena involves observing the dynamics of forces that are present and acting in each context; if reality is an ongoing process of change, science should not freeze it but study things by changing them and seeing their effects (Lewin, 1951). Lewin believed that scientific theories and transformative practice should and could intertwine in a fertile reciprocal process in which hypotheses drive actions, and those actions stimulate and change the knowledge itself. To this end, researchers and community members can usefully cooperate and share needs, expertise, and resources. In this way, the role of the researcher is profoundly revisited: from being an extraneous and distanced actor, he or she becomes a participating and involved subject (Greenwood, Levin, 1998). The democratic movements of 1968 gave significant impetus to the spread and development of action–research while critical theories provided new insights into action–research approaches with explicit emancipatory intentions. After Lewin's death, the principles of research intervention were deepened and variously applied both by his direct U.S. collaborators and by some groups of European researchers.

Among the European groups, the Tavistock Institute of Human Relations in London should be mentioned for the relevance of their research.

Since the early 1970s, social research in developing countries in the southern hemisphere (namely, Latin America, South Asia, and Africa) has focused on the manifest problems of inequality, poverty, exploitation, and oppression. The dissemination and wide use of action–research in these areas has allowed for an emphasis on the central idea of this model: to involve, from the outset, in the research processes, the recipients of the intervention and/or change process. This has given rise to a particular model of action–research: participatory action–research (PAR)

2.   What is PAR

PAR can be defined as a type of inquiry that, while having a cognitive purpose, promotes the active involvement of all significant stakeholders in the context to then collectively practice an activity on the reality being sought to know.

Local actors, the owners of local knowledge, are encouraged to actively involve themselves and collaborate with the experts conducting the research process (Maiter *et al.*, 2008; Kagan *et al.*, 2011) in the belief that the information they provide offers relevant insider perspectives and relates to issues that may remain inaccessible or invisible to academic researchers (Stern, 2019).

Each stage of the PAR (problem definition, subject recruitment, information gathering, data processing, data signification, final report writing, public restitution) is discussed and shared local people. The intent pursued through such modalities is not to obtain full agreement but rather to explore, through dialogue, the complexity of relationships among participants. As Rappaport (1990, p. 54) effectively points out, "When people collaborate, they engage in a process of mutual influence by reviewing their views in light of the views of others." Self-reflection, collaborative solution-seeking, and responsibility for change depower delegating or dependent attitudes while fostering empowerment and confidence in one's abilities (Dworski-Riggs, Langhout, 2010; Dudgeon *et al.*, 2017). This has a clear effect in terms of the sustainability of implemented development efforts. Therefore, PAR is recognized as an empowering alternative to positivist research methodologies that reflect more traditional power relations in which the research process and results are guided by the "outside expert" (Braun *et al.*, 2014).

Some authors (Dworski-Riggs, Langhout, 2010), however, point out that to carry out a PAR process, researchers need to understand the level of awareness that community members have of the power relations present in the context of investigation. In an environment without a history of egalitarian participation, for example, social norms may hinder participatory processes by causing tension between the community's rules of coexistence and the egalitarian model proposed by PAR (Wallerstein, 1999). PAR may occasionally activate processes of resistance in the community that necessitate accommodations in the use of participatory modes and tolerance toward attitudes of delegation (Dworski-Riggs, Langhout, 2010). In communities where there are clear power asymmetries, PAR should be adapted to be responsive to community needs and norms and be relevant in that specific context (Harper *et al.*, 2004; Tomai *et al.*, 2017).

2.1.   Characteristics of PAR and areas of application The distinguishing features of PAR, fully embraced by community psychology, are:

- The nonlinear but circular relationship between theory and practice through which continuous processes of transformation are generated;

- Be intentionally designed to change the field of inquiry as it is studied;

- Be designed and conducted collectively with the active participation of technicians and people from the community. This means that it is not the psychologist or researcher or practitioner who first research then communicate the results and eventually the intervention. The diagnosis is made together , simultaneously changes begin to take place, always collectively intervention strategies are studied. The community is both subject and object of knowledge and social change, and the technician is not the holder of knowledge but is a resource available to the community.

These elements allow us to grasp the democratic perspective that guides action–research, which always takes an emancipatory approach that aims to improve the social and democratic participation of people in society and to establish equality and social justice (Boog, 2003). Such an approach has, as can be guessed, privileged areas of application. In agreement with Cohen and Manion (1984), practical and emancipatory approaches seem to be most suitable when:

- The definition of a problem is relatively open;

- Participation is pluralistic;

- The purpose is to generate and promote processes of social reform or transformation;

- Research methods are transparent and understandable to participants;

- The aims and methods of the change project are shared by all participants;

- Consultants are not unduly compromised by their institutional base.

The connection between theory and practice, the circular and systemic view of scientific progress, and the stimulus to citizen participation and involvement still appear to be among the most valuable and innovative legacies of the Lewisian legacy.

1. A procedural model of participatory research-intervention

Cunningham (1976) proposes a procedural model that effectively summarizes the principles of participatory research intervention. This model describes the development of a working group that, within an organization, plans, executes, and evaluates the intervention. We follow Cunningham's scheme, which is divided into three points: group development, research, and intervention (Fig. 1).

• *Beginning*. Intervention research begins based on a motivation, in terms of a need or problem to be solved, by members of an organization. The perception of the problem is discussed and shared in its broad outlines by the people interested in addressing it.

• *Group constitution*. Within the reference context (organization or community), the working group or committee that will conduct the research-intervention is identified, while not neglecting to inform other members about the steps taken and the results achieved. Membership is voluntary, but it is necessary for the group to possess skills and influencing capacity, within the organization, to facilitate the implementation of the experiments.

• *Specification of goals*. The group needs to agree and define goals of common interest that are realistic, flexible, and not long-term but articulated *step by step*.

• *Group training*. Training refers not only to the technical training that can be given, but also to the positive evolution of the group in the sense of improving the skills of cohesion, interdependence, and creative cooperation.

At this point the group has formed and is able to get into the thick of the investigation.

•*Formulation of hypotheses*. Considering knowledge of the overall needs of the context, the group must now discuss, identify and operationally define the research hypotheses.

• *Determining how to collect information*. The tools will have to be as valid and reliable as possible and, at the same time, appropriate for the type of context in which one is working; one must also carefully choose the people, the categories from which to acquire information, whether interviews or questionnaires are used or whether one relies on archival documents or other indicators. At this point, the group can embark on *data collection*.

•*Data analysis*. It is the group itself that analyzes and processes the data, trying to avoid protective and defensive mechanisms or internal oppositions that involvement with the problems explored might foster.

Figure 1 A procedural model of participatory research-intervention

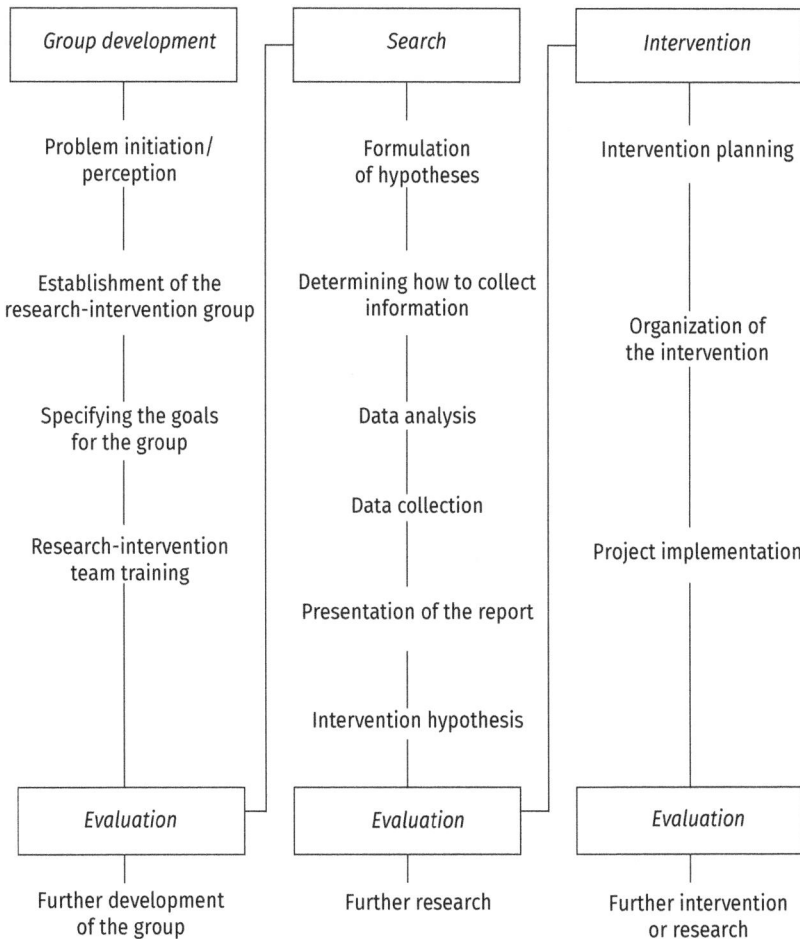

Source: Cunningham (1976), p. 218.

- *Presentation of the research report.* Research results are presented to other members of the organization or community for joint discussion, usually at an assembly time.

- *Action Hypotheses.* The information and knowledge gained in the research enables action hypotheses to be formulated as alternative solutions to organizational or community problems.

- *Intervention planning.* The planned intervention must be linked with research findings and have operational, realistic, observable goals.

- *Organization.* At this stage, the group establishes the resources to be employed, clearly assigns responsibilities and tasks, and schedules verification   deadlines.

- *Project implementation.* Finally, the project is fully implemented by the team, whose functions are now implementation and control. It is evident from this outline how at all stages of the cycle it is necessary to assemble different techniques and skills (the formation and coordination of working groups, the construction of survey instruments, the analysis and processing of data, the planning of interventions). Particularly important is the dissemination of scientific knowledge as well as the enhancement of community experience and "culture."

As made very clear by the graphic model, reflection can never be disconnected from action. Otherwise, we run the risk of alienation, utopianism, dogmatism, scientism, or fundamentalism (Boog, 2003).

Finally, we feel the need to emphasize the importance attached to this model in the evaluation process. As Figure 1 indicates, at the end of each phase, the group conducts an evaluation that serves as feedback to calibrate the progress of the work. The entire implementation phase is continuously monitored to check how the program is progressing; in fact, part of the effectiveness of the research intervention lies precisely in the cyclical nature of the process and the correction in progress. The process closes with an overall evaluation of the effectiveness of the intervention in which the quality indicators planned and included in the operational plan are verified.

3.   A comparative consideration of experimental and participatory research-intervention   approaches

From the description of experimental research and participatory research-intervention, it is possible to understand the depth of differences between the two methodologies, which also seek to combine inquiry with action in the community. The fundamental difference lies in the very purpose of the two approaches. Experimental research aims to test hypotheses, establish causal relationships, and determine the effect of treatments on representative samples to be able to generalize the results achieved or extend their validity and establish universal laws. In contrast, participatory research-intervention aims to achieve socially relevant changes, the applicability of which tends to be broadened. In this case, the guiding criterion, rather than the representativeness and validity of the results, is the actual and potential utility, the social relevance of the change introduced.

The other differences between the two approaches then seem to us to be related to these general different objectives. In experimental research, the subject of the investigation is simplified and "stopped" in time, the variables as isolated as possible, and the control group randomly selected to make *matching* with the experimental group reliable. In participatory research intervention, subject and object are often difficult to distinguish, and multiple variables are examined in their dynamic process; *matching* is almost always impossible because the intervention is implemented on a single group without control. Moreover, we must add that these considerations are related to the two types of research considered in the abstract as ideal types. In the concrete reality of social programs, much of the research fails to be as rigorous as the experimental method would like or to involve many citizens as participatory research intervention would demand.

And yet, there are various ways to implement "quasi-experimental" research interventions that can often serve as a satisfactory compromise – always to be verified and improved – between the demands of applicability and validity in the process of knowledge and social transformation. At this point, it seems even clearer why the model of participatory action research appears to be the most congruent with the fundamental premises of community psychology. As Amerio (2000) states, it is currently the tool that best captures and methodologically addresses the problems at the interface between the social and the individual, intervenes with real groups, directs action toward change (of the individual, the group, the organization), and produces social knowledge, which is intersubjective and co-constructed.

4. Critical evaluation of PAR: advantages and limitations

The work of Lewin and his epigones contributed to the definition of the participatory research-intervention methodology, which because of its specificities is the most coherent and appropriate research method for the community psychology approach and most closely reflects the guiding principles for a theory of technique in community psychology that we identified in chapter 3. A particular strength of this methodology is that it emphasizes the constructive role of action, understood as a process that articulates mental and practical activity, individual and social spheres, providing participants with the opportunity not only to adapt to contexts but also to change them.

Participatory research-intervention aims to stimulate the involvement of hypothetical intervention recipients by promoting their participation in the planning and implementation of all steps as well. Proposals born out of collaboration with at least some of the users are more likely to be accepted by the community and more effectively motivate change. This research model gives rise to a process a process in which "local knowledge" and "professional expertise" are integrated to promote social change. Communication between social scientists and community members, which can be considered the most innovative feature of participatory research-intervention, requires a not always easy and quick integration, with the need for researchers to find a suitable language to give training to community members.

Participatory action–research has the merit of being a particularly flexible and adaptable tool, useful for understanding and intervening in a specific ongoing process, effective in guiding problem-solving, decision-making and the development of action plans, and able to increase the awareness and responsibility of the group involved in the research. Shared knowledge (*shared expertise*) and participation in decision-making (*shared governance*) are the main takeaways of participatory action–research (Procentese, Marta, 2021): they facilitate the promotion of a sense of ownership because they increase the likelihood that people will invest more in the direction indicated by the shared decision.

Some limitations of the participatory actions are; it does not always provide opportunities to examine the historical roots of social problems, nor does it necessarily bring out the link between individual empowerment and social struggles. Narratives may or may not be included, but they are not a typical part of the methodology, so it will not always be possible in research-intervention to give voice to existing minority narratives that break the tacit consensus by which social actors accept the systems of convention in which they are immersed. In research-intervention much depends on the initial composition of the research group; if members of marginalized groups are included, different interpretations of problems are likely to emerge. Moreover, as already mentioned, even when it purports to be systematic, participatory research-intervention risks not achieving satisfactory scientific validity. Its objectives are often situational, the sample small and unrepresentative, control over interfering variables poor; the researchers' own involvement may limit and constrain observational capacity. Of course, there are fewer radical modes of implementation in which, for example, user participation in the research-intervention is limited to the initial phase, that is, to the gathering of queries and problem definition, and to the phase in which the researchers present the survey data. Even in this case, however, involvement can be significant where the discussion of the report becomes the basis for a problem-solving process and the formulation of a decisive action plan.

However, despite its limitations, PAR and participatory processes are among the most effective tools for addressing some of the most pressing social problems in the world today, such as migration flows, new forms of coexistence, poverty, and religious intolerance (Huutoniemi, 2015; Lake, Wendland, 2018).

15

Community psychology and climate change

1.   Expanding the ecological paradigm to include the natural environment

The natural physical environment has been little explored in both community psychology and environmental psychology (Riemer, Harré, 2017). Both disciplines emerged in the 1960s in the United States, adopting systemic ecological perspectives, but these focused on spatial and social features of physical environments inhabited and transformed by humans and rarely on natural environments. Community psychologists adopted the ecological constructs of biologists, but focused on social ecology rather than natural ecology. The low interest in the environment is also evident from the fact that few community psychology texts include chapters on environmental topics (Prilleltensky, Nelson, 2009; Bond, Serrano-García, Keys, 2017). Some community psychologists have considered the natural environment to be the main topic of environmental psychology, instead, until very recently, even environmental psychologists have been primarily concerned with built or human-inhabited environments (Bonaiuto *et al.*, 2016). This lack of attention to the natural environment, its biodiversity, and the thousands of sentient animals living in it has remained unchanged for decades, during which socioeconomic and cultural factors have pressed for an improvement in the quality of life and psycho-physical well-being of the planet's *human inhabitants* alone.

This selective focus on human inhabitants characterized the four main groupings of North American environmental psychologists in the 1960s-1970s. Followers of the cognitive approach explored *how humans come to understand* and use cognitive representations of the environment, also called images and schemas (Lynch, 1960)

Proponents of the behavioral approach examined *human behavior* more using behavioral mapping, a direct observation methodology that records people's movements and behaviors in specific locations to "identify uses of space as a factor in behavior" (Ittelson *et al.*, 1974, p. 232).

Environmental psychologists following the socioecological approach favored the "multidisciplinary study of the impact of physical and social environments *on human beings*" (Insel, Moos,1974, p. 9). They developed the concepts of behavioral setting (Barker 1968) and environmental pressure and constructed special scales to measure psychosocial environments in daycare centers, universities, and group homes.

A minority group of environmental psychologists with a psycho-political-pragmatic orientation aimed to pursue a better understanding of the environmental problems reported by Barry Commoner (1971). Wohlwill and Carson (1970) and Swan (1970) began early research on public attitudes toward pollution, "documenting how air pollution is most severe in poorer neighborhoods of large cities. Residents of these blighted areas, however, tend not to engage in anti-pollution campaigns because they feel they have no chance to influence decision-making processes and have pressing socioeconomic problems" (*ibid* p.74)

Others examined and criticized public housing projects that aimed solely to provide low-cost housing for low-income families, highlighting the lack of semi-public spaces and facilities for developing social relationships (Yancey,1971). Psychologists with a psycho-political-pragmatic orientation believed that the role of the environmental psychologist also involves participating as activists in political initiatives aimed at achieving legislative changes to improve environmental conditions. However, most environmental psychologists, in the subsequent decades, did not follow this socio-political approach and mostly focused on university contexts and politically moderate research, at most they supporting gradual reforms.

Environmental psychology and community psychology emerged in the sixties, a decade of the contestation of individualistic ideology, which produced great social inequalities, exacerbated by the racism and sexism of American society. Civil rights struggles spread, and the myth of the open society of equal opportunity was cracked. Some psychologists began a critique of those aspects of applied psychology that trace all behavior problems back to individual factors, indirectly relieving the social system of any responsibility for them.

1.1.　Early developments in environmental and community psychology in Italy

DonataFrancescato, as a doctoral student in Clinical Psychology at the

University of Houston, after publishing an article on the image that citizens of Rome and Houston have of their city (Francescato, Mebane, 1973), following Lynch's (1960) theories, devoted her doctoral thesis to the patterns and images of the city of Houston, but during her internship year (1970-71) in a mental health center in Boston she was attracted to community psychology. Back in Italy in 1972, she met Raffaello Misiti, who was very interested in environmental issues and offered her a job at the CNR, the Italian National Center for Research. He also asked her to meet with Professor Eraldo De Grada at Sapienza University, where the first graduate program in Psychology was created in 1970, to have her U.S. university degrees legally recognized. Instead, De Grada, examining her *curriculum vitae*, offered her a one-year contract to teach Personality Techniques for the 1973-74 academic year and advises her to write something about this new community psychology she told him about. Francescato chose the university contract but also published her environmentalist doctoral dissertation (Francescato, 1975), and in the foreword on early developments in the discipline in the United States; she wrote:

"A major limitation of the approaches of environmental psychologists lies in the failure to elaborate at the theoretical level the interaction between political and social systems and environmental phenomena. The absence of this comprehensive analysis means that even interesting attempts (e.g., to plan the environment of geriatric shelters to better meet the needs of residents) remain extremely sectoral interventions, limited and isolated from the broader social context. Psychologists and architects do not even consider the question of whether it is appropriate to enclose the elderly, to exclude them, albeit in gilded ghettos; it does not even occur to them to analyze the relationship between the sociopolitical context and the progressive marginalization of entire segments of the population. The same argument applies to those who study environmental behavior in other institutions such as schools, prisons, and psychiatric hospitals. Too often, a critique of the role of the institution is missing, and psychologists and other researchers take refuge, mythologizing it, in their role as experts. Even when, in the social-ecological approach, the authors insist on pronouncing a choice of values, thus positioning themselves politically, they end up only wishing for optimal human environments without at all specifying what interests currently stand in the way of creating better environments, how redistribution of resources, now unequally divided, etc., should take place (*ibid.*, p. 29).

Also, in Italy in the 1970s, environmental and community psychologists focused on the human inhabitants of the plan*et al*one. Marino Bonaiuto (2017), in an interesting analysis of the evolution of the discipline in Italy from the 1960s to 2015, defined environmental psychology as the psychological study of the relationships and interactions between people and the physical environment, although very often it is the physical-social environment and thus the disinterest of environmental and community psychology in the natural environment continued for decades.

As early as the 1960s, some environmental psychologists in Italy began to carry out various collaborative activities with professionals such as designers, architects, urban planners, and builders to design or redevelop places or buildings. Mirilia Bonnes initiated the first research projects on urban ecology in Rome. In the 1970s, the first Italian publications appeared (Francescato, 1975; Bagnara, Misiti, 1978; Bonnes, Secchiaroli, 1979).

The years from 1980 to 2000 also saw environmental and community psychologists engaged in formally introducing the two disciplines into academia, sometimes competing for scarce resources. While environmental psychology became rooted in social psychology, community psychology would be included in both social psychology and dynamic and clinical psychology (Palmonari, Zani, 1980; Francescato, Contesini, Dini, 1983; Bonnes, Secchiaroli, 1992; Francescato, Leone, Traversi, 1993).

Environmental and community psychologists have done research and interventions in the same areas: they have collaborated with architects and other professionals in creating new residential centers or renovating old ones, considering the psychological needs of the users (daycare centers, playgrounds, retirement homes, urban trails for children, etc.). They studied how residents use parks and other neighborhood services experiences relative to crowding, traffic, and technological risk. They explored the psychology and physiology of living, attachment to places, and emergency housing for earthquake victims (Caia, Ventimiglia, Maass, 2010).

Environmental psychologists have been examining work environments more often, improving perceptions of products, documenting that noise reduction increases not only satisfaction with the environment and work and mental and physical well-being but also identification with the company, and that architectural aspects also impact users at the level of cognitive processes: a court, hospital, or university building considered beautiful but threatening induces greater pessimism about the outcome of its specific functional activities than one not perceived as such.

In the 1990s, environmental psychologists conducted research on new communication technologies such as video conferencing, the redevelopment of degraded urban areas, the humanization of hospitals, and the beneficial effects of nature, particularly in terms of perception and enjoyment of greenery and psychological catering within various contexts. In the early decades of the 21st century, the rise of extreme weather events reminded some environmental and community psychologists that there is a natural environment that is not built only for humans but instead is damaged by human activities.

1.2. The proliferation of extreme weather events awakens the interest of some environmental and community psychologists in the natural environment

Environmental movements denounce the melting of glaciers, the degradation of air, oceans, and water quality, and the increase in extreme weather events and point to eco-sustainability. The natural environment has become more present among environmental psychologists but is always viewed from a predominantly human-centered perspective (Bonaiuto *et al.*, 2016).

Members of both disciplines have been involved in environmental education, and some community psychologists have developed innovative courses with environmental experts exploring both how best to apply some pivotal principles of community psychology (Dittmer, Riemer, 2012) and the role of negative and positive emotions in environmental attitudes, integrating aspects of social-affective and environmental education (Francescato, 2020).

Environmental psychologists have most often worked on developing quantitative psychosocial tools and methodologies for testing the effectiveness of environmental education programs conducted at primary and secondary schools, creating and evaluating questionnaires and scales (Bonnes *et al.*, 2006; De Dominicis, Schultz, Bonaiuto, 2017). More recently, they have explored how to measure and intervene in the psychology of sustainability, that is, how psychology can help the ecological transition for environmental sustainability (Bonaiuto, 2023).

Community psychologists have theorized the need to adopt multidisciplinary visions and have developed models of analysis and intervention such as those examined in several chapters (e.g., 5, 7, 11, 12, 14). Moreover, they denounce injustices and all types of discrimination that harm the physical and mental health of women, people of color, and the disabled; they have also explored the perceptions, responses, and resilience of people and communities affected by climate change, but also examined the rise of climate migrants, and the relationships between mental health, community well-being, and climate change (Manning, Clayton, 2018). However, Nielsen and coworkers (2020) have documented how psychologists have focused too much on changing individual behaviors that have low environmental impact, and have neglected to promote changes in organizations and social networks, and, most importantly, have neglected their role as activists, which is necessary to demand and achieve structural policy changes that can have a more lasting and significant environmental impact.

The COVID-19 pandemic increased interest in global climate issues: in particular, new members of the boards of the Italian (SIPCO) and European

(ECPA) associations of community psychologists, where younger members most threatened by climate change have been elected, have promoted webinars and publicized special issues on climate change in journals such as "Community Psychology" and "Community Psychology in Global Perspective" (see ECPA website).[13]

2.   Climate change, health, community, and COVID-19 are close relatives

The COVID-19 will be remembered as a great pandemic that disrupted the lives of billions of people and claimed millions of lives. However, it also gave us a great gift that we can best use to promote health and the environment in our communities: it showed us how much air pollution increased coronavirus mortality. The very areas in northern Italy where the highest mortality due to the virus has been recorded have had the highest levels of particulate matter (PM2.5; EEA, 2019) for years. Diseases resulting from local air pollution (ischemic heart disease, cerebrovascular disease, chronic obstructive pulmonary disease – COPD, lower respiratory tract infection, lung cancer) emerged among the main comorbidities of patients who died from coronavirus in Italy in a sample of 10,000 cases (Mebane, 2020). In the spring of 2020, the first national survey of long-term air pollution exposure and coronavirus mortality (in the United States) found that an increase of only one microgram per cubic meter in long-term average pm2.5 exposure was associated with a 15 percent increase in COVID-19 mortality rates (Wu *et al.*, 2020).

The COVID-19 pandemic has brutally reminded us of what we should have known long ago. The European Environment Agency estimates that there were 76,200 premature deaths in Italy, already in 2016, most of which (58,600) are attributable to PM2.5 (EEA, 2019). These pollutants are produced using fossil fuels. Mortality resulting from air pollution is caused by inhaling or ingesting fine particles that are as small as 2.5 micrometers in diameter, so fine enough to penetrate the lungs and bloodstream. These particles, which are called "particulate matter," can be emitted directly from combustion or formed indirectly from atmospheric reactions with sulfur dioxide or nitrogen oxides. High concentrations of particulate matter increase the frequency of four lethal diseases: stroke, chronic obstructive pulmonary disease, ischemic heart disease, and lung cancer, that is, the same diseases found in patients who died from coronavirus..

The world population has now surpassed 8 billion and is on track to reach 11 billion by 2100. This massive growth increases environmental pollution because it leads to a staggering increase in GDP and energy used to produce

13. See https://www.SIPCO.it/ and https://www.ECPA2024.it/.

more goods and services, housing, and transportation. Today, 90 percent of the energy consumed comes from fossil fuels (coal, oil, natural gas). Burning these materials produces energy, plus carbon monoxide and fine particles. Energy derived from fossil fuels is used, for example, by automobiles, domestic boilers, airplanes, and thermal power plants to produce electricity. But its wide use increases environmental pollution and damage to our health.

Outdoor air pollution from fossil fuels caused 4.2 million deaths, and indoor air pollution was estimated to be responsible for 2.9 million deaths in 2015, a total of 7 million deaths per year (IMF, 2019). In comparison, traffic accidents kill far fewer: 1.3 million people a year (Harari, 2019).

2.1. Our health is increasingly in danger Hurricanes, violent cloudbursts, floods, and fires cause deaths and injuries, but summer heat waves also increase mortality among the most vulnerable (the elderly, the poor, children, and people with mental illness), mainly affected by respiratory and cardiac crises. In addition, forest fires, by lowering air quality, increase respiratory and cardiovascular diseases, and higher temperatures raise ozone levels, damaging lung function, and aggravating asthma. The greatest impacts are evident in poor neighborhoods of large cities, where the urban heat island effect is greatest and access to health facilities is most difficult (Butera, 2021), as early environmental psychologists of the sociopolitical approach already argued (Francescato,1975).

In addition, our health is undermined by the spread of atmospheric aerosols caused by the burning of fossil coals in industry and transportation, the burning of biomass in intensive livestock farms, and by forest fires and desertification. Chemical and biological pollution caused by the spread of plastics, pesticides, pharmaceuticals, antibiotics, hormones, and chemical compounds adversely affect human and ecosystem health.

Rising global temperatures, moreover, due to lower agricultural production combined with floods and droughts, exacerbate the problem of food insecurity, which in turn increase migration and wars and promotes the growth of diseases caused by malnutrition and undernutrition.

2.2. Climate change makes a shift from fossil fuels to renewable energy more urgent Already, the planet is one °C warmer than in 1860, and environmental disasters are multiplying: unprecedented hurricanes, boiling summers, devastating wildfires, dying coral reefs, desertification, and melting glaciers. Rising temperatures due to global warming could also put viruses and bacteria, frozen in permafrost and glaciers for thousands of years, into circulation, with serious consequences for the health not only of our planet but of all living things. Changes in temperature, humidity, and soil conditions promote so-called "**species jumping**." Sixty percent of emerging infectious diseases are transmitted to humans from wild animals (SARS, Ebola, HIV).

Humans destroy thousands of ecosystems each year, and many species must migrate in search of new niches in which to survive.

These phenomena have become more frequent over the past fifty years, with a more pronounced increase from 2000 to the present. One billion two hundred thousand people are at risk of flooding, and 1.8 billion suffer the consequences of land degradation, desertification, and drought. The situation is expected to worsen in the coming decades, affecting especially people in developing countries, but also those living in Europe, the United States, and Australia. Hurricanes and heavy rains affect warmer and wetter areas with tropical climates: in India, 12 percent of the land area is at risk, in Nepal, 20 percent, in Bangladesh, 75 percent (Butera, 2021)

Sea level rise is already beginning to cause damage to low-lying lands and salinize coastal sources of fresh water. The global average sea level has risen by 3.6 millimeters per year from 2005 to 2015 and is projected to rise by nearly a meter by 2100 if emissions continue at current rates. But even if we can stabilize them, the level will continue to rise by 30-40 centimeters, enough to create serious problems, given that 10 percent of the world's population (about 750 million people) live in at-risk areas, particularly in Egypt, India, and Bangladesh. Already, inhabitants of some Polynesian atolls have been forced to migrate, encountering reintegration problems. If millions of people must migrate, host countries will have to manage enormous health, social, and economic challenges. In Africa and South America, droughts increasingly threaten crops and increase the intensity of wildfires in Australia is augmenting.

We are also close to implementing irreversible changes such as the loss of Arctic ice, the slowing of ocean water circulation, the thawing of permafrost, the disappearance of the Amazon rainforest and coral reefs, the melting of glaciers in West and East Antarctic Greenland, and the loss of biodiversity (*ibid.*; Thunberg, 2022). The latter factor influences the future of all inhabitants of the planet, as well as the survival of the human species and other life forms fundamental to us. Moreover, the future of the evolution of the entire Earth's ecosystem depends on universal biological laws governing its existence, which we humans do not know today except in a very limited way. Biologist Rob Dunn and the great ethologist and activist Jane Goodall remind us that, of the species that inhabit the planet, we know one in eight, and to avoid a collapse, we will have to learn to respect all life forms (Dunn, 2022)

3. Environmental problems are systemic: physical, chemical, biological, and social changes interact and reinforce each other

The Anthropocene, a term various scholars have proposed to refer to the current geological time interval, is characterized by many conditions and processes that have been altered by human impact (Raworth, 2017; Butera, 2021). We have increased erosion and sediment transport associated with urbanization and agriculture. We have changed the biochemical cycles of elements such as carbon, nitrogen, phosphorus, and various metals, along with various chemical compounds. Environmental changes promoted by these perturbations are global warming, sea level rise, ocean acidification, rapid changes in the biosphere both on land and in the sea, the explosive growth of domesticated animal populations, and the proliferation and global dispersion of many materials, including cement, fly ash, plastics, and a myriad of techno fossils.

To combat climate change, we need to implement a set of changes: zeroing out all climate-altering gas emissions and stopping the destruction of forests, grasslands, and wetlands, which absorb carbon dioxide by removing it from the atmosphere. To counter biodiversity loss, we must use fewer resources and exploit them better: climate change reinforces pandemics, and deforestation leads to mass extinctions of many life forms.

We need to give up hyper consumerism, the continuous flow of often unnecessary goods designed for rapid obsolescence, critically examine the basic assumptions of the market economy that dominates the world today and adopt a circular economy (Raworth, 2017).

3.1.   The circular economy model The circular economy is a production and consumption model that involves **sharing**, **reusing**, **repairing**, **reconditioning,** and **recycling materials** and products designed to last as long as possible. This extends the life cycle of products, helping to minimize waste because production is minimized. Once the product has completed its function, the materials from which it is made are reintroduced into the economic cycle wherever possible. Thus, they can be continuously reused within the production cycle, generating further value. The principles of the circular economy contrast with the traditional linear economic model, based on the typical "extract, produce, use, and throw away" pattern.

This classical model is self-sufficient and ahistorical because it postulates, according to Butera (2021), a circular flow between production and consumption without inputs or outputs, which takes place within a system in which a spontaneous and progressive increase in its structural, organizational, and functional diversity takes place, thanks to technological and social innovation that also expands through economic growth. An isolated system in which entropy spontaneously decreases. In essence, this model would work by contradicting the second principle of thermodynamics: the economic system has inputs, which are high entropy matter and energy in the natural environment, and it is through the transformation of low entropy flows into high entropy flows that the economic system functions, that is, through the **degradation of natural resources** and **nature's ability to absorb degraded**

flows. Low entropy natural resources, such as nonrenewable minerals, arable land, or the number of fish in the sea, impose the concept of a limit because they are not infinite. In addition, there is also a limit to the waste stream that the planet can absorb without damaging low-entropy resources (at the local community level, for example, water and land pollution reduces the availability of water or food resources; at the planetary level, climate-altering gas emissions reduce the availability of water and food or increase the number of catastrophic events that destroy resources). Most economists who flank politicians in state government have ignored the limits posed by the second principle of thermodynamics, which had not yet been enunciated in Adam Smith's time, embracing instead the scientific discoveries of telematics and information technology.

The classical economic model also fails to consider another limitation: time, which is no longer an independent variable. The rate at which a resource is used, that is, the rate at which it is degraded, cannot be greater than the rate at which this resource is regenerated, thanks to the only continuous, low-entropy flow available to the planet: solar energy, which keeps the water cycle alive and feeds ecosystems. So, there is a limit even to productivity, something many economists deny, always hoping that technological innovation will devise systems to increase.

In addition, large corporations incorporate small and medium-sized companies, often buying start-ups that could become their rivals. In fact, "10 companies control 70 percent of the home appliance market, five companies control 70 percent of the automotive market, five companies control 70 percent of personal computers, three companies control 50 percent of the smartphone market, and one company controls 70 percent of the social media market" (*ibid.*, p. 232). This concentration of production increases global inequality: in 2019, the 1 percent held more than twice the wealth of nearly 7 billion people. Moreover, the richest 10 percent contributed 45 percent of emissions to pollution, the poorest 50 percent only 13 percent, but it is the poorest who pollute the least that suffer the most severe consequences of climate apartheid. It is estimated that there is a different spillover of climate change as well: the **geopolitical South of the world**, while being responsible for only 13% of emissions, **will have to suffer 75% of the damage** (Butera, 2021).

According to Butera, the powerful multinational herbicide and textile industry corporations are among the main culprits in putting plastic into rivers and oceans. For years, statistics described ever-increasing economic growth as a victory over scarcity until it came to light that this growth was endangering the future of the planet and destroying more than it was creating.

There is a need to shift from a production-based economy to a maintenance-based economy to the development of mental and physical well-being in addition to economic well-being (O'Neil *et al.*, 2018; Hickel, 2019) and finally achieve the goals of the 2030 Agenda for Sustainable Development, which established an action agenda for people, planet and prosperity signed in September 2015 by the governments of the 193 member countries of the UN. It incorporates 17 goals for sustainable development:[14] defeating poverty and hunger; ensuring health and well-being, quality education, gender equality, clean water and sanitation, clean and affordable energy, decent work and economic growth; reducing inequality; creating sustainable cities and communities, responsible consumption and production; combating climate change; protecting life forms underwater and on land; promoting peace, justice, and strong institutions; and strengthening partnerships to achieve the Sustainable Development Goals. On climate, we are unfortunately far behind schedule, but we can still make it if we can make changes at the local, continental, and planetary levels.

3.2. Interventions to address changes It would be most urgent, for example as far as Italy is concerned, to multiply clean energy projects, which could be implemented by allocating additional resources to existing energy programs, with quick start-up and easier management. In addition, reducing local air pollution would mean decreasing fossil fuel and carbon dioxide emissions, and lowering the growth of pollution-related diseases. The actions are part of the energy-climate plan (PNIEC) already submitted to the European Commission, and funding for the initiative could include new European funds allocated after the pandemic.

Italy is already one of the leading countries for energy conservation and renewable energy and could offer many good new jobs. Many energy efficiency activities are labor-intensive, meaning a high number of people are employed per value added. Typical energy efficiency interventions include building upgrades (of apartment buildings, houses, schools, public buildings, and offices) and the integration of renewable energy sources in buildings, with commitment to the phase-out of heating oil. In the transport sector, the priorities indicated are the transformation of passenger mobility from private to collective and *smart mobility*, shifting freight transport from road to rail, and vehicle efficiency. Using available European and national funds well, we could simultaneously improve environmental conditions and public health and create new jobs for many young people, starting with those most affected by the COVID-19 pandemic. Europe has been a driving force in issuing directives to protect the environment and promote the circular economy but has suffered setbacks due to the rise of right-wing political

---

14. See https://unric.org/it/agenda-2030/ for information on progress and delays with respect to achieving these goals.

leaders and autocratic tendencies (e.g., in Poland and Hungary) who deny or downplay climate change and the health crisis due to the 2020-21COVID-19 pandemic

Cristina Figueres, a leading Costa Rican environmentalist, argues that the pandemic has caused tremendous suffering in the world, causing the climate issue to slide into the background. However, right now, we must. addressing the climate crisis with dogged optimism. We are in a transition phase, and in the present of transitions, there is always a co-presence of past and future. Now, we experience uncertainty, but we need to make an extra effort to improve the future without denying the crisis we are experiencing, we need to move. And to do that, we need a goal, a positive direction that mobilizes people and justifies the sacrifices needed every day to get there. If we think all is already lost, why should we engage? We need to spread the good news of progress, e.g., in 2020, 32 nations succeeded in dislodging economic growth from rising CO_2 emissions and demonstrating that it is possible to increase GDP, but not pollution (Figueres, 2020).

According to Figueres, we are at a crucial moment in human history. This decade will decide the future of the planet. We tend to think that global issues that overwhelm us are bigger than us. Instead, they are decided by people, by individuals, who have spaces to exercise responsibility, and we must decide where we stand. So, we can influence the governments that have to make the laws to protect biodiversity, but also the managers of big companies and multinational corporations that have the technology and capital to implement the changes and spur the managers of financial funds that are supposed to invest in the country's future. We citizens in civil society must influence CEOs and managers. Above all, elect capable and courageous politicians because the war in Ukraine, which began in 2022 and is still underway as we write, has made it even more complicated to achieve the goals for a sustainable world listed by the United Nations. After October 7,2023, when Israel was attacked by Hamas and 1200 Jews were murdered and more than 200 kidnapped, Israel has beingbeen fighting Hamas in Gaza to find the leaders of Hamas, who hide underground near schools, hospitals, so more than 45000 men women and children have been killed and thousands wounded. Moreover, Hezbollah in Lebanon has attacked Israel, supported by Iran. **All these wars have used funds that could be employed for fighting climate change and poverty and instead are being used to bebuy weapons. That why is more important to listen to ecofeminists.**

3.3. The contribution of ecofeminist politicians Ecofeminists (Marcomi, Cima, 2018; d'Eaubonne, 2022) argue that we need to vote more women out and support them in their careers to get them to the top. Women, in general,

are more supportive of funding health, education, and environmental projects. Men are more willing to favor business and arms spending. In our research with local and regional parliamentarians and politicians, we also found the same gender differences (Francescato, Mebane, Vecchione, 2017).

Seeing how important it can be to have women in leadership when dealing with a crisis was another of the "gifts" that the COVID-19 pandemic brought with it. The countries with the fewest deaths were New Zealand, Iceland, Germany, South Korea, Norway, and Finland, all nations led by women. The crisis claimed the most lives in the United States, Great Britain, and Brazil, male-led nations that have minimized health risks to aid economic recovery. Trump, Johnson, and Bolsonaro are authoritarian, narcissistic men with outsized egos who have conceived the fight against the coronavirus as an opportunity to dominate nature and flaunt their manhood by refusing to wear masks. These are our adversaries, claim the ecofeminists, and not all males. It is no coincidence that they are the ones who detest environmentalists and deny that there is climate change favored by pollution produced by us humans. Bolsonaro has given the green light to the brutal exploitation of the Amazon rainforest, heedless of the rights of Indians. Trump, on the other hand, has eliminated much of the environmental legislation, reopened coal mines, and encouraged the construction of new coal-fired power plants. His successor, Biden, has resumed an environmental policy, but Trump had said he wanted to run again, and after his reelection in 2024 he will close climate change programs.

Fortunately, a woman at the top of the European Union, Ursula von der Leyen, has presented the most forward-looking green blueprint. The European Green Deal (2020) comprises a set of European Commission initiatives aimed at achieving climate neutrality in Europe in 2050. The plan includes penalties for countries that do not cut their harmful emissions and promote a circular economy and environmentally friendly agriculture. Unfortunately, the wars in Ukraine and in Gaza have diminished the funds for fighting climate change. So we are not reaching our target and our progress may prove too slow to address the growing environmental problems, as Fridays for Future activists argue.

3.4.   Criticism from the Fridays for Future movement Activists who, with Greta Thunberg, make up the Fridays for Future movement support the European Green Deal ,but cry out that this package of initiatives falls short of the problems: "We have only eight years to make drastic changes and prevent global warming from leading to irreversible damage" (Fridays for Future Italy, 2020). Now we have fewer years, and cannot waste precious time. FfF demand that governments declare a state of emergency and "tell the truth" about the environmental situation. They want politicians to take extraordinary measures to get to zero emissions by boosting renewable energy,

energy efficiency, waste recycling, and the circular economy. Adopting a circular economy requires radical changes.

Moreover, Fridays for Future activists fight for planetary environmental justice. Developed countries have created most environmental problems, but the most serious consequences already weigh heavily today and will burden even more heavily the future on the poor countries of the Global South, and this environmental injustice must be remedied. Fridays for Future denounces the rise of environmental conflicts in many countries; land acquisition to produce palm oil (*land grabbing*; Honduras, Colombia, Mexico, Indonesia, and Burma); dam construction (Brazil and Africa); large water-polluting mines (Latin America and Africa); fossil fuel extraction by extreme means such as *fracking* (Arctic); waste management (Gaia); illegal sand mining (India).

Fridays for Future support the movement that seeks to promote fair distribution of environmental benefits and harms and equitable access to decision-making on environmental measures. They denounce racism and discrimination that have led to the placement of waste management facilities and hazardous industrial plants in poorer neighborhoods. They protest and demonstrate with adherents of civil rights movements against racism and homophobia. They march in the streets with feminists against violence against women and with exploited migrants in the countryside around the world. As early as 15 years old, the movement founder Greta Thunberg understood that the environmental problem to be solved requires a struggle against all the inequalities that plague our planet (Francescato, 2019; Thunberg, 2019). Fridays for Future argue that environmental problems are planetary problems. Climate change, ecological degradation, climate migration, conflicts over water, intensive livestock farms, huge inequalities, and predatory trade practices threaten the health of many, particularly the poorest on the planet, as we have seen with the COVID-19 pandemic.

Not only do Fridays for Future think that environmental crisis and social crisis are interconnected, but also a group of scientists from various disciplines and system analysts, members of the nongovernmental association called the Club of Rome, who for two years worked together to create two models of possible futures (Dixson-Declève *et al.*, 2022). The first model, called "Too Little Too Late," which is still the dominant model today, causes increased social instability, conflict, and greater environmental deterioration. The second, called "The Big Leap," identified five very effective goals: 1. abolish poverty; 2. decrease extreme inequality; 3. achieve gender equity; 4. develop a food system that produces healthy and sustainable food; and 5. produce clean energy. The scholars also calculated what measures might be effective for each of the objectives identified by the second model. We refer to their book for a thorough description of all the variables examined, and here we simply report a few examples:

- To eradicate poverty, the International Monetary Fund should lend 3 trillion a year for green jobs (understood as both environmentally related professions and environmental interventions), write off the debts of the poorest nations, protect their industries, and promote trade among states in the Global South;

- To reduce inequalities, interventions should be made to increase taxes on those belonging to the richest 10% of the global population and to establish a Citizens' Fund that shares the profits and dividends from investments among all citizens;

- To promote gender equity, educational opportunities should be provided for all girls and women, equality in work and politics should be increased, and adequate pensions should be ensured;

- To have better food for all, it would be appropriate to enact laws against food loss and waste, promote healthy diets, and increase economic incentives for regenerative agriculture;

- To have clean energy, they propose to eliminate fossil fuels now and increase renewables, triple annual investment in new renewables, and invest in energy storage.

The authors say the program they drafted would not cost a huge amount of money, for example, improving food and clean energy would cost only 2 to 3 percent of the world's GDP. Compared to the 1.6 trillion spent in 2022, it would come to about 3 trillion a year.

The authors also point out another important aspect concerning women: they make up 51 percent of the population, and because they live often and in different places in the world under oppressive conditions, for this reason, they are probably more interested in solving various crises.

Finally, they believe that eradicating poverty would be a smart way to "upgrade" capitalism. They argue that the commons, such as **air, water, soil, minerals, and oceans, belong to everyone**, and the entrepreneurs who use them should pay a fee to care about the asset and the people who work with it. The authors propose higher taxes for large corporations and a decent minimum wage; however, they argue that they do not aim to create a socialist system but rather a more inclusive capitalism for all stakeholders.

4. What can we do as psychologists?

4.1. First, support community-oriented public health As some physicians at a hospital in Bergamo write:

Western healthcare systems are built by putting the sick person at the center, but an epidemic requires a shift toward an approach that puts the community at the center... solutions are needed for the entire population, not just hospitals. Home care and mobile clinics avoid unnecessary travel and relieve pressure on hospitals. Early oxygen therapy and adequate supplies can be provided at home to patients with mild symptoms or in convalescence. A widespread surveillance system needs to be established to ensure adequate isolation of patients using telemedicine. This approach would limit hospitalization to a targeted group of severely ill patients, thereby decreasing contagion (Misuraca, 2020, p. 2).

We were able to see what a community-based approach can mean by assessing what happened in Veneto and Lombardy. The differences between the two regions' healthcare systems are striking. Veneto has a decentralized territorial model with many local public units, while Lombardy is based on the privatized hospital model with fewer decentralized public facilities. The number of coronavirus cases per 100,000 residents was 373 in Veneto and 771 in Lombardy (about twice as many), and the number of deaths (also per 100,000 residents) was 31 and 141, respectively, more than four times higher in Lombardy. Putting the community at the center (Zani, 2012; Francescato, Mebane, 2018) means that public health interventions following the COVID-19 pandemic should promote new employment to strengthen the sector in terms of prevention and management of epidemics that will occur in the coming years. It also means prioritizing the promotion of collective well-being in its environmental, economic, psychological, and social components.

4.2. Promoting individual and collective psychological well-being After the crisis, many people who have faced family bereavement without being able to be near the sick or hold funerals suffer from emotional disorders. Usually, human beings comfort themselves by being close, hugging, and kissing each other. The COVID-19 pandemic, on the other hand, required distance. So many still need psychological counseling or mutual-help groups to process the griefs they could not fully experience and lessen the guilt and regrets of not being able to give a last goodbye to a loved one who desperately wanted it, as we learned from social workers, who tried everything to connect the sick and family members, sometimes unsuccessfully A great many others have experienced high doses of negative emotions such as fear, anxiety, disappointment, and anger for months, and especially concern for their future and that of their children, whose lives were on hold, with no secure prospects for work or study. Staying at home, including working and/or taking distance learning classes, improved couple and family relationships that were already good enough but have worsened already conflictual relationships between partners and between parents and children. To prevent the increase of domestic violence on women and children and the growth of depression and psychosomatic disorders, it is necessary to immediately strengthen territorial psychological counseling services but also to create online services that offer opportunities for listening and sharing. It is necessary to offer psychological support for all social workers who have been dealing with the coronavirus emergency, who are at risk of having nightmares, insomnia, and other psychosomatic disorders, as well as problems with children and partners whom they have "neglected" for months. So far, few interventions have been prepared: there has been talk of establishing community-based territorial houses, but funds have not been allocated; the psychologist bonus has been created with some free sessions; unfortunately, only after numerous physical and verbal attacks on operators and citizens in emergency rooms were police garrisons set up in some hospitals.

4.3.   Facilitating mental well-being through cascade training We need to triple the number of family counseling centers and women's homes, which have been greatly reduced in recent years. Multiply the number of meetings and gathering centers and offer psychological counseling to families, condominiums, schools, and workplaces to pass on skills in conflict resolution, as is done, for example, in family mediation courses. We have been very good in Italy at offering a wide range of psychological help to private citizens who can pay for individual, family, and group therapies. The very high number of psychologists competently doing clinical interventions reflects the training opportunities in this field, both at the level of master's degrees and especially in the high number of specialization schools existing in our country. Rarer, on the other hand, are the cascade training offering knowledge and know-how in a variety of places from condominiums to neighborhoods, from communities of interest to nonprofit and voluntary organizations and associations. These different organizations networking together can also build pathways to preserving mental health by succeeding in implementing a set of circuits of well-being and increasing social capital, trust and mutual aid (see chapters 16,17).

The best volunteer organizations are already sharing experiences, trying to address crucial questions. After the pandemic, everyone, including the Pope and the President of Italy talk of the need for more community care. . Many health and social workers are asking: How do you work on individual situations and personalized educational projects from a community perspective? How do you work in a community center, daycare center, educational, or therapeutic community from the community perspective? How do we initiate animation and participation processes in the regeneration of an urban peripheries? (https://www.animazionesociale.it/it-schede-3020-energie_dicommunity, 2021)

The environmental problems are so serious that we need to stop tending to our little hives, fighting over scarce financial resources, and putting up fences, we community psychologists need to argue that the community perspective is not a surplus but a different way of intervening, centered on making those we "help" participate and empower, eradicating feelings of distrust and helplessness in the face of problems they think they cannot affect.

4.4.   Health and affective education and active citizenship projectsg Above all, it is urgent to give more space to prevention, and mobilizing positive energies, and the widespread desire to contribute to the rebirth of a better Italy. Health education and active citizenship projects should be encouraged for all segments of the population that promote interpersonal skills, mutual respect, a sense of community, care for places, mutual help among neighbors, and **intergenerational solidarity**. Respect and a sense of community are learned in the family, but they need to be promoted in all the contexts in which we live, from apartment buildings to schools to workplaces and online, where we spend an average of six hours a day.

These projects that use a variety of online and face-to-face techniques are particularly effective when participants can take advantage of both in-person and online learning opportunities with a collaborative pedagogical model (CSCL, *Computer Supported Collaborative Learning*; Francescato, Mebane, 2015). They are even more effective when, in addition to learning collaboratively, participants must also develop a project to implement, thereby increasing their empowerment, that is, their belief that they can at least partially affect the development of their lives and their local, national, and international communities (see the training experiences described in chapters 11 and 12).

By involving thousands of citizens in these health education and active citizenship projects, we can improve the health and mental well-being of many and implement small, medium, and large projects that take care also of ecological problems of the community. An empowered community becomes a more empowering place for its inhabitants, creating a virtuous circle. In many campaigns to promote health, active citizenship, and emotional and environmental education projects, community psychologists have for decades used the circle time method, wherein circle participants explore positive and negative emotions aroused by the issue being addressed. To promote change, it is also necessary to consider the emotions they provoke, as circle time-based social-affective education theorists argue. This empowering method was introduced in Italy in the book *Star bene insieme a scuola,* published in

1986 (by Francescato, Putton, and Cudini) and reprinted many times until the new edition of 2022. TThis new edition (Francescato Putton, 2022) documents how this methodology can promote the well-being of students, teachers, and parents, educate in living a sexuality that respects feelings, and help counter social issues such as school dropout, cyberbullying, and drug use, and above all improve environmental improvements.

Circle times methods promote also individual and group sociopolitical empowerment, eradicating feelings of distrust and helplessness in the face of problems and experimenting with nonviolent modes of conflict resolution. In addition, the Francescato and Putton (2022) illustrates how the method has been used in training projects in various territorial contexts, such as family counseling centers, SERT, and depressed areas. In the latter, the use of circle time has supported youth entrepreneurship, fostering the entry into the world of work of young people who had dropped out of school and learned to work in networks, creating community welfare from below (see also chapter 13).

By addressing complex issues (e.g., environmental degradation and climate change) with an empowering methodology such as social-affective education, participants choose small changes they can undertake in their daily lives and explore the contexts, organizations, and movements they can join to fight for change locally and internationally.

Empowering orientation and training courses require five or six days of face-to-face or online training (Francescato, Tomai, Mebane, 2004). However, in a time of climate emergency, some secondary school teachers have requested to set up shorter interventions to motivate their students to become interested and participate in environmental movements. We, therefore, tested a pilot intervention that can be done not only in schools but in sports centers, third-sector associations, senior centers, and other civic gathering places to promote awareness of environmental issues and increase individuals' willingness to take action to address them.

The brief outreach ecological intervention (minimum 2 hours extendable up to 8) is usually implemented by two people: an expert in environmental issues and a psychologist experienced in social-affective education and empowering methodologies. An outreach intervention involves four phases. In the first, the two trainers introduce themselves and tell how they became environmental activists. This first phase serves to create a climate of trust and curiosity, which is crucial for stimulating interest in climate change. In the second phase, the environmental expert explains the key concepts of climate change and explains various aspects of the current environmental crisis. In the third phase, the psychologist facilitates the identification and recognition of negative and positive emotions elicited by climate change with individual, pair, and small group exercises. The final fourth phase explores what participants can do as individuals, as a class, and as a school if they are students or as a group if they belong to other organizations. It also examines which environmental movements can be contacted.

These interventions were also implemented online during the pandemic; children worked alone, in pairs, or in small groups (3-5) and followed debates, watched films, and listened to experts in large groups of students from various schools (Mebane, Benedetti, 2022b; Mebane, Francescato, 2022). Initial results have shown promise in both in-presence and online and in *blended* courses, which each have specific strengths and problem areas. Notable is the creativity of the projects developed to improve school environments and address local environmental issues.

There are many small actions that each of us can take to overcome feelings of indifference or helplessness in the face of big problems. As Greta Thunberg often repeats, "No one is too young or too old; everyone can do their part, and if there are many of us, things can change" (Thunberg, 2019). Here is a summary of the ideas that emerged in the environmental emotional education courses.

At the individual and household level, we can consume less energy by buying more efficient (certified) household appliances, separating waste, using greener means of transportation (let's walk and bike, do *car sharing*), and public transportation. Above all, we must spread and practice the circular economy, which is based on the "four Rs": reduce, reuse, repair, and recycle.

If we work in a school or have school-age children, we can ask the school administration to have an energy audit done to estimate the appropriateness of energy upgrading (and greenhouse gas reduction) through interventions on school buildings, explore the concept of the circular economy as applied to school consumption (reduce, reuse, repair, recycle); discuss the concept of sustainable development and its goals; and follow the developments of the European Green Deal

In addition, each of us can request an energy diagnosis for each building where we work, play sports, or do other activities. In the years 2020-22, the government has proposed a tax rebate of 110 percent of the cost of the intervention aimed at energy saving, applicable to public and private buildings.

We can also vote for parties and politicians who defend the environment, support environmental associations such as WWF, Environmental League, Greenpeace, and Fridays for Future, and follow the Environmental Justice Atlas (EJATLAS). Moreover we can visit websites on the internet, which collects battles, failures, and successes of the environmental justice movement.

5. How to combine the biophysical well-being of the planet and social well-being by developing a planetary sense of community

Major social processes are creating both a sense of planetary belonging and national and local movements. This transformation is tortuous and conflictual because the forces that unite the world also generate the forces that divide it. Integration actions trigger fragmentation reactions, geopolitical activism provokes nationalist isolationism, and economic globalization triggers localist *backlash* (Falk, 1994).

The words "planetary sense of community" appears on the NASA website (in 2013): "A mission to our nearest neighbor that offers the best opportunity for collaboration will highlight our common interest and lead us to a planetary sense of community." However, many authors in fields as diverse as environmental sciences, international law, citizenship studies, and religions have in recent decades pointed to the need for a planetary sense of community (Hayes *et al.*, 2006; Francescato, 2020).

So, the concept of a planetary sense of community has multiple roots. It has commonalities with the concept of a sense of community developed by community psychologists such as Sarason (1974), McMillan, Chavis (1986), Cicognani, Albanesi, and Zani (2012); it includes beliefs of belonging, influence, connection, and emotional support among inhabitants; however, it differs in terms of the space occupied by the community, from the local/territorial level to planet Earth. Community is not identified with a specific geographic place because the world has become a place. A second important difference is in the two types of solidarity that a planetary community requires: worldly solidarity between like-minded and sublime solidarity between different.

"Worldly" solidarity develops among like-minded people (identity politics), unites like-minded people, and affirms similarity (the *idem sentire,* or "the same feeling"). But in the "sublime" register of solidarity, a group or individual pushes its gaze outside, beyond similarity to something broader. A solidarity that aims at transformational change requires that we not only recognize and sympathize with the fate of others but also join them as equals, moving beyond differences without erasing them. Solidarity, in its sublime form, breaks down the boundaries of identity, connecting us with others, even when we are not alike (Hayes *et al.*, 2006). Sublime solidarity and environmental, economic, and cultural justice underpin a planetary sense of community. Pope Francis proclaimed during Environment Day (2021): "We need to reestablish equitable social relations by restoring freedom and the commons and forgiving each other's debts. We must not forget the exploitation of the global South that has resulted in huge ecological debt, fundamentally due to the plundering of resources and overuse of the common environmental space as a dumping ground for waste. It is time for restorative climate justice: we must cancel the debt of the most vulnerable countries, recognizing the serious impact of the health, social, and economic crises they are facing due to COVID-19" (L'Osservatore Romano," clxi, 125, June 5, 2021, p. 10).

5.1.   The challenge for community psychologists The challenge in which community psychologists must feel involved is to foster a planetary sense of community that facilitates the implementation of feasible and necessary changes.

Everyone on the planet must feel responsible for the survival of the Earth because there is no "plan B." If environmental injustice in its broadest sense continues, even the "winners" will live mutilated lives, prisoners in their luxury ghettos, protected by their bodyguards. We can still live better lives together if we increase our efforts to understand the scope of current and future environmental issues

Using European funds, we could improve the environment and public health and create new jobs for many young people in poor areas of each European country.. It will be very difficult to prevent the melting of polar glaciers with the consequent rise in sea levels. We will not be able to stop the intensification of extreme events, which will worsen the food security of the planet. We can, however, try to stop greenhouse gas emissions, processes of biodiversity loss, and land use. We can reduce emissions in transportation and construction by replacing fossil sources with renewables and improving equipment efficiency, transform the agricultural production model by abandoning the industrial one and activating agroecology, close to cities, and consume less meat.

According to Butera (2021), we need to produce less in developed countries and better in developing countries by realizing the circular economy following the second principle of thermodynamics, transferring resources from rich countries to poor countries. Above all, in our view, new values need to be promoted by creating an ethics of sustainable development, which requires valuing cooperation, sharing of goods and services, solidarity, equity, and sobriety. We will need to change the law, now too much based on an economic model centered on the prevalence of private property over the commons. From the Indigenous Peoples of the Global South rises the demand that environmental defense should be among the basic values of a state, such as freedom and justice. Some argue that there is also a need to develop forms of ecological democracy (*ibid*., p. 287), which will have to change models of democratic representativeness to overcome the growing incompatibility between the length of electoral terms and the need, as far as environmental sustainability policies are concerned, to refer to the medium to long term. It will be necessary to attribute new and different roles to representative assemblies using even the drawing of lots for part of them, as proposed by various scholars (Bourg, Whiteside, 2010; Pluchino *et al.*, 2011).

Indicators need to be created in addition to GDP; the time is ripe to shift the emphasis from measuring economic output to measuring people's well-being. And measures of well-being should be placed in the context of sustainability. Some measures have already been constructed. For example, the **BEI** (*Better Life Index*), developed in 1980 by the OECD (Organization for Economic Cooperation and Development), considers certain *material living conditions* (housing, income, work) and *quality of life conditions* (social relations, education, environment, governance, health, personal satisfaction, security, work-life balance) as essential elements of well-being. The HDI (*Human Development Index*) was introduced in the *First* United Nations *Human Development Report* in 1990 as a new tool for measuring the development of the world's nations. This tool was promoted by Pakistani economist Mahbub ul Haq in 1990 with the explicit purpose of shifting the focus from GDP economic development to development policies for people. Haq believed that to get a true idea of a people's welfare, it was necessary to assess not only economic development but also people's development. The HDI is defined as the geometric mean of three basic indices: life expectancy, educational attainment, and income.

In 2020, an index of environmental pressures (PHDI, *Planetary Pressures-Adjusted*) was added on an experimental basis. The HPI (*Happy Planet Index*) is a measure of a nation's well-being and environmental efficiency, introduced by the New Economics Foundation (NEF) in July 2006. This index considers life expectancy, subjective life satisfaction, and a measure of environmental costs to also assess global sustainability and is weighted to give progressively higher scores to nations with lower ecological footprints. The index is not a measure of happiness but rather a measure of the environmental efficiency of supporting well-being in each country.

The **Genuine Welfare Index** measures three types of welfare: market welfare (with economic indicators such as income inequality and economic subsidies), nonmarket welfare (where we find services to increase human capital and social capital), and environmental costs (which include costs of reducing natural capital, pollution of economic activities). Finally, the **Fair and Sustainable Welfare Index** includes the labor non-participation rate, average disposable income, CO_2 and other climate-altering gas emissions, and the inequality index.

However, developing indices of well-being that are not only economic is necessary but not sufficient. A cultural paradigm change is essential to promote some of the cardinal values of community psychology that nurture psychological well-being, two in particular, cooperation and solidarity, which underlie the development of a sense of community (see chapter 5).

Santinello and co-workers (2018) and Arciadiacono and co-workers (2021) describe experiences that can promote a sense of community in residential buildings and neighborhood streets, including using online platforms linked to specific urban spaces, which facilitate the development of mutual help and solidarity.

We also need to develop a new form of globalization, no longer guided by the view of exploiting nature and the most marginal workers to have maximum production with maximum profits. We can be inspired by the millennia-old traditions of Indigenous Peoples, who conceive of themselves as part of nature and value the commons, as critical community, liberation, and decolonization psychologists remind us, as we discussed in chapter 3.

We European psychologists must also recognize that our nations have inflicted damage on the countries we colonized in the past and that we have created environmental waste and pollution in recent decades. Therefore, we must pursue climate justice by supporting the implementation of the proposal made at COP 26, in November 2022, and COP 28 in November 2024 to provide financial aid to low-income countries to cope with the climate crisis.

We have a difficult but exciting task ahead of us: we community psychologists have developed several intervention tools that increase cooperation and solidarity. So, we can make a small but significant contribution to addressing the complex problems of environmental pollution and climate change. It will not be an easy path after the war in Ukraine, Israel, Gaza, Lebanon, and Sudan, and the presence of rival political groups aspiring to become the dominant autocratic government in several Middle East and African countries confirm that firms producing weapons will continue to make huge profits. Moreover, the rise of competitive tensions for a new world order between the United States and Europe on the one hand, and China and Russia on the other, the growth of economic inequality, sexism and racism will continue to increase conflicts. But we community psychologists think that times of crisis also offer opportunities for positive change and that hope and desire for happiness, combined with our ability to value a sense of "we" in a society that has glorified individualism and egotism, will help us grow a planetary sense of community in enough world citizens to be able to enact small and large positive changes (see the experiences fostered by community psychologists described in chapters 16-17).

Part 3
Community psychologists tell their experiences

16

Freelance clinical community psychologists and practitioners in public services

1. Clinical community psychologists in public services

1.1. "Knowing how to be" a community psychologist: twenty years of work in Calabria. Paola Giacco (community psychologist, family psychotherapist; 45 years old) My story as a community psychologist was born out of a deep "crisis," and crises always have potential for improvement, they are precursors to change. On changes, we have some scope for redefinition if we have developed within ourselves a good concept of empowerment.

Halfway through the second year of my degree in Clinical and Community Psychology, in 1999, I entered a deep academic crisis; I could no longer find the meaning in my studies or the motivation. I had enrolled in the faculty, like most of us, to resolve some personal issues, but the faculty is certainly not a path of 'psychotherapy.''psychotherapy.' I was a student living away from home, and being in Rome gave me a sense of profound loneliness during those years.

A phone call with my mother at the time was very restraining; faced with my idea of wanting to change faculties or, in the worst-case scenario, drop out of my studies, she told me to go ahead with my exams and finish the year anyway. Supported by an "internal" boundary because I knew my parents would never tolerate such a kind of choice, I looked around for something that would give me back a sense of *concreteness*; the whole "clinical" part didn't make much sense to me. Studying Freud and the whole psychoanalytic movement destabilized me.

In the curriculum, I had community psychology as a core subject. I, therefore, decided to attend it and was lucky enough to meet Professor

Donata Francescato; the lectures opened a world to me. I still remember the emotions she was able to convey in class. There was a lot of structure, but a huge load of emotion, tenacity, and desire to change contexts and to concretely help people through practical tools passed through. Francescato brought herself and all her humanity.

She not only "knew" and "knew how to," but "knew how to be" a community psychologist. The more I listened to her, the more I studied, and the more I participated in EPGs (guided practice experiences, i.e., workshops where one could learn various tools through working groups and practical exercises), the more I understood the direction to take. I was not only faced with clinical theories, but I had the opportunity to learn concrete tools that could give me definite answers on how to help contexts evolve. It was not the world of occupational psychology, perhaps overly structured for me, but it was a world where "people" could relate to "contexts." The concepts of crisis and empowerment, the power of groups, and the "revolution" from below were being translated into tools that integrated the clinical and practical aspects. I fell in love with community profiles. I still remember that looking for material on the ground helped me to better experience the neighborhood in Rome, where I resided as a student, and the city itself became less "dark" and more familiar.

I had decided that was my path; I had found a purpose, a "sufficiently structured" but not "rigid" structure. I therefore decided to apply for the thesis and join the research groups that had been formed. I enrolled in the group on identity and belonging (after all, these were concepts very dear to me) and proposed the thesis on the *Sense of Identity in Workers in International Government Organizations*. The thesis set out to study how much of the "local" survived in a world that was becoming increasingly "globalized." For reasons of security in 2001, after the attack on the Twin Towers in New York, the head of the Foreign Department told me that he could no longer give availability for interviews with his employees. I therefore chose to change and direct my research to NGO workers. This was also a wonderful experience. One hundred and ten interviews around Italy allowed me to meet a lot of people and to have postgraduate internship offers (from the International Red Cross to Amnesty International). Those who knew me were impressed by my tenacity in requesting interviews (sometimes, I would tail people for days just to get their yes). I graduated on June 21, 2002, and my thesis co-rapporteur was Professor Andolfi. Someone on the thesis committee said he didn't think such complex work was "truthful." However, the "fieldwork" I had done in those months, together with my research group, had increased my sense of empowerment, and I was ready to answer that I was confident in my experience. I took the highest score and was proud of it. Do you want to know what was the outcome of the interviews? The concept of "glocalism" emerged: that is, people who traveled globally retained within them a strong local identity that they carried inside them forever, opening to new worlds but making movement and change new parts of their identity while increasingly strengthening their roots. They felt and were "citizens of the world."

After my exams, I also finished my master's degree in Community Psychology and Formative Processes, taught by Professor Francescato, where I learned everything, I needed to do training, and left for my internship experience in Romagna, a community for drug addicts of CEIS (Italian Solidarity Center). The tools I acquired during the master's program were a big help in fitting me in professionally. Through the "network" I had built and acquired as a tool of community psychology, in addition to the internship, I tried my hand as a trainer and had my first experiences in schools. I had the opportunity to experience social-affective education in the school-work paths of difficult children in the province of Ravenna. At the end of the internship year, I received an unexpected phone call from Professor Francescato. She was proposing that I return to Calabria, my region of birth: here, a training organization in Rome, through funds from the Ministry of the Interior, had launched a project coordinated by Francescato to combat school dropout in the three Italian provinces found to be at greatest risk (it was the year 2003): in addition to Catania and Cagliari, there was also Vibo Valentia. A thud to the heart-another "crisis," what to do? Stay or return? Return.

1.1.1.   The experience in Calabria The beginning of my experience in Calabria was anything but simple. The work group was composed of people my age (around 25) from diverse backgrounds, mostly in the humanities, who had been trained to be 'network operators.' There were initial clashes, and the months were not easy. My ideal reality clashed with a harsh reality. Within the group, some believed in the work we were doing, while others were less convinced. Despite the difficulties and a group that was certainly not motivated to the same extent, I continued to do what was asked of me: focus groups, interviews, building networks, and community profiles. The coordinator, Francescato, had assigned us 'zones' (small inland municipalities) and the opportunity to contact 'key individuals' whom we were to interview and ask to hold focus groups. I had the chance to meet a young priest from a small town; he was only 33 and believed he could change the world. I thus entered his small yet significant universe, filled with youth groups and many young people, and I began proposing activities for the youth and their families. The young priest was from Milan. After about two years, as the project was coming to an end, I found myself in crisis again. Being in Calabria meant 'fighting' in a challenging, but simultaneously 'virgin' territory (circa 2005-06). No one was aware of socio-affective education or national operational planning (PON), there was much potential but also immense difficulty in putting it into practice. I received a phone call from Ravenna, from the community where I had done my internship. They asked me to return because a position as a psychologist had opened within the community. At the same time, I had been contacted by a training company to work with them, thus returning to Rome; additionally, the priest I had met during the project proposed that we open a social cooperative and work in Calabria. I accepted this last proposal, and after twenty years, I am still here. I also became a family psychotherapist and a trainer at the Roman School of Family Psychotherapy. I teach colleagues from the offices in Rome, Cagliari, and Naples. I am a consultant in schools and collaborate with the court in Vibo Valentia. One must always believe in it. One must believe in oneself while also having the humility to feel like it's the first day of school, always ready to learn.

Working today in Calabria is possible; Calabria needs prepared individuals who are willing to engage. It needs colleagues who bring socio-affective education into schools, who train teachers, and who meet with families. To make your way in Calabria, you must 'be' a community psychologist because only by entering the territory with these tools can you seize the opportunities that the territory itself offers and connect with people who 'think like you.' Working well and with professionalism, creating networks, developing empowerment, and experimenting – this is the secret.

Am I happy? Yes, with what I have done and the changes I have brought to my land. Thinking of being an alternative for the development of the potential of many people gives me a sense of who I am and what I have become.

1.1.2.   Projects in Calabria. Below is a chronological list of the projects I followed in Calabria in the years 2002-22.

- **Social Affairs Research Center** and Community Psychology Chair project on school dropout and meeting with a key person in the area, a young priest, during focus group interviews with youth in the Maierato (VV) area (September 2002).

- **Birth of the cooperative** and work in the territorial sphere (management of the social center Safe Paths, in Maierato, whose president was the priest already mentioned, 2006-16, born out of the study of community profiles, which had shown the need to create structures for the territory of the province of Vibo (September 2005).

Sponsors: activities offered in the center were financed by membership fees paid by members. The areas covered: social (with groups for parents, minors, workshops for young people, mutual-help groups, and groups over 50). The design was conceived from a network perspective involving the whole area through a set of volunteers. The activities stemmed from a needs analysis of the territory itself in collaboration with the parish and the municipality.

Goals: cultural promotion and land development.

Difficulties: the difficulties encountered in the long years of working in this setting were mostly concentrated at the beginning of the journey. It is not easy to work in small communities. They are often characterized by such a strong sense of belonging that it is difficult to be accepted if your thinking is "divergent." The community center was recognized at first only by those who revolved around the church; it was harder to be recognized by a part of the community that was far from this kind of thinking. Other difficulties included boundaries with people, who tended to consider you part of their family; recognition, when the village priest devoted his energy to other activities and no longer to the center; political election periods because the tendency was to instrumentalize the center's work; and relationships within the cooperative's working group. One might wonder if there were aspects of 'Ndrangheta (the mafia of Calabria) pressure in contexts like these. In my years of work in Calabria, I have never had these kinds of problems. Social work does not move large amounts of capital, and the 'Ndrangheta does little or nothing with human capital. Instead, there is great recognition for things done well and for contexts that restore confidence, hope, and competence, and this has universal value.

Results: Today, the children who attended the center all have their own work and family identities. The group intended for the over-50s continues to operate in the area and, in turn, founded the University of the Third Age in 2021.

A curiosity: the opening of the center in 2006 was attended by Cesare Moreno of the nonprofit organization Maestri di Strada. For musical entertainment, a singer who was young and unknown at the time, Dario Brunori (his brother was a citizen of the town of Maierato). The center's mission was to provide space for new talent.

- **Profiles of community of Cassano allo Jonio** (2007).

Client: bishop of Cassano allo Jonio. The parish priest of the town where the community center was located was the conduit for the work proposal.

Sponsor: the bishop of Cassano allo Jonio.

Goals: through community profiles, draw a picture of the area to design interventions with volunteers. To build a participatory planning group.

Difficulties: the greatest difficulty encountered in the working group was related to mentality. It was not easy at first to make the group understand the purpose of the intervention. People were very doubtful about the possible changes, and it was difficult for them to understand the meaning of the profiles. The phrase that resonated most often was, "Doctor, here there is nothing, and nothing can be changed" (Cassano allo Jonio is one of the municipalities with the highest rate of organized crime in Calabria). The most challenging aspect was being able to "think" about change and being able to "imagine" what to change. The trump card turned out to be thinking of small changes possible: creating groups, organizing small initiatives, and, in short, thinking small and concrete; empowering training produced active, actionable thinking. It restored hope in concrete action, starting with each of the participants, about 30 or so people. Of course, to enter and be accepted in some contexts, you are more facilitated if you talk about it and know the languages and culture.

Results: through participatory planning, training paths and volunteer association groups were initiated. Some examples of paths activated, starting from the resources present, were the birth of cultural groups, film paths, and cooking workshop paths.

- **PMOA for a company in Fuscaldo** (2011).

Sponsor: business owner (participated in the parent group proposed by the center).

Goals: to improve the business climate and help the family business make the transition to more structured management.

Difficulties: the main difficulties we encountered in the training were maintaining the boundaries between professional behavior and Calabrian hospitality. Calabrian hospitality is a characteristic that reflects and distinguishes somewhat all contexts. In this family business, the air at first (even among employees) was very "unstructured." The word "boundaries" was like ant repellent: simply suggesting that participants refrain from indulging in the lavish "breakfasts" with goods generously offered by employees during training sessions sparked resounding disappointment and feelings of "suspicion" and "distrust." It quickly became clear that to be accepted—and more importantly, not perceived as "outsiders" or agents of "change" that staff feared could be "catastrophic"—it was necessary to embrace the ritual of sharing bread and salami alongside the training. However, this wasn't ideal for fostering focus. To gain their trust and cooperation, a compromise was struck: bread and salami, yes, but only during breaks. In the end, it became a shared pleasure, savoring a taste of Calabria together.

Another difficulty was eliciting not only the experiences of the employees but also those of the owners. The company had hired us to assist in transitioning from a family-run identity to a more corporate one. The management consisted of the four owners, who were also brothers, each recalled to work in the company after the death of their father, but with skills that did not fully align with production needs. The major challenge was navigating the shift from a "family" to a "corporate" identity, especially with some of the hired staff coming from the old family-run management. This transition was intricately tied to the family's "culture," the relationships between the brothers, and the three-generational history of the family. Over nearly a hundred years, the company had experienced cyclical patterns: periods of great expansion and productivity were followed by significant "downfalls," as the brothers in charge struggled to reconcile success with family dynamics, often buckling under the weight of success itself.

The psychodynamic dimension, therefore, helped us to overcome this kind of resistance. The construction of images that each represented themselves as a role within the company, the placing of their own "symbol" on the company map constructed by them, the graphical drawing the colored lines regarding the type of communication (using different colors belonging to each component depending on the role), the quality of communication, the interruptions, redefining the map graphically, from one real to another real passing from the ideal change, and all this done through "images," colors, sheets, maps (a whole series of graphic "traits" and "traces" that served to lower resistance); we were not talking about them, employees or owners, but about trees, dry stone walls, sun, ladders, buckets, pens, lamps; each object represented self and at the same time was easier to use in the construction of fables and scripts talking about "other," in a as if. This, I would say, proved to be the trump card.

Results: through the analysis of the four dimensions, an improvement in working group dynamics was achieved; improvement in the relationship of leadership (consisting of siblings), greater growth and development based on greater definition of roles and boundaries; and greater awareness of the three-generational history of the owners and their differentiation in different sectors. At the end of the first four meetings, whose purpose was to analyze the company, we were able to derive a picture related mainly to the company climate and internal dynamics (which were most needed to get to the goal required by the client) and condensed into eight pictures that encapsulated:

1.   the departure of the company (related to the three-generation history);

2.   the "gray areas" of the tri-generational story;

3.   family identity fused and confused with corporate identity;

4.   the family/business myth that characterized the birth of the company itself and "unknowingly" traced its trajectory over the years;

5.   power management over the years;

6.   the proposed "challenge" for change;

7.   the strengths and weaknesses of the company;

8.   the path of the change process with the risks and opportunities of the same.

Tough challenge because transforming a family reality into a business reality also meant having to change the relationship and internal modes of communication between person and ownership, as well as between siblings (who represented the management), and "create" their model that encapsulated both aspects, rethinking themselves different from their ancestors.

Master's degree in empowering training and social-affective education (2011-13). Directed at social and health workers, which mostly encapsulated training models related to the methodology of socio-affective education. The tools presented in social-affective education, i.e., the integrated method, became training sessions with insights and experiments.

Sponsors: private operators who self-funded training.

Goals: create a network of practitioners in the region; create empowerment and network professionals from all over the country.

Difficulties: the difficulties that emerged were, in the beginning, to make ourselves known in and around Vibo credibly. To overcome the initial mistrust of the practitioners, who usually trained outside Calabria, we contacted most of the colleagues that each of us knew from outside the region and who were recognized as capable of treating the topic in depth. I, personally, after years of training that took place outside the region, did nothing more than contact colleagues whose contact details I had noted on the famous "address book of potentially useful addresses for networking," which Professor Francescato had spoken so much about in class, and which turned out to be a very useful tool. Today, the network is vast and is spread throughout the country as well as regionally. We often contact each other for job opportunities, supervision, and/or simply to arrange trips and share free time.

Results: one hundred practitioners, including teachers, educators, social workers, psychologists, and prison educators, brought the community psychology model back to various work settings and formed a network of professionals with whom to create interventions. Multi-year interventions in schools are still in place in some.

- **PON in schools from a networking perspective**

Sponsor: national operational program (PON) of MIUR.

Objectives: presentation of projects on social-affective education from a networking perspective.

Outcomes: activation of groups for parents; training for teachers on social-affective education; classroom-based social-affective education interventions for children. Structuring of networking to ensure in the project hours (e.g., 30 per year) the greatest possible involvement of all school participants and reach the greatest number of pupils.

To date, the school that has the most applied community work internally from a networking perspective is the Istituto Comprensivo Murmura in

Vibo Valentia. Thanks to the foresight of the school principal, it has been possible to propose and develop parenting support groups, training groups for teachers, and listening desks; to build a network with services in the area, to offer social-emotional education in the classrooms, and to start a research on the sense of security and emotional experience following the COVID-19 pandemic.

- **Prison self-help groups with sex offenders: prison odyssey project** (2014-16).

The center got the assignment through a prison educator who had taken our training courses; together, we structured the project.

Sponsor: Vibonese prison system.

Goals: to form a self-help group within the prison system.

Difficulties: entering a correctional institution emotionally is not easy at all. Then, confronting a group of sex offenders is even less so. The first impact within the walls of the Vibo correctional institution was strong, harsh, and raw. Hearing the doors close one after the other behind you, being subjected to constant scrutiny, waiting outside the gates with the many families waiting, these are just some of the difficulties one encounters.

The prison institution is mostly made up of big rigidities, and the educator, Dr. Barbara Laganà, was the "key" that made it possible to carry out the project in an area of sex offenders, which is delicate for everyone. The prison director was not very sure that he could come up with changes for this type of user, but Barbara was very convincing, and even the smallest reluctance was overcome. The topics put on the table concerned the selection of participants (in fact an interview had been done to see if the type of users could join the project), the description of the course and possible positive developments (less friction in the ward between inmates, the possibility of improving their mood, and increasing the probability for some not to re-offend). That said, our adventure began. I remember some rather significant anecdotes with both prison police and inmates. Once at the control, I had to take out all the colors in the case I had brought with me one by one (there were about a hundred colors), and we spent most of the time allocated to the group checking them (the next time I showed up with only ten ballpoint pen); another time I was alone with the group of ten inmates (without Barbara) and with the prison guard outside the door (locked of course), and at one point the policeman looked at me through the skylight to see if everything was all right; the inmates, noticing the tension, looked at me sensing the guard's concern about them, and to see if I was scared too. With a toothy smile, which I learned was very useful in such situations, I said smiling, "Don't worry, they're just afraid you're going to kill me." It all ended with a laugh that dampened the tension. It also works a little like that in prison. The inmates must not feel that they are afraid because otherwise, they get afraid, too, and fear turns into anger and aggression as an element of defense. Once, an inmate in prison for violently beating his wife said to me, "Doctor, you treat us as people and not as animals, so we have respect for you. We must thank you and Dr. Lagana if I have learned that a bulldozer is something that can be used to build and level and not only to destroy, but I have also only the third grade, and here I have been given back the dignity to exist. I learned the value of culture." I came out of the experience of that humanity changed. My eyes glazed over, and a weight on my heart; how much of the prison experience can be rehabilitative and restore dignity, and how much is destructive and debasing?

Results: the founding of self-help groups. To begin forming the self-help group, we started with getting to know the people within the group beyond their offense. In order not to be confused by the stories in the prisoners' records, I asked Barbara not to let me read the files; I needed to meet the men, not their crimes, and especially to get to know them for what they were there at that moment and not what they had done. It worked because they were men afraid of judgment (someone said he felt free to speak because he had not felt judged and seen as a person and not as a monster).

- **Project Chance, my second chance** (2013-15).

Sponsor: ministry of the interior. Obtained through the network created through the formation of the cooperative.

Goals: improvement of the dynamics of family contexts and accompaniment in the reintegration of boys with precautionary measures.

Difficulties: the difficulties encountered in the course were getting the families of the boys and the boys themselves to understand the meaning of the group and the meetings. Initial mistrust, thinking that the content that emerged in the groups could be used against them, and doing the path because it was part of the "testing" were not easy elements to deal with. Here again, the possibility of entering into a "therapeutic" relationship made up of acceptance of the other without judgment helped us practitioners to move congruently, but above all, to create a restraining and supportive network so much to create spaces for reflection and a rereading of the conviction as a new beginning, evaluating together with the family what had worked and the family elements that had instead facilitated the exit from legality.

Results: Thirty families were supported in their paths of reintegration and rehabilitation. I remember a rather complicated case that struck me for the tenacity and strength of a mother. It involved a family of five people – three children, all adopted, and two parents. The oldest, a minor at the time of the offenses, had been reported for possession of drugs and was thus placed in the community and probation system. The process was, in fact, a continuation of his exit from the community. It was not easy to work with Raffaella (the colleague who collaborates with me on all the projects) on this family: on the expectations of the adoptive parents, who had welcomed the three children idealizing them without considering their cultural complexity and age differences, on the guilt felt by both the adoptive parents and the adopted children; and on the image linked to the past of the kids (a past that felt to them like a 'brand' that could never change, no matter what they did). We followed the family for a long time, guiding them along the path of empathy, becoming conscious parents, and "seeing" the children.

1.2.   Being a community psychologist in complex organizational contexts. Rita Porcelli and Maria Andò (community psychologists; ages 48 and 47) When you enroll in psychology school, usually the image you have of yourself in the future is that of a professional who is locked in a study and relates with one person or with a small group of people and their inner world. The dimension of help and social usefulness pervades everything and makes you relish already, just thinking about it, the gratification you will get from it. Then you go on with the course and come across subjects you don't know much about, or rather, from the title you imagine one thing and then discover a world you didn't know. One of these is community psychology.

The first time you walk into the classroom or open the textbook, you say to yourself, "It's going to be about therapeutic communities for the treatment of addiction."; even today, people, when they hear that you have a degree in Clinical and Community Psychology, respond, between concerned and admiring, "So you work in facilities for addicts!" But no, or at least not only: We work in the world, we work wherever there are people relating to each other and the environment because Community Psychology was born as a primarily empirical and applied discipline and gives strong relevance to the social context in understanding individual psychological functioning. It is the belief of community psychologists that people's behavior and well-being are strongly interdependent with their social settings and contexts (Francescato, Ghirelli, 1988).

You realize this immediately when you hear or read the three words that always return in the books, articles, and stories you hear: groups, organizations, and communities. Working with people means confronting everything in which they are immersed.

As you take the community psychology course, you realize that what you are studying is not aimed at passing the exam but is getting inside you, so you choose to make this world your own and decide that discipline will be the subject of your dissertation. Everything, every idea, can be dissected and then reassembled through the lens of community psychology. In the late 1990s, there were numerous dissertations assigned by the chair, all on pivotal themes and all strictly experimental, but each one offered a particular point of view. Each construct and methodology was declined in the most varied and imaginative contexts and circumstances.

When we finished college, we felt we had accumulated a lot of knowledge, but in our case, the know-how we had acquired mostly through participation in the community psychology course and the countless initiatives (then called "guided practical experiences") offered by the chair. Knowledge of community psychology techniques was the real practical baggage that we brought as a dowry to the job market. Hence, the decision to continue to enrich this baggage through the internship at Professor Francescato's chair and enrollment in the Ecopoiesis network's master's program in community psychology and training processes, which allowed us to experience the application of the different techniques in multiple contexts, to develop skills in the design, organization, and implementation of educational activities, and to experiment in research activities that led to the publication of several contributions. Here, we were able to experience firsthand the "power" of empowering education. The need to face ever-new challenges in the application of techniques, the possibility of experimenting in real contexts, the awareness of being able to do so, and the habit of observing the world and trying to make the most of all opportunities allowed us to increase our skills but at the same time to work on the emergence and enhancement of our resources, in a nutshell on our sense of personal empowerment.

Looking at our educational journey today, which led us to embark on the Ph.D. program, we realized that our life project was built *in itinere* thanks to our ability to "seize" the opportunities that the professorship offered us and to "embrace" and face challenges, even complex ones, by mobilizing our desiring energy. All this has enabled us to take control of our lives and gain the feeling of protagonism offered by the condition of being able to choose among different possibilities. Today, we feel we can say that we have experienced what Bruscaglioni (2007) calls the "thinkability of change" as an increase in possibilities, choices, and, therefore, freedom.

We had thus become community psychologists!

Being a community psychologist is not a profession or even a mission; it is a way of seeing the world and acting within it, it means having a privileged lens of observation of contexts and life stories, as it is based on a holistic, contextual, multifactorial and therefore complex reading of social phenomena. Such a reading enables a responsible, critical, and proactive approach to life choices. It is a *forma mentis* in which professional, as well as personal, action is dictated by necessity, by the instinctive impulse to analyze, know, and capitalize on the context as a whole, with a gaze that, like concentric waves, widens its scope more and more and reviews resources and constraints of the person, of one's groups, of the organizations and communities of reference; seeks to identify as many solutions or strategies as possible, assessing advantages and criticalities from time to time and connecting them so that the former can absorb the impact of the latter and perhaps evolve them into strengths. It is a spiral evolution that is enriched with new elements of knowledge and possibilities for action. To be a community psychologist means, moreover, to believe in the power of knowledge and to consider the evolution of the person as a continuous process that takes place throughout life; the results of this development, since they result from the joint function of individual and context, also depend heavily on the environmental changes associated with changes in contests with time. The study of the contaminations between individual and context has been the common thread of our training and has ferried us into the world of work, we have capitalized on the specific knowledge gained in the implementation of our research projects (collaborative learning in face-to-face and online educational settings and the variables of school failure) and turned it into marketable expertise.

Our professional paths have seen us engaged in different organizational contexts: public organizations, private companies, nonprofit organizations, cultural associations-and have been characterized by more than a decade of precariousness, marked, however, by a continuous work commitment; our wealth of skills has allowed us to go from contract to contract until we have stabilized in those same organizations in which we have invested for decades. Precariousness usually brings with it a strong experience of not being rooted, of not belonging to work contexts, a strong limitation to future planning, and, in some cases, to the construction of a professional identity. As community psychologists, we have been able to make the best of even what may seem a critical aspect of our path: despite the undeniable limitations that precariousness entails, for us, it has been an opportunity for professional growth, forcing us to experiment in the most diverse contexts and offering us a plurality of experiences.

For us, community psychology has been the "safe base" from which to explore the world, allowing us to build a dynamic, flexible, responsible as well as autonomous professional identity, but above all, one capable of continuous and new adaptations, the result of an experiential puzzle in which the combination of the different pieces has been cemented by the awareness and mastery of our training and professionalization path. In some ways, the precariat also allowed us to fully understand and experience the importance of sociopolitical engagement for the protection of rights. Participation in labor organizations with a constructive attitude based on goal orientation, a strong sense of hope and trust, and interest in the inalienable rights of workers was a way of expressing our sense of belonging to those organizations in which we had invested so much and in which we wanted to continue to invest, with the idea or hope that we could, through them, contribute to the development of the community.

The ability to share knowledge has led our paths, though different, to intersect and enrich each other; our doctoral dissertations have become a common heritage:

Investigation of the determinants of school failure, which particularly analyzed the correlation that exists with cognitive, emotional, and behavioral variables, formed part of the background that enabled it to develop or contribute to intervention projects at the national and international levels, applying the methods of research-intervention, empowering training, and socio-affective education to the issues of youth distress, gender differences, and school-to-work transition. This then contributed to the development of orientation models and tools aimed at different types of targets (youth, employed and unemployed adults). These models and tools have been tested in schools and employment services throughout the country by specially trained guidance workers;

- **Experimentation on the evaluation of the effectiveness of collaborative learning in face-to-face and online educational settings** demonstrated how professional knowledge and skills can be learned equally effectively through the two different modalities and formed the basis for experimenting at a distance with the empowering training approach on different topics and with different targets (university students, professionals in education, training and guidance).

Twenty years after experimentation, we were faced, due to the necessities imposed by the pandemic, with a momentous push in the use of the collaborative learning methodologies pioneered by the Chair of Community Psychology back in the 2000s. Re-reading what we wrote then and realizing how "terribly" current all this is brings with it satisfaction but also a great feeling of frustration. The satisfaction comes from being aware that the learnings gained, thanks to the principles and methods of community psychology, are still relevant today; the frustration, on the other hand, arises from the fact that, as scholars of these issues, we must perhaps, more influence the capacitation of systems to make the results of the experiments rights and practices "enforceable" for all citizens of our country.

Empowering orientation and training, which have been our workhorses, complement some aspects of the global interdisciplinary model placed at the center of the interventions promoted by community psychology, which we can summarize as follows:

1. Implement in different contexts the spaces for exploration of positive emotions that promote empowerment and negative emotions that may hinder desirable changes;

2. Use a historical approach that emphasizes the importance of

understanding one's past and the social part of the group one belongs to in envisioning one's future;

3. To enhance the interaction between individual factors and social variables by paying close attention to individual stories and collective narratives in bringing out desires, needs, and values;

4. Use the small group as a tool for individual and collective empowerment and as a generator of new knowledge;

5. Increase personal and political awareness through the development of skills in reading the group, organizational, and community contexts in which participants are embedded;

6. To encourage critical reflection of dominant personal and social interpretations;

7. Foster processes of constructing new personal narratives to give voice to minority social narratives that generate possible visions of feasible personal and collective change.

In our personal and professional experiences, these principles have become a *modus operandi*.

How have we exercised this approach? By grafting into the projects and training interventions carried out in our 20 years of professional experience, the principles synthetically reconstructed, but above all, by approaching the work in our respective organizations (INAPP and ANPAL Servizi),[15] with a critical attitude and action, open to confrontation and proactive. Specifically, the principles and tools related to networking, "working groups and teamwork," AOMP(participatory multidimensional organizational analysis), community profiles, and social-affective education were included in pills in all the actions implemented. The train-the-trainer methodologies tested in the master's program in Community Psychology and Training Processes (Ecopoiesis network) can be reread in the ways we design and co-lead. We often hear about people-centeredness: active employment policies (whether understood as training and/or guidance activities) make this construct a workhorse. But **how much and in what ways can citizens be made free and able to choose**? Real freedom is defined by two conditions: the subject's ability to choose and the opportunities and means available to him or her. Community psychology provides psychologists working in various public and private contexts with the possibility of fostering learning processes aimed at increasing empowerment and thus enhancing all those resources indispensable to being conscious actors in one's personal and professional life story. Our professional efforts have been aimed at building tools, including operational tools, capable of expanding people's possibilities

---

15. INAP - National Institute for Public Policy Analysis (formerly ISFOL), a public research organization; ANPAL Services, an instrumental entity of the National Active Employment Policies Agency.

of reading contexts and understanding their complexity (e.g., tools for describing and systematizing the contents of work and training).

This approach, by fostering the integration of emotional-affective components with cognitive ones, aims not only to enable people to better understand themselves but, more importantly, to provide tools for reading social contexts to implement the individual's propensity and abilities for meaningful use of the opportunities offered by the environment. By placing great emphasis on feelings, introspection, and the utilization of the full potential of the human brain, it leads people to question and observe themselves in their interaction with others in the context, requiring a strong commitment and generating, at least initially, resistance to getting involved. Being willing to activate oneself in lifelong learning and updating processes, being open to listening, and having an attitude inclined toward interaction and integration with and in the groups in which each person finds himself/herself inserted are not innate skills but must be learned, nurtured through practice and cared for. This is even more true in complex organizations that exhibit a certain level of "rigidity" from a structural and functional point of view and consequently also psycho-environmental and psychodynamic. With regard to such organizations, which are characterized by highly structured models where communication, interaction, and actions are confined within the formalities of roles, significant cultural changes have been observed over the years. These include a greater emphasis on understanding reference contexts, fostering teamwork across different organizational structures, and prioritizing not only the professional but, in some cases, the personal growth of employees. Together, these elements signal a meaningful process of organizational empowerment.

1.3. Community psychology between training and work experience Mirella Cleri (healthcare executive and university lecturer; 58 years old) I present my journey through the various stages by which it has been marked.

1.3.1. First phase. Undergraduate and postgraduate: among the pioneers

I began studying psychology in 1985 at Sapienza University of Rome. One of my first great passions was social psychology: studies of individual and group beliefs and behaviors, attitudes, coping strategies, attribution theories, and studies of health and behavior modification. There was a copious body of studies that emphasized the great potential of social psychology on issues of health and illness. I collaborated for about two years on the research projects of the Chair of Social Psychology, graduating with an experimental thesis with Professor Eraldo De Grada and Professor Lucia Mannetti. During my third year, my first encounter with community psychology occurred: teaching was one of the core exams of the applied major I was enrolled in. The community psychology exam opened the world to me and brought me closer to applied psychology out of the box. Certain aspects immediately struck me: the study of people in their life contexts, the focus on the construction of well-being, and the multifaceted professional profile of the psychologist as an agent of change. This area of psychology proposed an approach away from fascinating theoretical speculations, embraced a systemic and multidisciplinary approach and a perspective of intervention ranging from single-person care to health promotion projects in communities, with a firm theoretical concept that proposed a clear model and guided the implementation and verification of interventions.

Starting with an initial focus on stakeholders, this model used consultancy and participatory design strategies and always contemplated project evaluation, offering not only viewpoints or theories but also empirical data to compare against. The model knew how to systematically combine analysis, participation, design, evaluation, research, and intervention. One studied the problem of the effectiveness and sustainability of an intervention before it was even started in the design phase. It did not limit the work to psychotherapy but was open to new ways of dealing with health problems.

Community psychology presented all these themes and went beyond the systems approach through an ecological view of human beings. This approach was appealing and fulfilled my desire for research and practice in the field. Social psychology boasted great studies, for example, on social representations or individual and group behaviors. However, it was difficult to apply and did not propose practical interventions that could produce change and results, whereas community psychology, through the study of people in life contexts, offered documented and effective tools and operational methods.

It was within the postgraduate course in health psychology that I was able to delve into the community psychology approach and begin fieldwork through the direct supervision of Professor Donata Francescato, specifically training with the integrated method for socio-affective and sexual education. Among the experiences made available in the course, for me, this was the most formative: the work experience lasted for a year beyond the training through participation in the design and implementation of a research intervention in socio-affective education in the classrooms of some Roman schools in which the administration of psychological tests, statistical processing of data and fieldwork were planned. It was 1992, and the Chair of Community Psychology was testing the integrated method; our research intervention was about testing this method in a project on educational continuity at some primary and secondary schools. My first fieldwork experience was conducting circle time in class groups of primary and secondary schools. I remember the enthusiasm and support that came from the constant support and supervision of the work done by Professor Francescato and the working group.

The experience was so engaging that I had no doubts about continuing my master's degree training in community psychology at Ecopoiesis. Passion was immediately sparked, that extra element that makes us put aside doubts and reasoning and take it into our own hands to live it, embracing risks and potential. Training and working with community psychology for me has been and still is a great passion. Looking back, I am certain that I made the best choice. They have been years of intense study and work, without respite, but also of great satisfaction and great joy.

And a new phase of my life as a student and as a psychologist, the beginnings of my professional career began.

1.3.2. Second phase. Master's degree in community psychology The master's training model, marked by a multidisciplinary perspective, included classroom learning and field experiences for each module: participatory planning, the implementation of interventions and training courses for social-affective and sexual education, using the integrated method, on network building, on the analysis of the profiles of a community, on the analysis of an organization through the method of participatory multidisciplinary organizational analysis, on peer training and the management of self-help groups. The complexity of the phenomena requires different levels of analysis and different types of intervention, starting from the examination of needs, including through epidemiological knowledge and the ability to read the various systems involved: families, work contexts, and economic and social systems. It is an approach that considers the public and political perspective by overcoming the individualistic dimension and focusing on the interdependence between the individual and the social structures. Training in community psychology enables the construction of a professional profile with eclectic and versatile skills that are not only clinical. It proposes a professional identity that makes the psychologist capable of counseling individuals and communities: he or she is not just an expert but a **promoter of change that enables individual and community empowerment and shares psychological knowledge**.

I carried out my first organizational analysis in a social cooperative in Arezzo, SCASA, built the community profile of the Orvieto area, and began designing socio-affective education programs in elementary schools in Umbria. At the same time, I started projects in Lazio and Abruzzo. I also participated in the project of the municipality of Torino di Sangro (Chieti), which was carried out with funds for the prevention of drug addiction and housed within it a large socio-affective education project aimed at teachers and parents of schools in the area, to work not only on information but also on the empowerment of individuals and the community. Along this path, I was a lecturer for socio-affective education courses aimed at parents and teachers. These projects were carried out by accessing funds and intercepting funding that at that time was provided to schools or municipalities and working as a freelance psychologist with occasional services; I had the activities included in the annual training programs approved by the schools' teaching boards and school boards.

Of all the work experiences during the two-year master's program, the most extraordinary was participation in 1994, first as a tutor and then as a lecturer, in basic motivational courses with Professor Francescato. The Sviluppo Italia project was carried out with European funds and was aimed at promoting youth entrepreneurship in disadvantaged regions (project for the Promotion of Youth Entrepreneurship, Law No. 44 of February 28, 1986). Of the many trips made, the first one, to Calabria, deserves to be shared, as for the first time, I was able to see the impact of a training program in the field. The program included guided hands-on experiences and face-to-face lectures using projective techniques, including the work career film, as well as narratives, such as reconstructing the work narratives of families of origin, developing a community profile, and learning how to read an organization. I could see for myself the impact on the perception of events and the potential for change that group training in the field could bring. In a class, a working group had produced a film at the first meeting titled *Pietro li Pompi (Pietro's Funeral Homes)*, in which the protagonist, a certain Pietro, who was the same age as the students, chose to work in funeral services due to the high number of 'murdered' people in the town where he lived. By the end of the project, the title of the film (and the story of the protagonist) had changed to *Pietro Lavanda,* meaning a man who opened a laundry in his hometown, where many other commercial activities had flourished.

In addition to the training, of this project, networking with schools, institutions, and local businesses, meeting with key figures, interviewing, and building an alternative network in the area in which schools and students were included and involved was really interesting. I remember meeting the leader of a group of young people trekking in Aspromonte and an entrepreneur of a well-known mineral water brand who was trying to develop a business in their land, people who cultivated alternative values to those of the 'Ndrangheta and who courageously fought for a reality of life where a different future was possible.

1.3.3. Third phase. Four-year training school in clinical and community psychology, and psychologist at the family counseling center in Orvieto I began my clinical specialization by enrolling in the four-year ASPIC (Association for the Psychological Development of the Individual and the Community) school in Rome. The start of this training path coincided with the beginning of my work as a psychologist in the family counseling center of the Orvieto district. With my hiring at the end of 1994, another phase of my professional life began, which would add a piece to my professional identity as a manager and public employee – an identity for which I was completely unprepared, as at that time, there was no specific training for this field, and I had no relevant experiences in my family background.

My life experience has been shaped by fieldwork, which I can distinguish into three phases: the second half of the 1990s, after 2000, and from 2010 to the present. The family counseling center is a 'service for family and maternity assistance.' Due to the existence of regional laws implementing Law No. 405 of July 29, 1975, which establish staff levels, service areas, and more, each counseling center is unique; thus, there are countless counseling centers, each with its history, evolution, and unique development.

It was necessary to achieve the goals of healthcare reform, translating them into behaviors and professional actions while responding to a specific mandate that also included health education, which called for an innovative role that went beyond purely clinical-ambulatory psychological work done with individuals or families. There was a demand to design alternative interventions in the community aimed at minors, families with disabilities or elderly members, and in schools for sexual education. As a community psychologist, I spoke a language appropriate to the institutional mandate of the health service in which I worked, but it felt as though I was talking about something that did not exist; everything was still to be built, especially for the psychological profession. At that time, community psychology was not well known, and in health services, people would say about me that 'I did not have the vocation to be a psychologist' because my professional profile did not align with the clinical work of diagnosis and individual or family interviews.

Alongside those who did not understand the 'different' work from traditional clinical work, I also encountered open and willing individuals with whom to interact, allowing me to build many projects within the community. I still work today as the only psychologist in a team composed of an obstetrician, a social worker, and three midwives. Training in community psychology has enabled me, from the beginning of my professional experience, to pay special attention to the organization of work through careful analysis of user demand community, and internal service resources.

Psychological activity included a reception and orientation response of users and a "basic psychology" activity with the intake of situations relevant to the areas of intervention of the counseling service of a specialized type for sexuality. By the end of the first year of work, psychological activity constituted a substantial area of clinical work. It was necessary to reconcile psychotherapy activity with the mandate of prevention, health promotion, and health education. Initiative projects needed to be implemented to overcome the risk of working only with those who already had resources and sought help by spontaneously orienting themselves. Thus, through a demand and needs analysis, the importance emerged of avoiding irrelevant intakes and lowering the risk of "psycho-therapization," that is, responding to requests only with clinical psychotherapy interviews. The eclectic and integrated training was of great help to me in this, it allowed the construction of a working model that, from the *analysis of the demand,* allowed a *differentiation of the* appropriate and effective *psychological response.* The data from this work are collected in my master's thesis (Cleri, 1999).

This has allowed me to develop a working model that integrates clinical skills with prevention and promotion skills, where the psychologist's professionalism finds its professional specificity in the *operational diagnosis* phase, which is the central moment of the intervention, which is followed by psychotherapy or group empowerment and social support interventions in the service, or a referral to another relevant service of the company. Operational diagnosis identifies not only the patient's pathological structure but also healthy resources (with emphasis on desires, coping, and problem-solving skills), family structure, the network of systems in which he or she is embedded, economic and cultural resources, service resources, and the possibility of intake. I have worked significantly in building referral networks between territorial services, particularly for psychiatric referrals in mental health and drug addiction services, but also with the hospital for abortions. The psychologist with integrated training can build differentiated interventions and responses: brief counseling, support, guidance and motivational interviews, individual and group psychotherapy, psycho-diagnostic interviews, but also community interventions through health promotion projects and community profiling, network building, participatory goal-directed planning, and verification through the involvement of recipients.

I have organized various types of interventions: *Interdisciplinary projects* were linked to the work of the team and involved welcoming users for all areas of service intervention, from the "birth path" and sexual counseling to the "youth counseling" at the headquarters and in schools. As a psychologist, in the "birth path," in addition to individual interviews at the headquarters, I was also involved in puerperium support at home and in conducting self-help groups for new mothers. Among the most relevant interdisciplinary projects, I remember *childbirth preparation.* The activity in the birth accompaniment courses includes three meetings, in which in addition to theoretical lectures there are guided practical experiences in groups and pairs for mothers and fathers with the circle time mode; *Clinical specialist services,* which included specialist reception, psycho-diagnostic counseling, sex counseling, couples, and family counseling, counseling for the implementation of Law No. 194 of May 22, 1978, and parenting support counseling; *Specific psychology projects*, which involved two target populations, adolescents and parents, through the "adolescent group" and the "parent continuing education group." These activities were designed according to the methodological principles of community psychology, and the integrated method of social-affective education was used in both groups.

Regarding sex education, as a psychologist, I have been participating in service projects continuously for 30 years now, going through many vicissitudes and changes. The activity has been carried out in schools of all levels on pupils and then through training for teachers and parents. The sex education activity was also carried out in collaboration and with special attention to operators and parents with children with disabilities, with the mental health service for children and together with a social cooperative that ran daycare centers aimed at boys and girls with disabilities. The socio-affective education method had the advantage of having already overcome the view of only informational programs in the area of sexuality compared to other models of sex education. The model argued for the need to implement relational-affective education by training teachers and not just having the psychologist as an expert enter the classroom directly, working on relationships in groups and not just information. Particularly not insignificant is the awareness that, to date, there is no law in Italy for sex education in schools: working with this method allowed for soft interaction in the contexts through the involvement of all actors, making the projects included in the school training plans and allowing a great awareness and mediation in the community.

Around 2010, what I call the "third phase of my fieldwork," when regulations allowed for the construction of the corporate health promotion operating units, it was possible to bring activities into the system to provide all schools in the district with the same opportunities, to avoid inequalities in the services provided. This was the starting point of a larger work, which allowed the construction of a network of all schools in the district, with the signing of a memorandum of understanding between the health company and schools of all levels. It is very recent – it dates to June 2022 – the implementation of a regional memorandum of understanding that defines new guidelines for us health workers and schools about health promotion activities. At this stage, as a psychologist, I have served as head of the "network of schools that promote wellness" in the Orvieto district for more than six years, and I have been the scientific head of courses for social-affective education and peer education. In the counseling center, minors represent a significant clientele that deserves special attention not only in waiting activities but also in initiative activities. In our experiences in the counseling center, we found that children, although they did not have pathologies, had relational difficulties related to the stage of their life cycle. To respond to these requests, a group was activated in the office that provided cross-access from all secondary schools. It was a strategy that made it possible to respond to an increasing number of users, cutting down the waiting list of individual interviews when there was no clinical pathology, with a response offered through the group and participation in a psychologically led social-affective education learning program. The same strategic motivation was behind the design of the socio-affective education group aimed at parents, which enabled, through psycho-educational pathways, to respond to requests appropriately and effectively. *The ongoing training* activity *for parents* also began on an experimental basis as a response to a growing demand for requests for help for those who had adolescent children and served to break down the waiting lists of requests for support on parenting.

The socio-affective education activity aimed at parents was structured on three levels:

- Level 1: Awareness;
- Level 2: Training course using the integrated method;
- Level 3: in-depth theme course.

The three levels had different modes of access and different structural characteristics, such as numerosity. All groups included an entry questionnaire for the detection of needs and knowledge and the exit administration of satisfaction and learning questionnaires. What was surprising, noted consistently, was that parents who participated claimed to be unfamiliar with social-affective education, but by the third meeting, they were already using new modes of communication with their children (Cleri, 2022). This continuing education experience dedicated to parents has developed with an articulated path of nearly two decades, which has gone through three different phases:

- Groups in counseling;
- Reception project groups in area schools;
- Groups in the network of schools promoting wellness.

Over time, annual interventions designed to raise awareness among parents and families on certain aspects of psychological growth and affective education have been carried out. Class representatives and teachers have been involved, and this counseling activity has covered an increasingly large segment of the population.

Even today, counseling centers in the National Health Service are directly accessible and free of charge, offering counseling and health promotion to families and young people. Their openness to health promotion activities makes them more relevant than ever, as we know that foreseeable transitions and crises are sensitive phases in which motivation to overcome difficulties facilitates change, representing very important gateways to work on lifestyles and individual and community empowerment. All these issues have been advocated by community psychology since its origins and are more relevant today than ever before.

Initiating online interventions training in community psychology has also prepared me to manage *crisis intervention* and work in the field in *emergencies*. In 30 years of work, I have dealt with two earthquakes and a pandemic. The latter experience was a great test and allowed me to break new ground. For example, it made it possible for the health service to transfer services and activities online, which was unimaginable under normal conditions. This transformation, which made a virtue out of necessity, enabled the successful use of clinical interviews and the conduct of online groups. Like all healthcare facilities, at the time of the outbreak of the COVID-19 pandemic, the Orvieto consultation center, where I serve, was caring for numerous pregnant patients. For all of us who remained in the field, maintaining continuity of care during the spread of the virus was a great challenge. Psychological counseling activities for pregnant and postpartum women were recognized as imperative by government and regional guidelines; therefore, many services were reorganized with telephone contacts and video conferencing. I increased the listening space so that the group could also have an opportunity to rework information, which was very conflicting and caused disorientation and confusion, unnecessary alarmism, and anxiety.

I have taught and continue teaching on the topics of communication for safety in the workplace (Legislative Decree No. 81 of April 9, 2008). My most significant university experience is teaching occupational psychology at the Faculty of Medicine, University of Perugia. Within the course, I developed a program that, starting from a systemic view of the work experience, advocates the importance of integrated training and the development of so-called transversal skills, emphasizing personal values and ethical choices (Cleri, 2021).

1.3.4. Conclusions

It is not always possible to choose one's work, yet when this possibility occurs, it can be a source of well-being and gratitude. I feel my educational experiences are closely intertwined with my work experiences. I owe to working in public administration an awareness of the importance of equity, sustainability, and appropriateness of interventions. I am in a profession that fulfills my tireless curiosity and allows me continuous growth and learning.

Still worth mentioning is the secondary benefit I enjoyed, thanks to my training in community psychology, in being able to read organizations in their structural, functional, psychodynamic, and psycho-environmental aspects. This skill has been a key support in adapting to the management role I have held in the company in which I have worked, allowing for work well-being and quality experiences in daily work. I believe this training, which is often overlooked, can be useful to anyone working in complex organizations, school personnel, healthcare personnel, as well as all people who are employed in businesses. Community psychology offers models and methods that can help build a more benevolent and resilient society in which people can live and thrive through good relationships with humans and the natural environment. I feel immense gratitude for those who have helped me grow in "science and consciousness," the teachers, patients, students, colleagues, school faculty, and parents who have shared so many extraordinary experiences with me and with whom I have walked side by side on a stretch of life.

1.4.   Social and community psychology in Friuli. Orietta Pagnutti (community psychologist; 69 years old) The story of my relationship with community psychology has very distant roots: I had landed at the Faculty of Psychology in Padua fortunately in the early 1970s. I had wanted to study medicine, but this path being too long, I had diverted my interest to psychology despite being accompanied by a chorus of disapproval that said I would be doomed to unemployment. However, I was convinced that even through this discipline, I could come to a better understanding of myself and, above all, contribute to changing the conditions of life in my world, a Friuli that offered no prospects, no possibilities for change for those like me who had come into the world in a working-class family and a father lost too early.

Those were difficult years, in which the heated political struggle had seen Those were difficult years, in which the heated political struggle had seen the emergence of violent and blind terrorism that ran through the days like a chill and went on to strike people unpredictably (once, on my way home by train, the convoy in which I was riding was stopped in the middle of the Venetian countryside and we passengers were forced to leave the train and take refuge in a small station, waiting for some explosive device to be inspected and detected). But at the same time, they were exciting years because full of ideal aimed at building a more just, less authoritarian, and more participatory society - in which the distances between social classes would be reduced, everyone would be given access to knowledge and university, workers would have rights, women would have the same opportunities as men and the possibility of deciding about their bodies - were perceived as something concrete and possible. Padua and Rome then were the only two locations in Italy for the Faculty of Psychology, and I had landed in Padua.

In 1976, a powerful earthquake devastated Friuli, which appeared to the world in its poverty and immobility: it was a region that had no hope, bent over by work that barely ensured subsistence, whose language was a sign of minority for those who spoke it, known by most only because they had done their community service in this remote place. My country, even my home, had been damaged. So, I interrupted attendance in my senior year of college and began working (after a brief period of volunteering) in the population aid organization. I would graduate the following year with a research thesis on the condition of the elderly in a small village in Friuli.

During that period, my work choices were intertwined and had ample room for action in synergy with the events related to the reconstruction in Friuli. The earthquake, which had destroyed a good part of middle Friuli and the Piedmont area and caused many victims, had the power to mobilize an extraordinary response in terms of both economic aid, the presence of sympathetic volunteers, and affective closeness from all of Italy and from many foreign nations. Thus, employed by the municipality of Martignacco (Udine), I was able to design and organize territorial social services for almost two decades. Martignacco became a laboratory of innovation and experimentation in the social and healthcare field, so much so that, from this experience, the first regional framework law on social assistance took inspiration.

The principles that inspired my action then were:

- *Participation*, i.e., soliciting the active presence of citizens called upon to express themselves at institutionally scheduled moments on the quality and appropriateness of services offered;

- *Social promotion of the community* as a response to the social, cultural, and economic marginalities then present and particularly acute. Dialogue, knowledge, and animation both through experiencing the community and through research activities aimed at knowing and measuring its needs and resources.

The first research work, conducted between 1976 and 1977, focused on the condition of the elderly population in Martignacco and led concurrently to the establishment, among the first in the region, of the home care service. That work made it possible to highlight the frailties of the elderly at that time, when deprived of a family network, lacking income (at that time, not all elderly people enjoyed a pension), in precarious socioeconomic conditions or with inadequate housing, in a particularly poor living environment. To encourage and organize the participation of the population, the annual assembly of the elderly was established, with the task of evaluating and approving the work program of social services and the final balance of the activities carried out, and the council of the elderly, as a group to support and promote social work.

In 1978, the first summer stay was organized in the municipality of Carnia, Prato Carnico. Over the years, the offerings became very extensive, allowing for the inclusion of non-self-sufficient or disabled individuals, always to build connections and create moments of meeting and knowledge with the places where we were hosted. A few years later, I conducted a similar study for the Municipality of Prato Carnico on the condition of the elderly population in Val Pesarina, which aimed to promote the initiation of social services in that area and was published a few years later. An intense program of cultural and socialization activities began (meetings, outings, conferences, parties) for the elderly, aimed at intertwining and strengthening bonds among them and getting to know Italy and Europe. In the town's capital, I then promoted the opening of a self-managed recreational center, which for many years served as a meeting and gathering place, especially for women (as the men still went to the tavern).

As part of the activities aimed at valuing elderly individuals, a participatory history project was launched in collaboration with some teachers from the local first-grade secondary school, where the students were able to dialogue with a small group of elderly people who recounted their experiences during the war and the Resistance, as well as elderly women who shared their perspectives on wartime.

Long before the establishment of social welfare areas (inter-municipal bodies for managing services), I initiated a collaboration with the neighboring municipality of Pasian di Prato aimed at ensuring continuity of service through a system for substituting service managers. In 1980, the 'Martignacco model' was presented as one of the most innovative during the International Year of the Elderly. This organizational model remained structured for about twenty years.

1.4.1. The family counseling center In 1980, I was commissioned to collaborate in the opening of the first family counseling center in the province. On February 17, 1981, the family counseling center in Martignacco began its activities with a broad multidisciplinary team in which I held the role of psychologist. It was born amid much controversy and conflicting expectations and took its steps in a climate of great uncertainty. They were times of strong opposition, and the fact that the health personnel were non-objectors and thus also applied the provisions of Law 194/1978 on abortion, was a source of exasperating attacks and controversy. In the short time that the counseling center was under the City Council, the first childbirth preparation courses, the first women's cancer screenings, and courses on child development were initiated. We dealt with responsible procreation, contraception, adoptions, and separations. In 1982, I promoted and supervised the first research activity on the condition of women, carried out by three psychology undergraduates.

As part of the counseling activities, a close collaboration with schools was initiated, which continued independently of the events at the counseling center, which a few years later came under the administration of the then-local health authority (USL) of Udine. This was followed by a long period of depletion of activities and resources until the headquarters was transferred to Udine, and the activities were concentrated in the city.

1.4.2. Working with schools The work in the area led me to meet with primary and secondary school teachers and listen to the problems they reported. In the 1980s, in our places, schools were still selective, and struggling families had no tools, other than punitive ones, to intervene against undiligent or unproductive children. The rule was that everyone had to make it on their strength, and no help should be asked or given. The school was a duty, and this the boys were expected to do. And so it happened that those who lived in an environment that was culturally poor, or too simple to understand, or too burdened by the vicissitudes of life to see what was happening to their child were sooner than others rejected from school as well. Starting from the spaces for action opened to me by the counseling center, I tried to apply some of the tools that psychology had given me. Knowing distress in detail allowed me to identify spaces for action with a view to inclusion, prevention of school failure, and youth maladjustment, shared with both the school leadership and the local authority. Alongside the assessment and construction of support projects for individual children, I proposed initiating an activity aimed at the reasoned construction of classes. Thus, came to life, in the phase of transition from primary to secondary school, the activity of *class formation*, based on the assumption that a heterogeneous context in terms of gender, territorial and educational background, social skills, and productivity would create better preconditions for teaching and learning activities. Criteria and tools would be shared each year with school leadership and parents both in preparation and later as restitution. The activity lasted for about ten years, involving around sixty young people each year, until it became impossible to continue because of declining birth rates.

The relationship with teachers and parents highlighted the difficulty of many in relating to adolescent boys, with whom dialogue appeared stunted and too superficial. The difficulty in communication and confrontation manifested by the boys was pointed out by the school as a primary problem, so much to limit their potential both on the school level and on the broader level of self-affirmation/self-realization. And so, to deepen the knowledge of these adolescents and to facilitate dialogue among them, parents and teachers, in 1989, I promoted (with the collaboration of a colleague) the research *Adolescence: analysis of a dynamic process*, aimed at acquiring information about adolescents' attitudes toward their affective world and interpersonal relationships. The instrument used was the "semantic differential," the scores of which were subjected to factor analysis according to Thurstone's method. The research involved 233 boys attending the second grade of secondary school and belonging to three different schools: Udine, Martignacco, and Lestizza.

1.4.3. Since 1990 aggregation of social services In the early 1990s, the Friuli-Venezia Giulia region intervened in social services, substantially changing its organizational structure: from that time on, services would no longer be managed by individual municipalities but by municipalities aggregated in contiguous territorial areas and with a population then of about 20,000. So, I suddenly found myself at the top of one of these areas and having to perform a very different function from the ones I had held up to that point. I realized that I didn't have all the necessary skills, so I started looking for a school or courses where I could acquire the missing knowledge. And then I needed to get out of my region, which I felt no longer had the innovative and constructive drive it once had. So, I chose to attend a master's program in Personal, Group, Organizational, and Community Empowerment, which was held in Rome under the direction of Professor Donata Francescato. *I know for a fact that it was the turning point of my life because it gave me new ideas and provided me with keys to understanding organizations* that opened me to possibilities that I otherwise would not have even been able to think of. It also enabled me to find powerful tools to do what I have always enjoyed: that is, to make others see the possibilities they possess, to manage groups by encouraging and promoting their members, to read organizations with the tool of organizational analysis to overcome impediments to doing, to design and grow resources within a community, etc.

I would like to say that it was from there that my second life began, in which I began to think of myself as an independent professional, a consultant, and not just as a part-though with the ample freedoms of action that I had had of the public administration.

The first workshop I attended was the one with Professor Anna Putton on social-affective education, and I knew right away that that would be the right tool to offer to kids and teachers and help overcome those social and cultural limitations that too many times I had seen mark the fate of so many young people. Later in my vocabulary, during another workshop given by Professor Francescato, entered the words *empowerment* and *generativity*, which would be my revolution and the design goals of all my subsequent actions.

1.4.4. Learning and wellbeing together in school In the spring of 1995, as a natural evolution of the many activities carried out in support of compulsory schooling, with the financial support of the Province of Udine and with the support of the relevant educational directorate, I launched the project *Star bene insieme a scuola*, to give tools to teachers, parents, school administrators, and non-teaching staff for the construction in the school of a relational environment concretely favoring learning, inclusion and growth of children and young people.

The project also south to enhance educational collaboration between school, family and community.

The first training course, aimed at a group of primary and secondary school teachers from the municipalities of Martignacco and Pasian di Prato, was held by Professor Putton and covered personal, group, and community empowerment, the social-affective method, and effective small group management. After that first training course, several teachers undertook to apply the tools acquired in their classrooms. The experimentation was a success, so much so that the project shortly involved all schools, preschool, primary and secondary, in the two municipalities. Being in charge at the time of a social service area that included two other municipalities, the project was soon extended to the educational circles of Campoformido and Pozzuolo del Friuli, and its activity developed for almost a decade, until 2004. The aims were to work for the educational success of all children, to build the conditions for real well-being at school of students, parents, and teachers, to create shared knowledge by placing the collaborative presence of the student at the center, to give life to an educational alliance between family, school, and community to encourage the overcoming of prejudices, suffering and mutual distrust.

It was necessary, however, to start with an analysis of the needs, resources, and weaknesses of the different schools. Therefore, my first action was to initiate AOMP (analysis organizational multidimensionald and participatory, see chapter 7) intended for a group consisting of teachers, parents, school staff, and principals. The AOMP turned out to be a powerful tool to bring out the interdependencies between the different parts of the school organization, to listen to and recognize the various visions of teachers, parents, janitors, administrative staff, and principals, to put in place small but significant changes already during the development of the AOMP path. Eventually, it was possible to articulate and plan the project's actions over time, summarized in a memorandum of understanding with a multi-year scope signed by the two educational directorates and the four municipal governments involved, which pledged to support the project financially and organizationally. From that time, I organized and held, at different times, the training of teachers and parents, the supervision of teachers applying the content and methods learned in their classrooms, and the evaluation of the results achieved. During its decade of existence, the project trained 429 teachers and over 700 parents.

So, in December 2003, school leaders and local administrators presented the *Star bene insieme a scuola* project to the Friuli-Venezia Giulia Region as a model both for the results achieved in the relations between school, family, and territory and for the psycho-pedagogical contents that proved to be effective tools for the exercise of their respective and interdependent educational roles. All of us felt that we had acquired sufficient experience and skills to support other schools in acquiring tools to facilitate school success and being well at school as well. Our interlocutors liked very much what we presented; a regional councilor proposed a law that seemed to establish the services to support school autonomy. Subsequently, the region avowed to itself the continuation of the project; the school leaders who had meanwhile taken the place of the promoters did not seem, in fact, interested, and slowly, the experience died out. However, for a few years, in the Upper Tagliamento Valley, in six small municipalities located in a large mountainous area, it was possible for me as a consultant to initiate a similar project, with initial PMOA work and later with the implementation of the training paths for both teachers and parents; the project went on for a few years.

1.4.5. LAFORIT (Laboratory for training, research, transformation) My experience over the years has made me realize that to make an area grow or to offer other opportunities to the people who live there, it was necessary to make certain strategic knowledge for the development of oneself, and one's community become usable, particularly to managers, decision-makers, and people in significant roles in services, schools, healthcare, and work. I had grasped how community work was relevant to enhance processes of social inclusion, participatory democracy, and recognition of the many identities of which it is composed. So, with two colleagues in 1998, I gave birth to LAFORIT, which in its six years of activity offered three editions of a two-year master's degree in Social Empowerment, allowing the trainees to come into contact with leading experts in their respective subjects, such as, in addition to the aforementioned Professors Francescato and Putton, Professors Massimo Bruscaglioni, Piero Amerio, Anna Ravenna, Bruna Zani, and Miretta Prezza.

1.4.6. Nursing homes The last part of my professional career entailed working in nursing homes. At the beginning of my working career, the promotion of territorial services meant building a real alternative to the institutionalization of people, and for many years, the principle of allowing people to be cared for in their homes was strong. In the meantime, the world was changing. Degenerative diseases of the central nervous system, dementia, and particularly Alzheimer's disease, were increasing exponentially, families had shrunk and became less stable, the young and adult generations more mobile, work more precarious, making it virtually impossible for children to care for their parents, even if they wanted to. Regionalized healthcare and social work were responding very differently to the care needs caused by long diseases and behavioral disorders that were increasingly less manageable at home, where care was entrusted to caregivers, usually foreigners, who were not always prepared to handle these complexities. In the absence of a satisfactory response to non-self-sufficiency, the regional territory quickly was filled with nursing homes.

In 2007, following a deteriorating relationship with a public administration for which I worked, I began working with a large local nursing home that housed and still houses about 450 dependent elderly. Initially, I had a great fear of being harnessed within such rigid gears, but I could no longer work in the territory where an unstoppable process of despoliation and bureaucratization of services and mortification of the professionalism working there was underway. Therefore, I began my work inside a "total" institution, I started to deal with the key figures of the facility (which at that time had more than 250 employees), read about nursing homes and their legal nature, degenerative diseases (the first cause of hospitalization in the facility); made contact and visited the facilities that offered more advanced models of care in Trieste and outside the region.

Eventually, I identified several critical points on which to act:

- Models of care that were hospital-based oriented more toward caring for the body than taking care of the person;

- Training of staff, who did not have the tools for cooperative group work or interaction with people with dementia and behavioral disorders;

- The presence of unorganized volunteerism;

- Family members.

Along these lines, I developed my work.

1.   Models of care: the psychosocial reactivation service carried out inside the nursing home. It was urgent to bring life inside the facility, adding to the healthcare intake everything else and opening it to the territory. So, I began to plan and organize different and simple activities aimed at socialization. The activities grew quickly also because they were liked by both the people and the health workers, who saw their load lightened somewhat. Eventually, the psychosocial reactivation service, so called because its task was to rebuild in the facility the relational dimension of the person and act to maintain the skills of the residents, consisted of a group of ten professionals, including psychologists and professional educators.

2.   *Animation activities within very frail elders.* These are dedicated to the people most affected by illness, organized by the animators according to a shared monthly program and carried out both directly by them and through the contribution of volunteers.

3.   *Animation activities in the general animation room.* After a few years of operation, I had obtained to have available, properly furnished space that was intended for animation but had remained unused for years. It would later be equipped with an internet connection and a big screen. Other activities

would be organized in that space, tailored to people with greater autonomy and good cognitive abilities and to enhance their skills, history, and interests. Over the years, many activities have been tried out: workshops, pet therapy, art therapy, *laughter yoga*, film and screening cycles, carpentry workshops, etc.

4. *Library.* Exploring the facility, I discovered the existence of many books, the result of bequests from private individuals and institutions. After an initial reconnaissance, I arranged for the oldest books to be placed in the historical library, intended for enthusiasts, and the newest ones in the library for residents, in the general entertainment room so that all residents could enjoy them.

5. *General entertainment activitie*s that also involve people who are not residents of the nursing home. Under this heading, I grouped all those activities intended for the generality of residents, also open to family members and often organized with the contribution of resources from the city or territory. My relationship with the territory always involved seeking its resources and collaborations at both the individual and institutional levels, *primarily* with organizations and educational institutions. Here found a place for moments of total animation of the structure:

a) At Christmas and early summer, which ended with a collective lunch with family members;

b) A concert season organized thanks to the willingness of the local high school of music that featured the students themselves;

c) A season of poetry, with readings by poets or amateur writers in the area and realized through the collaboration of one of them.

Courses on communication, group management, and dementia have been implemented, seeking to transfer the tools, value framework, and principles of empowerment and community psychology to that context. Over the years, the training has also included operators of healthcare facilities and small nursing homes in the area.

I like to remember the time when we could have, as a trainer, in an event built together with the most important nursing home in Trieste and attended by more than 270 operators, Naomi Feil, author of the Validation method for relating with people with dementia, which still represents one of the most effective nonpharmacological therapies in this field.

Networking among nursing homes, volunteers, community service, and family members. I built and shared with various volunteer organizations a protocol in which they defined their respective roles and responsibilities, the protection of volunteers and residents, and how to interact with the services and apparatuses of the facility and, with the psychosocial reactivation service, which had the task of coordinating and integrating their activities with those in the facility. I also organized training courses for volunteers who want to work in nursing homes.

I tried to involve the civil service to open another space for intergenerational dialogue. After a year of bureaucratic work, our project was approved by the Prime Minister's Office. Since 2012, six civil service youth have been attending our nursing home every year.

Family members, whether children, husbands, wives, or siblings, with great suffering and effort, decide to place their relatives in the facility, experiencing this transition with guilt, sadness, and often with the difficulty of managing a physical as well as an emotional separation. One of the first actions with family members was to promote research work conducted by one of my staff members, aimed at learning about their experiences, perceptions, and expectations toward the nursing home.

I then proposed working on two fronts: with training moments aimed at caregivers regarding taking charge and others aimed at family members, with the establishment of a collegial moment called the Participatory Individualized Care Plan, in which the core team motivates and explains the goals of taking charge of the person and family members express their expectations, fears, and differences of view. The meeting, led by the psychologist, is aimed at building shared caretaking with sustainable care goals and building a nonetheless meaningful presence and an effective relationship with family members.

Finally, it resulted in the creation of a specific training course for nursing home operators.

1.4.7.   Retired but always empowered The highlight of my journey came just days before my retirement, when I had the opportunity to organize the final training event at the nursing home. I was fortunate to have my teacher and professor, Donata Francescato, lead the session, focusing on "promoting empowerment at the individual, organizational, and community levels." Today, the tools of community psychology and empowerment have become my forma mentis, allowing me to approach life and the world proactively, centering on solutions and acting in the present.

2.   Freelance community psychologists

2.1.   Training in community psychology with new technological tools and in virtual spaces. Marcella Autiero (psychologist, doctoral student; 29 years old) and Emanuele Esempio (doctoral student; 31 years old)

2.1.1.   Educational and vocational paths I am Marcella Autiero, a psychologist with a master's degree in clinical and developmental psychology from the

Federico II University of Naples. After graduation, I did my internship first at the chair of Community Psychology in the Department of Humanistic Studies at Federico II and then at the clinical psychology service of ASL Napoli 1 Centro, always on the issues of male violence against women. My attention and interest in this phenomenon continued with further training in other services (Protocollo Napoli project) and then online violence (European project CTRL+ALT+DEL). In these experiences, I began to combine clinical and community expertise with acquisitions inherent to social media management; the expertise in conducting groups acquired with community psychology became the driving force of didactic training and preventive actions with the organization of webinars, videoconference events aimed at colleagues and professionals, the city territory and the global world.

I am Emanuele Esempio, my educational background in the field of psychology can be divided mainly into two moments. The first is constituted by an initial interest in the purely clinical perspective, in which the *vis-à-vis* encounter with the other and the deep and personal individual relationship dominate the scene. By enriching my perspective of analysis on community places through comparison with different European cities visited over time, I was able to identify in community psychology a deeply concrete perspective, strong in tools capable of uniting theory and practice and stimulating the direct action of psychology on living contexts. I therefore chose to follow a logic that I would soon encounter in other professional fields as well, namely a classic "funnel" or *funnel* logic.

I borrowed the term *funnel* from my other profession, that of copywriter, *new media* manager, and online communication manager, which accompanied my studies as a psychologist. A profession that allowed me to learn about additional tools useful for community psychology (and beyond) in a professional mixture that made me a perhaps hybrid, but probably necessary, figure to operate in social contexts in line with current needs and instrumentation. Experimenting in the field of online communication has helped me to actively employ new communication and information technologies as a useful psychological tool to get in touch with users and operate in the full spirit of action–research that cannot ignore the use of new media today.

Adopting a "funnel" perspective in the psychological field means arriving at the empowerment of the individual and the improvement of the individual's perceived well-being by acting *first and foremost* on the broader contexts. By imagining the individual at the narrowing end of the cone and the contexts at the wider mouth, we can act on multiple levels by systematizing interventions from the general to the particular. I find myself to this day being able to operate daily following this logic, acting for the well-being of the individual

– which had approached me in the beginning, to psychology – from a broader contextual action that becomes all the more all-encompassing and "possible" the more we psychologists become able to enrich our toolbox with new tools, ideas, inspirations, and techniques that we will tell you about in these very pages.

2.1.2. Toward new career paths for community psychology: social design and virtual spaces During our postgraduate internship, we encountered the world of planning in the context of community psychology with the European project VIDACS (Violent *Dads in Child Shoes*; scientific referee Professor Caterina Arcidiacono). The product of a fruitful collaboration between the chair of Community Psychology in the Department of Humanities at Federico II University and the Beyond Violence desk of the ASL Napoli 1 Centro, the project brought our professional backgrounds and mutual expertise together.

Knowing how to design through community psychology means knowing how to deal with individuals in a broader sense, always keeping in mind their relationship with contexts, whether near or far. Community psychology teaches that individuals, as well as contexts, talk and ask; they carry questions and needs, but these are often not intercepted. So, how do you intercept these questions and needs? The planning psychologist does not wait for individuals to seek him out but goes to places and interrogates contexts, trying to give clear answers to concrete needs and setting general and specific goals. Then, it identifies problems recognized as salient by local, municipal, national, or supranational bodies and thinks of medium- to long-term solutions to address them. This all takes place within boundaries and rules that require important *skills* and interdisciplinary teamwork: *first and foremost,* the ability to thoroughly analyze and understand the problems one addresses as a psychologist; then, having good research skills to identify useful material already produced in the scientific community, thanks to which one can better identify the problem, possible solutions or hypothetical paths of work; and, finally, systematizing and putting down on paper the ideas developed during the planning process, which, as it is typically conducted in teams, requires skills in organizing and negotiating the contributions pooled in the working group, for the realization of feasible and sustainable interventions.

In this mode of work, the psychological intervention must be structured within patterns that may initially seem to limit to the psychologist but instead allow the creation of a solid structure of thought and action within which to build pragmatic and realistic interventions capable of improving the condition of end users.

This is what we did precisely in the context of the VIDACS project, which was created to counter men's violence against women. Starting from the OMS(WHO 2012) and ISTAT findings that this type of violence is particularly prevalent in family relationships, we proposed a theoretical-operational model useful to identify pragmatic and cohesive measures to counter violence against women and, in particular, domestic violence (Arcidiacono, Di Napoli, 2021; Di Napoli *et al.*, 2019a).

This project moved by following an ecological perspective (Di Napoli *et al.*, 2019a) to understand and actively intervene in domestic violence by overcoming the existing fragmentation of current measures to counter it. We started from the collective level, within which we suggested wide-ranging interventions aimed at promoting a consistent change in the current cultural system, change based on valuing gender differences and, at the same time, on equality and equity in terms of rights and opportunities. At this level, the role of new technologies was clear and valuable, which, as we shall see, allowed us to address broader social and cultural systems by acting as a megaphone for the voice of our team and our project. A level that, moreover, intended to work on individual and social representations and slowly deconstruct the collective perception of violence, still steeped in prejudice, stereotypes, and defenses, promoting, as far as possible, self-reflection in both the recipients and promoters of the intervention.

The second level, the organizational level, allowed us to explore a double plane: on the one hand, it led us to analyze how services, professionals, and public safety, as well as training and gathering places such as schools, parishes, etc., interacted with each other in the contrast and signification of domestic violence; on the other hand, it allowed us to explore the representations and emotions of practitioners that inevitably affect, positively or negatively, the quality of the contrast intervention (Autiero *et al.*, 2020).

Lastly, at the relational level, since domestic violence always occurs within a relationship, we investigated how the protagonists of family systems themselves signified and experienced violence to obtain a clearer conceptual and understanding framework and to identify new courses of action to counter it also thanks to new communication technologies (Arcidiacono, Di Napoli, 2012; Di Napoli *et al.*, 2019b).

2.1.3.   The ViDaCS: New technologies for community psychology **The VIDACS project** aimed to promote the creation of alternative (mostly virtual) realities that would serve *as new places of thinkability*, recognition, and awareness on the topic of domestic violence with particular reference to family contexts. The core of the project was the construction of a *serious game* in which fathers, perpetrators or not, experienced a domestic violence scene by putting themselves in the shoes of their children through a virtual immersive experience. This experience made possible using *virtual reality*, allowed the creation of an interactive scenario designed to stimulate reflection on domestic violence and the impact it has on the child who witnesses it. We used new media and web tools and social networks to create virtual spaces of thinkability through which we sought not only to inform and raise awareness among the public but also to promote confrontation and dialogue with the public and to create *active participation*. We aimed to establish new democratic and easily accessible meeting places for the direct and indirect recipients of the project to promote truly inclusive, active, and participatory action on domestic violence issues. We then built, through a site designed for the purpose, a bridge between institutions and users, inviting the latter to explore the issues they were reporting to us online through telephone counseling (when not possible in person) with psychologists and psychologists working in institutions who were part of the VIDACS team.

First, it was understood how it was necessary to distinguish the users into two macro-categories: the general population, to whom the existence of a project such as VIDACS and thus the concrete activity of psychology in combating domestic violence should be communicated, and the population directly involved in the phenomenon of domestic violence in the family environment.

Secondly, it was necessary to profile the population directly involved in the project into two additional micro-categories: users capable of increasing the *engagement of* the communicative facility and users that we can define as "conversion" users (and on whom we will focus later).

Starting reaching as wide a generic population as possible, therefore, constituted the first step in the communication management of the entire VIDACS project. Communicating with a wide audience allows professionals and, consequently, the projects in which they are involved to alert the population about a given social issue. Achieving these preliminary goals was thus one of the starting points of the entire communication framework of the VIDACS project. In addition, this communication to a wide audience enabled the so-called primary prevention process because it allowed us to raise awareness of the average population rather than just the specific population that may have domestic violence problems through web channels. Thanks to social media, it was possible to maintain contact with the target audience even during the lockdown that affected the whole of Italy during the COVID-19 pandemic. Even with the project at a standstill, the target audience remained connected with VIDACS, which allowed us not to lose the results achieved in months and months of strategic communication. In addition, we achieved an additional outcome that was not directly budgeted for during the design, which was to be able to help family members contacted during the pandemic to renegotiate the use of family spaces.

The goal was to position the project and the VIDACS team as approachable and insightful communicators, capable of fostering meaningful reflection on the issue of domestic violence. By presenting themselves as accessible – just a screen, click, or glance away – the team aimed to be readily available not only in critical moments when fathers recognized the need for help in managing their emotions but also during everyday moments of life. This ensured continuous engagement, making the project both useful and relatable.

As VIDACS is a project aimed primarily at fathers and consequently at family systems, Mother's Day and Father's Day were particularly useful times to engage families through the VIDACS Facebook page we built in a virtuous circle, with posts aimed at positive *engagement*. For Father's Day, videos were produced in which dads themselves told us about how they experienced the family environment in quarantine. These contributions were intended to bring fathers closer to the topic we proposed and, at the same time, to create a space for shared reflection about the issues, emerging and otherwise, of fatherhood in the time of the coronavirus. We were positively struck by the interactions we obtained, as if to underscore, from a psychological point of view, the need fathers had to talk about their parental function, which was sometimes steeped in taboos and hidden by a domination of the maternal function within households and public spaces.

Instead, for Mother's Day, posts were created in which children, with the help of their fathers, shared their fantasies and wishes about their parents' relationship. Since VIDACS focuses on countering the effects of witnessing violence, we felt it was important to give children a voice during a time when they were also affected by the restrictions of quarantine and lockdown. Among the posts that resonated with us the most was one from a child who openly expressed a desire for his father to be more involved in domestic life, particularly regarding household chores. This initiative not only allowed a male child to voice his wish and challenge certain family stereotypes but also prompted his father to reflect on those same stereotypes.

The use of the social world has thus become one of the pillars of the VIDACS project and has led to a previously unprecedented involvement of the population with whom to interact both formally and informally regarding domestic violence, secondary victimization, and, above all, respect to ways of combating these serious phenomena. In this, the user reclaims their communicative power, embarking on a two-way dialogue with projects that know how to use this very important tool. No longer, therefore, a monologue from institutions to the population, but rather a constant and mutual exchange that leads to a better and deeper construction of meaning for all parties called upon. In particular, the webinar experience enabled reflection on the implicit premises and gender stereotypes that substantiate violence, opening a dialogue with masculinity too often excluded from discourse on the topic.

In this perspective, the figure of the psychologist is evolving and is characterized by new capacities for action and new tools that are inextricably linked to the progress of the society in which we live. New media thus become new interfaces within which to construct psychological intervention and promote change in a different sharing space constituted by cyberspace, which creates new social and cultural dynamics. Mastering online tools, therefore, means offering new points of contact to users, generating a virtuous circle whereby professionals and users can mutually enrich each other's perspectives and experiences. It means, too, enriching ourselves with new resources while constantly keeping in mind that cyberspace can become a crossroad where we can meet each other, get to know each other and act together for change. In concrete terms, this must be done in a deeply pragmatic way, studying, understanding, and mastering the logic and changing balances of the constantly evolving Web 2.0. Skills in content production and sponsorship, website creation, social network management, but also web-oriented copywriting and writing skills, as well as the use of platforms for working and meeting remotely (such as Zoom, Meet, etc.), become central to a hybrid professional figure in step with a liquid, hypermodern society.

2.1.4. The VIDACS project tool: the use of new media in the field of psychology

We used social networks as a building block for the first level of the *funnel*, namely, to generate **awareness** of the phenomenon of male violence against women in the undifferentiated population mentioned above. Social, combined with the website and webinars conducted in collaboration with the Beyond Violence desk, then helped us generate interest in the users about the phenomenon. Particularly useful were the creation of *ad hoc* campaigns and messages aimed at meeting the public on a common platform where they could discuss mutual representations and meanings of the phenomenon, as well as their positioning in this regard, and the use of the aforementioned webinars as focus groups to engage the male audience, interested through *dissemination* work, to activate a discussion among men, mediated by two psychologists, about their views on the issues proposed during the campaigns.

Capturing a strong intrinsic demand of the users, we also decided to partially follow the natural **interest** of the people with whom we came in contact, who explicitly requested to reflect on their emotionality in relationships, their relational and family roles, and a range of important contents as they are present in the constellation of relationships and violence in domestic contexts.

The third step required users to **decide**, namely how and whether to position themselves regarding the phenomenon in question and what to do about it. A step, this one, in which many men decided to collaborate by contributing video posts and discussing the salient issues proposed by the project. The last step of actual conversion is certainly the most arduous in any *funnel* project, as it aims to get the target audience to act in a well-defined way. In the case of VIDACS, the final conversion should have been about the spontaneous trial of the *serious game*: an emotionally and pragmatically very strong request as it would have led the target audience to get involved in the concrete, investing time to go to ASL, making an effort to confront an emotionally strong content and facing a deep reflection on their role as a man, parent, relational subject. Conversion became impossible due to the social restriction measures that occurred in recent years but were replaced by a strong movement of participation in online discussion platforms, webinars, and events held in cyberspace, that is, in that place that allowed us a meeting outside the, unfortunately then limited, physical places.

Finally, it is important to point out that all the tools chosen were evaluated according to their ability to engage the target population. In the context of the VIDACS model, therefore, we identified a segment of the population characterized by sociodemographic variables especially touched by Facebook and certain communication modes that guided the design and production of textual, visual, and video content.

The social media environment is extremely competitive, rapid, unstable, and characterized by *background noise* that often desensitizes and alienates the user. *First and foremost,* it is necessary to learn how to construct *ad hoc* communication implants for the target audience, trying to understand which linguistic, emotional, and narrative registers are most appropriate and effective for a given intervention. In the case of VIDACS, for example, we replaced an overly formal communication implant, which risked alienating users, with one that was more direct, engaging, and informal. A more social-friendly and engaging register while remaining in perfect balance with the need to communicate sensitive and complex issues with the right and unfailing degree of seriousness, clarity, and reliability required by our profession.

Finally, we again call attention to virtual reality as a tool for raising awareness on the one hand and for primary prevention on the other, as well as an element certainly in the making in relation to which it is possible to begin thinking about projects and interventions in new and technologically innovative ways. The involvement, through *dissemination* social and *vis-à-vis*, of the general public and male operators, who work in various capacities in the different services where it is possible to interface with situations of domestic violence, has led to two orders of results: first, it has allowed to inform about the project, prevent and raise awareness among the male public; second, it has allowed to open a space of questions and thinkability within the services themselves, too often still collusive with respect to the typical dynamics of male violence against women. Although the different activities had heterogeneous specific goals, the entire VIDACS project was guided by the overall goal of engaging men, their question, and their words within the domestic violence debate. By moving to the other side, the side of the perpetrator, the participant was able to access, in a protected context, the dark side, too often expelled outside himself as an individual and as a member of a larger community and was at least able to hold the two positions with respect to violence together. What used to be a divergence now finds its way to a possible convergence that makes the two vertices of observation true simultaneously without them being mutually exclusive.

These results were made possible by starting with initial user involvement, which would have been impossible without an online hookup. Too often, community psychology (and other) projects and actions go unnoticed and underwhelming, not seen and picked up by the groups and individuals who could most benefit from them in both direct and indirect ways. By increasing the communicative reach of the VIDACS project, on the other hand, it has been possible for us to have a real and broad impact on target populations and to make the project itself visible to the public. In this way, we were able to concretely reach our target audiences, producing concrete change.

2.1.5. Conclusions

Thus, the role of the psychologist is enriched with new challenges, new paths, tools, and potential that are now also linked to the development and use of new technologies and social planning as a creative, lively, and interesting professional outlet.

Being community psychologists, therefore, means being able to meaningfully and creatively understand the physical, virtual, social, mental, educational, and professional contexts in which we live to travel roads that are sparsely traveled to constantly improve our capacity for action. This allows us to accumulate tools in a toolbox that must never become static but must continually renew itself according to the real needs that, as psychologists, we will face in ever-new and often very complex ways that will enable us to act in the contexts we have chosen to study, protect and improve.

2.2.   Listening to oneself and others to promote relational well-being in school settings Alfredo Ferrajoli (clinical and community psychologist, psychotherapist; 65 years old) I graduated with a degree in Psychology back in 1980, supervisor Professor Donata Francescato, discussing an experimental thesis on friendship and the importance of mutual aid in human relationships in collaboration with Marquette University in Milwaukee, Wisconsin. Before graduation, reading a book particularly excited me: *Open Families, the commune*, which Professor Francescato had published with her sister Grace. Those were the years of social commitment and the dream of changing the world by a young generation, mine, dreaming of "infinite paradises," unrealizable on earth, but which we then tended to live and feel within ourselves, abetted by the revolt embodied by protest songs and a certain kind of music, all embedded in the protest and revolutionary movements of hippy inspiration that hoped for epochal

changes in human relations and ways of living and being together. It was from the set of these experiences and lived experiences that had matured in that context that I was increasingly discovering the will, interest, and importance of being well together, where the individual and collective spheres were supposed to find healthy balances and living spaces of common satisfaction. "Networking," as community psychologists emphasize, pooling experiences, sharing, valuing each other's energies and availability, listening to each other and listening, and studying and understanding how all this could be achieved and put into practice, constituted an important part of my focus, while the other part was training in various types of psychotherapy: psychosomatic psychotherapy, bioenergetic analysis, psychoanalytic psychotherapy, and body psychotherapy.

Freelancing as a psychotherapist, coupled with my teaching work in schools, prompted me to use Donata Francescato and Anna Putton's text *Star bene insieme a scuola*, which integrates constructs from humanistic psychology and community psychology, for the implementation of training and refresher projects for teachers and students, described below.

2.2.1. The sharing and communicating project: a workshop for interaction It was carried out with the participation of boys and girls attending the ISISS Pacifici and De Magistris of Sezze and the secondary school of Priverno (in the province of Latina). The Workshop proposed to young secondary school students was characterized not only by the treatment of some theoretical aspects but also by group experiences on the emotions and feelings of the participating boys and girls to improve the knowledge of those deep and ever-changing aspects of their personalities and to promote solidarity and mutual help. The setting of the activities, also characterized by working in small groups, allowed socialization and sharing of everyone's experiences and fostered, in some cases, communication and contact with themselves. Through an interactive methodology capable of urging participants to open about themselves, it was possible to interact in a meaningful way, creating a climate of welcoming, listening, and understanding.

Overall, a picture emerged of a youthful condition tested by some of life's experiences: authoritarian upbringing, bereavements and disappointments of a sentimental nature that caused intense frustration, and feelings of bitterness about lost or terminated friendships. In particular, the participating subjects report that the Workshop experience enabled them to reflect on themselves, listen, question themselves, socialize, and open in a pleasant and serene atmosphere of acceptance. For some of them, attending the Workshop served to understand that both "good" and "bad" feelings, or those deemed as such, can be shared with other people. At a particular age, such as adolescence, when personal identity is being formed, understanding the importance of opening to others and socializing in-depth, including through the expression of emotional content that would like to "stay inside," is already a source of help, of exchange and deepening of the sense of "us."

Other participants experienced the experiences made in the Workshop as a time of personal enrichment and growth, capable of providing help through discussion with the group, accepting themselves more, and loving themselves more. In a circle time on traveling, some participants also reported that they wanted to take books with them since they considered them "irreplaceable travel companions"; most would have liked to take their romantic partners with them, without whom they would not leave and would not go anywhere. Others, however, preferred their best friend as a traveling companion, along with the inseparable cell phone chosen by many, which expresses both the attachment to productivism and the consumerist values of our society and, on a deeper level, the difficulty in leaving and being separated, even for short periods, from the world of affection. Some participants then reported a desire to leave with their cat or their favorite stuffed toy (teddy bear, some said), which, on a deep analytical reading, reveals an aspect related to relationship needs and requests for more "warm and soft" reassuring contacts.

From the constant monitoring carried out through open discussions, reports, and the administration of questionnaires, it was found that the Workshop found much interest in the participants, who requested more hours and insights. For one girl, the Lab is a place where "everyone has something to communicate and give, even a simple look or smile: simple gestures that make one understand that life can become beautiful, if enjoyed with fullness in every moment and if one can accomplish ordinary things always with extraordinary vigor and commitment." The Workshop was able to provide this girl, as well as others, with a valuable container for her moments of anxiety, which threatened to affect her schooling. After some time, one participant wrote:

"I participated in the Lab in the 2000s. I was a student in the high school of social sciences and participated with my desk mate, who has always been a psychology enthusiast, from the first day of school. I remember approaching that space and experiencing it timidly. I was a girl bored by the outside world, in which I could not find adequate, interesting stimuli. I was born and lived, for most of my life, in a town of about 25,000 souls in which it was easy to get bored or "lost." The Lab immediately appeared as something particularly challenging. It was difficult at first for me to participate in it since I was a closed, unmotivated, and quite sad girl. In that space, I met people who are still a part of my life today: that experience bonded and bound us inextricably and, at the same time, made my friendship with my desk mate tighten even more. In that Lab, I discovered my limitations and, at the same time, my potential. I learned to improve myself and accept the peculiarities that characterize me without judging myself or the people around me. It gave me the necessary inputs to start doing work, constantly, on myself, and to this day, those acquired tools guide me in some moments of my life."

Another participant:

I attended the Lab for five years, and it was a profound experience that significantly helped me during the often bizarre period of adolescence. I truly believe that, during that transitional time, attending the Lab was highly beneficial. It allowed me to better understand myself, deepen personal relationships, and discern what was truly mine in the perceptions I had of others.

More testimony from a participant:

Many times, in life, one finds oneself going through experiences without fully understanding the meaning they may have in the moment one is living them, this is what happened to me when I started participating in some groups when I was a teenager within the Lab. For more than ten years to date, I have been part of other groups, both for educational and work reasons, but few have managed to make me experience and feel emotions as I did during that time... I remember the joy, anger, and pain but also the self-awareness that came out of the closet at each meeting. Afternoons spent at school even though no one forced us to do it, and we were the ones who chose to do it because even though everyone in their way had the teenage torment inside, we felt it was better to live it and share it. Today, I consider myself very fortunate to have been able to live these experiences, and in my work at the listening desks within schools, I try to reintroduce these group meetings in a historical era in which teens seem to be increasingly distracted by living virtual relationships that promote distance from their emotional world, but at the same time have a desperate need to feel welcomed and accepted.

2.2.2. The Emotional Literacy Project: education in feelings, affectivity, and sociality as prevention of phenomena related to bullying This project was presented to teachers from schools in the Sabaudia teaching circle, as well as in Sezze, Priverno, and Rome, involving nursery school workers and teachers. We began by identifying the training needs of the groups. In the initial phase, urgent concerns were expressed, mainly centered around managing daily work challenges and addressing discomforts. These concerns, at least in part, reflected the difficulty of providing relational responses that adequately considered the emotional needs of their students. This allowed us to set training priorities, define clear objectives, and ultimately achieve various outcomes.

Aims and objectives
• Activate a research process in those involved;

- Train in the acquisition of communication skills that facilitate learning processes and improve the quality of relationships;
- Enable people to discover, use, and improve their skills;
- Fostering empowerment;
- Create a network of relationships in which each is a partner to the other;
- Breaking the isolation of participating subjects.

Goals

Reflect on the importance of relational dynamics and the emotional aspects that can arise in relationships with pupils through meetings aimed at:

- Improve and facilitate communication with pupils;
- Enhance educational skills;
- To provide an opportunity to question oneself, one's modes of action, and one's reactions, both emotional and behavioral;
- To foster and broaden awareness of one's way of communicating and the effects it can have on pupils;
- Acquire the ability to listen empathetically to empathize with pupils;
- Create a space for expressing, thinking, listening, questioning, processing, reflecting;
- Promote primary prevention, student empowerment, and mutual aid.

Some results

From the data obtained from the feedback received, the following emerged in the words of the participants themselves, including teachers:

- "The knowledge that I have carried out an extraordinary, meaningful experience, full of positive and new emotions for me."
- "Through group work and discussion, I realized the validity of my doubts; they belong to all of us, and together we can find the best strategies."
- "Surprised that so many different people who do not know each other inwardly can express emotions, thoughts, and feelings."
- "Unusual experience that prompts reflection on the profession."
- "Today's meeting helped me to feel understood as a teacher. Constantly questioning oneself, the famous 'I know, I don't know' is the dominant motif of this meeting. One should never pretend to know with certainty and security what the student has inside him or her."

Some participants expressed appreciation regarding the methodology, which fostered, in addition to working on emotions, the sharing of reflections among teachers. These methods promoted the cohesion of the group itself and fostered mutual help and a sense of community among professional teachers. Becoming aware of the emotional complexity of a child can be very difficult or even impossible if we do not relate to the "child part" that is inside each of us; when this empathy-identification operation is implemented, thanks to internal emotional resources, it can be physiological that this is experienced in a way that is, at times, painful and anxious: "What happens, or what will colleagues think, if I say what I think, what I feel, what I feel in a group context?"

The experiences made it possible for participants to get in touch with deep elements of their personalities, express content, and process thoughts and experiences. All this contributed to the personal and professional growth of all teachers and to creating a willingness to network between school and territorial organizations to create an educating community.

2.3. Knowing, knowing how to do, knowing how to be a community psychologist. Maura Benedetti (psychologist specializing in integrated humanistic psychotherapy and community psychology; 48 years old)

One becomes what one is. Sometimes, then, one is lucky enough to find in one's chosen profession, in my case, that of clinical and community psychologists, the tools to express aptitudes, skills, and competencies. Perhaps this was so for me if it is true that even as a child, crossing the streets of the neighborhood or the place where I spent my summer vacations, I loved to greet people - even those I did not know - and to entertain myself with them. I sensed that there is a beauty of people and a strength of relationships that somehow open the full meaning of life. What came later - the studies, the thousands of experiences in volunteer work as well as in the profession - gave substance and awareness to that small seed that I perhaps already carried within me.

The growing-up years, before and contemporaneous with the university education itself, were an important training ground for relationships and participation, especially within the neighborhood community. Belonging to a community educates one to exchange within and between generations; one learns the meaning of gratuitousness and sharing, one focuses on one's values, and one trains oneself to discover what one will find beyond that boundary. Inside the community, I also had my first volunteer experiences, which, as is often the case, opened my eyes to the different forms of human fragility and poverty, of which economic poverty is often the most visible aspect but which frequently hides a fracture in the very fabric of relationships, family, and society. At the same time, my encounter with initially, mainly through volunteering in the services the city offers for people in precarious conditions, also trained me to see the beauty and richness in people who survive life's travails that help us, too, to orient ourselves toward what is essential. The themes of poverty and fragility that often must do precisely with traumatic disruptions of relational fabrics would return many years later within projects in which I have participated and am participating, such as those dedicated to mental health or to building healthy and supportive neighborhood networks within especially large urban contexts. This is why, after initially choosing to study medicine at the Sapienza University of Rome, I soon turned, with intention, toward the study of psychology and, as my perspective was refined, toward community psychology, stimulated in this in a very strong way by the scientific proposal of Professor Donata Francescato.

I took the community psychology course for two consecutive years, animated by having found a theoretical model that I was immediately passionate about. The study, research, and experiences of operational interventions in different contexts were gradually an accelerator of desire and, at the same time, of awareness, nourishing on the one hand, the passion, the desire to do, and on the other hand, equipping it with theoretical and methodological knowledge and all those tools that are necessary to give form and substance to the things we feel and wish to follow. Those years, and the first years after undergraduate education, were exciting, just as when one feels that one has somehow found one's professional path dense with meaningful experiences such as the doctoral program in which I was involved in research on the beneficial effects of humor on health considered within the bio-psycho-social model.

The dimension of university research and teaching in collaboration with Francescato's Chair of Community Psychology and, following that, the specialization in integrated humanistic psychotherapy and community psychology were fundamental. They introduced me to new work contexts and created connections with international organizations working in the field of development cooperation, migration, and rights for inclusion inside and outside our country. Thus, I was able to acquire new tools for accompanying individual but especially group pathways, a fundamental setting for community psychologists and psychologists. For the enhancement of personal resources, each person is always the bearer. Also, from this point of view, the centrality of the empowerment category has been and still is a lintel of my work.

My professional path has gradually deepened, also thanks to a certain interdisciplinary openness, generating once again stimulating experiences, such as the one shared within the **Inter-University Research Center** for Sustainable Development at Sapienza University of Rome. In the specifics of my functions and competencies as a community psychologist, I have had the opportunity to collaborate with professionals from other scientific fields, with whom we have carried out projects that had as their focus the development of technological competencies at the service of paths of inclusion and empowerment aimed at individuals and groups. This vision and these skills were expressed in projects in different fields.

I have been involved in training **within the prison context** for the promotion of the possibility of successful reintegration in their contexts of reference after the period of imprisonment. I have been involved in projects to promote sustainability--environmental, economic, and social--in prisons as well as in organizations that care for the employment of disadvantaged individuals, working operationally and doing research to monitor and promote minority rights in national and international contexts. This is another pivotal element of community psychology that I have had the opportunity to promote through my professional and civic engagement: to try, through my work as a community psychologist, to give voice through research and training to the rights of people who are not included in decision-making processes, nor in policies to promote welfare, seen in the individual, family, organizational and community dimensions. My task has always been to accompany, monitor, and support the processes of personal and community change underlying the development actions envisaged within the projects carried out.

However, the dimension of belonging and engagement within my community of origin has never failed. I mention, among the many experiences I had, the creation of an initiative we called **Do-hands of solidarity**, an active citizenship project carried out in the Appio Latino neighborhood of Rome between 2002 and 2011. The objective was twofold: on the one hand, to raise citizens' awareness around issues concerning human rights, social justice, inclusion, the relationship between ethics and economics, and critical consumption through a series of periodic public meetings with experts from different areas of interest (teachers, journalists, administrators, third sector representatives, etc.); on the other hand, to foster citizens' commitment and direct participation through concrete actions in the social sphere, inside and outside the neighborhood perimeter. Volunteer groups have been supported in this way, collaborating with shelters for the homeless of Caritas of Rome and participating in the initiatives of the Food Bank, Fair Trade, with the Di Liegro Foundation's cooperation projects in Albania. Fundraisers were also organized to support development projects in the Global South, always connected to an action of promoting awareness of the issues underlying the situations of departure. At the same time, within the neighborhood, the network of volunteers ensured support for people with difficulties, for example, the elderly who were alone or to whom families were unable to offer the necessary support. The idea that a more aware and supportive territorial community is the first major resource for combating marginalization and loneliness, especially in large urban contexts, was among the project's cornerstones. To do this, as a community psychologist, I promoted the activation of people's resources and the social network of the territorial community.

Also, within the perimeter of the territory of my municipality, I then

supported political participation initiatives in support of projects and services for citizens. This testifies to the fact that the community psychologist finds its constitutive element in the deep sense of the civic dimension: participation in the life of the "polis," as well as a strong sense of the public (the value of services for all, from health to education) is also a beacon that has always accompanied me. Engagement in the areas of associationism and volunteerism can also become dangerous if it diverts attention from the need for a public dimension of services that is up to its task, which must instead always be defended.

In the last few years, I have been dealing with another crucial context in personal growth, **which is school**. I work as a school psychologist at the Newton Scientific High School in Rome. Here, I am responsible for the school's listening desk aimed at male and female students, teachers, and parents. I organize socio-affective education activities with class groups, parenting training and accompaniment, and active citizenship paths for students, working in a network with local services. An extraordinarily stimulating challenge and an enlightening experience in many ways in the post-pandemic time we are living in. Schools represent an extraordinary observatory of everything that moves within society, in terms of resources but also critical issues. It is precisely schools that are, today more than ever, a fundamental area of work for the community psychologist, an opportunity for the prevention of the discomfort rampant among preadolescents and adolescents, and for accompanying the adult generation that deals with them. Our commitment and professional skills as community psychologists, as well as the tools for working with groups, can implement the strengthening of togetherness as an element of health and well-being. To do this, we need to give and guarantee the continuity and quality of the professional presence of psychologists, even within the school context, a listening and supportive presence that anticipates and goes beyond the purely emergency logic (the psychologist called to intervene in the face of individual or relationship emergencies within groups).

At school, I first found myself going through the affair of the COVID-19 pandemic. How can we be a point of reference in such a delicate phase - critical, in the sense of a rupture and at the same time a possibility of discovery - not being able to count on the proximity and strength of physical presence? The same challenge is faced by teachers. Just as extremely delicate and compelling was – and to some extent still is – the task of accompanying students, but also family members and teachers, in the next phase, that of returning to school and processing what had meanwhile happened. In all this, as well as in the performance of my role, my outlook has always been that of a community psychologist.

The school's first resource is the school: it is an intergenerational space where an educational community can be developed. The community psychologist can promote among students, parents, teachers, and school staff the development of a sense of community, motivation and personalized methods of learning, respect for differences, and the value of inclusion of all, as well as the fundamental ability to navigate conflict and encourage interdependence. The perspective focuses on promoting participation, preventing mental health issues, and fostering a welcoming and attentive environment for individuals and groups. This approach is not evaluative or therapeutic but centers on community development, encouraging active participation within local networks and opening the school to the broader community. The community psychology approach aims to create an empowered school by nurturing the quality of relationships both within the school and the surrounding community, helping to strengthen the social fabric that can be mobilized collaboratively to address emerging needs.

The last professional experience I cherish is my participation in the **BuonAbitare project and association**,[16] conceived and coordinated by Raffaello Martini, another professional of reference for my training as a community psychologist. With the working group of the association, I feel I share a vision of relationships in which at the center is care for the bond that unites the individual to the community. In this sense, neighborhood communities are an excellent response to increasingly emerging needs for sociality and bonding, especially in the often "depersonalizing" contexts of large urban centers, but not only. As professionals involved in the project, we work for the promotion of supportive neighborhoods through the tool of BuonAbitare circles, in which people who live in the same apartment building or block choose to establish themselves as a group that, voluntarily, takes care of the physical and relational space of their neighborhood context. We thus intend to promote a virtuous circularity among people who live in the same context by trying together, through formal and informal initiatives, to make it a community in which they feel they are not alone and thus can count on a network of belonging. We are convinced that these kinds of generative relational processes should be accompanied by the specific professional skills that the community psychologist brings. Once people are accompanied in the pursuit of this intent, it will be possible for them to connect with other nodes in the network of the territory, in which we regain the ability to give, to contribute, to make our resources available through a pact of mutual responsibility and common belonging. With the BuonAbitare association, we want to promote the psychological well-being of the individual, which we understand to be linked in a double strand to the ecosystem within which he or she is born and develops, at each stage of his or her journey, including the more mature age, which is often exposed instead to the risks of loneliness

16. https://buonabitare.com/.

and marginality. Once again, it also means acting in terms of prevention and continuity, rather than on the emergency.

Beginning in 2006, I pursued these goals at the Di Liegro Foundation, where I contributed to the development of the **Families in a Network project**. The initiative aimed to promote mental health, support family members of individuals with mental distress, and combat stigma. The vision here, as elsewhere, is to imagine that the community and its network play a fundamental role in the quality of life and well-being of those affected by mental distress. The community can fulfill this potential when it is prepared and aware of its role as a resource, rejecting the logic of protection and exclusion. This vision aligns with one of the core principles of Franco Basaglia's philosophy – the constructive element of the reform that still largely awaits realization.

I close by mentioning an experience that, for family reasons, I was able to follow closely. That of a small community in the province of Bergamo, in the Valle di Scalve, which over the past two years has worked hard to open a path that would somehow honor the scenic beauty of the place and its cultural richness. The initiative is owed to a group of volunteers from the local subsection of the Italian Alpine Club (CAI), who from the very beginning thought of a path - born, moreover, from the networking of paths that all pre-existed - that could stitch together not only the different localities that dot this territory but also their different souls, united by a strong sense of belonging (also fueled by the fact that this is an area visibly bordered by the surrounding mountains) but also held back in part by the logic of small, large parochialisms. The aim of the work was not only to inaugurate the new path – which has taken the name Via Decia – but also to promote it through a series of tools and initiatives that can make this path a generative space over time through the exchange between the people who will arrive and those who will welcome them; through training experiences that schools, associations or companies can take advantage of; through the promotion of cultural events.

The interesting fact is that during the months of working on the project, the number of people who, although not part of the CAI, chose to make themselves available grew, each bringing their contribution: experts in local history, institutions, businesses, workers for the placement of signage, translators, photography enthusiasts, mountain bike guides, referents of tourist offices, hotel operators, etc. It was as if everyone had gradually really recognized in that path the red thread that could unite people. And at the same time, the investment made in caring for connections represents the greatest resource a community can have.

I am very grateful to the people I met, the "we" I experienced, family, friends, colleagues, and patients.

17
Community psychologists innovate: in training, the third sector, public administration, and the web

1. Researchers and community social psychologists

1.1. The road to becoming a researcher Galeotti Tommaso (recent graduate; 26 years old). "It is clear that there may be different skills involved, but you have to understand that a community psychologist is ultimately an ... all-rounder." These are the words that Professor Massimo Santinello used in the first lecture of the course on Intervention Models in Community Psychology, and if I am honest, they are words that I made my own.

What does it mean to be an all-rounder, though? Generally, this term has a negative connotation, referring to the person who improvises himself as an expert on everything without knowing anything, but in my opinion, the concept can be reversed. Being an all-rounder means being flexible, attentive, and able to work in different contexts, with multiple people, and with multiple objectives; but, most of all, it means being able to treasure any situation, finding in every encounter something to learn and to keep. I say this because it is my journey that testifies to the importance of such skills.

I enrolled in college in 2015, in the faculty of biomedical engineering in Padua, but after two years of study, I realized that that was not my path. Sure, I was able to pass the exams, and with classmates and fellow students, I was getting along well, but there was something that made me turn my nose up at it, a question that kept running through my head: "Once I graduate what will I do?" Indeed, I had realized that it would not be enough for me to find a job framed in an economic model that every day shows its limitations, I wanted something more, something different. That something came by starting over and beginning a new degree program in Psychology.

Although I had chosen a neuroscience-focused major, I had the opportunity to engage with the discipline as early as my three-year degree, thanks to Professor Alessio Vieno's Community Psychology and Wellness Promotion course, which changed the way I looked at psychology. Three concepts have stuck with me: the ecological metaphor, the focus on prevention, and the research-action paradigm. To someone used to thinking about the scientific process exclusively within a laboratory, first with engineering, then with neuroscience, the idea of working with people in contexts related to reality seemed almost like science fiction. Moreover, being able to improve the well-being of individuals not only by responding to their calls for help and by dealing with their sources of distress but also by trying to intervene in the context to prevent them fulfills the vocation of anyone who would want to be involved in public health. Finally, action research makes it possible to contribute to the scientific process without being confined to academia, rather than sharing one's knowledge with people and communities who can then put it to good use.

These characteristics made me choose the master's degree program in Community Psychology, which is distinguished by the possibility of following a highly experiential *curriculum* with various opportunities to test myself through group work and field experiences. Therefore, early on, I was offered the chance to test my skills in different areas, allowing me to experience multiple activities, almost only online due to the COVID-19 pandemic. During the Psychology of Sexuality course taught by Professor Marta Panzeri, we were allowed to work on a research project by the professor. The study focused on how the pandemic affected the sexuality of long-distance couples. The work plan called for holding focus groups with people from the identified target audience, divided into males and females, with moderators belonging to the participants' gender to decrease embarrassment in the face of sensitive issues. I found myself catapulted into running focus groups without having much experience with them, but already after the first meeting, I realized how constructive this experience could be. By the end of the session, all participants agreed that they had found a welcoming place that had put them at ease, allowing them to experience a moment of community far beyond what they thought possible.

If, for the experience just recounted, the online environment proved to be a positive factor, the same cannot be said for the *mentoring* one. Within my curriculum, there was the possibility of participating in the *Mentor-Up* project, shadowing a boy or girl between the ages of 8 and 12, but right from the start, the project changed due to the difficulty of finding schools able to participate in the initiative given the pandemic conditions. With some difficulties, I was assigned a mentee residing in a cooperative for unaccompanied minors, but halfway through, our relationship had to stop as the boy, having turned 18, moved to another city.

During my internship at Lab-id, I was allowed to design and manage a project entirely from scratch aimed at investigating how the use of Instagram impacts our body image and willingness to undergo cosmetic surgery. Being part of Lab-id also allowed me to organize and take part in an international meeting such as HBSC (*Health Behaviour in School-aged Children*), held in Padua in June 2022. The environment at the meeting has always been one of friendliness, and my position as a newcomer has never been made a burden on me; on the contrary, I have been repeatedly encouraged to put myself out there and be heard, taking part in discussions and meetings.

In July, I graduated. Along with the affection of the people who were with me that day, I am mostly left with the satisfaction of having achieved a goal that only five years earlier, when I was leaving engineering, seemed distant and, at times, unattainable. Since September, I have started an internal internship at the university to research so that I can contribute to the lab's publications and try again to gain access to the doctoral program. I know that the road is not easy, and that the academic environment is often marked by a great deal of internal competition, but I believe that with the right commitment and a little luck, I can achieve my goal.

1.2. Community psychology in the university Terri Mannarini (researcher; 56)

1.2.1. Training in community psychology My encounter with community psychology is related to the casual concatenation of a series of circumstances. Immediately after graduating (in Philosophy), I began to collaborate with the Women's Observatory Study Center of the University of Salento (at that time still the University of Lecce), founded and directed by Bianca Gelli. At the Study Center, we were dealing with gender issues and doing research mainly in the areas of health and politics, but from an interdisciplinary perspective. My first contact with empirical research was mediated by the approach and techniques of sociology.

At the instigation of Bianca Gelli, the first PhD in Community Psychology in Italy had started precisely at that time - the second half of the 1990s - in consortium with the Sapienza University of Rome (with Donata Francescato) and the University of Turin (with Piero Amerio): these were the years when the first community psychology networks in Italy and Europe were being consolidated. Thus, thanks to this unexpected opportunity of the doctorate, my academic path was born, not without difficulties and an initial sense of strangeness towards a disciplinary field that I did not fully understand. I did have to catch up in knowledge and skills but I found that I enjoyed field research, and that community psychology offered its avenue for this. However, by training and by nature, I have always remained eclectic in methods and approaches, and this has helped to create my version and vision of community psychology, which, over time, has become increasingly intertwined with social, cultural, and political psychology.

The doctoral path coincided with my first participation in European conferences and my first international contacts. I remember in this regard a summer school of European community psychology in Salento in 2001, which of community, the first in Italy, promoted by Bianca Gelli all the people then present cannot forget because it saw us together, dismayed watching the images of September 11.

1.2.2. How I became a researcher: strengths and difficulties encountered After a doctoral thesis on gender studies – a result of my work at the Women's Observatory Study Center – my research interests gradually shifted to themes typical of community psychology. For several years, I worked on mentoring programs and then on participation processes, both bottom-up and top-down, which led me to write my first monograph in 2004 and subsequently other works.

Meanwhile, it was becoming increasingly clear to me as a young researcher that the Italian academic and scientific culture was changing and opening to international scope and higher standards of research, gradually pushing me to publish more and more in English and scientific journals accredited in the international reference community. Going along with this changing context required commitment because the academic microenvironment in which I had grown up had a different setting and, in some respects, was not fully equipped for younger people to make the transition, but the direction was set and had to be pursued. Over time, this orientation – taken for granted for those beginning an academic path – has helped to make my work known abroad.

1.2.3. In the United States, as a visiting professor at UMBC An important step in the development of international relations, which propelled me out of Europe, was the choice in 2009 to spend a *visiting* period at the University of Maryland, Baltimore Country (UMBC) in the United States of America. I wanted to experience abroad out of curiosity, a desire to break away from routine and challenge myself, and to add to my *resume* a piece that was missing. My companion in this adventure was my friend and colleague Angela Fedi from the University of Turin, a community psychologist in the soul, with whom I had shared a doctoral path and several research studies on social movements - the most important, the one on the No TAV protest in Val di Susa.[17]

The experience at UMBC was unforgettable more on a personal level than on an academic and scientific one because of the friendships that arose from it (with Ken Maton first, Anne Brodsky later) and that only a few years later turned into scientific collaborations and those that were inextricably welded (with my friend Angela). The encounter with a different cultural and institutional reality was extremely formative on a personal level.

A few years later, in 2012, Anne Brodsky came to Italy for a *visiting* period that made her interact with multiple research groups, and an Italy-United States collaboration based on an affinity of interests and methods was initiated with her and her team. The first study we did together, and from which subsequent work has since budded, focused on analyzing the acculturation experience of certain migrant groups (Albanians and Moroccans in Italy, Latinos in the United States). An in-depth look, made possible using qualitative research methods, subsequently led us to investigate the role of a sense of community and processes of empowerment and resilience in migration pathways.

1.2.4. Relationships with the European network of community psychology Initially, and for a long phase, contacts in Europe were difficult for me to cultivate, because of a generational distance from the founders of ENCP (European Network of Community Psychology-now European Community Psychology Association, ECPA) and an approach to the discipline, my own, that was more research-oriented than intervention-oriented.

In 2016, I was co-opted to the ECPA board as *president-elect* (the figure who succeeds the incumbent president at the end of their term), however, after a little more than a year, I resigned and relinquished the role I was soon to fill. Coordination problems and poor attunement, along with my assessment of

---

17. No TAV is a movement in Italy that opposes the construction of the high-speed train line (TAV) between Turin and Lyon. The movement emerged in the 1990s and has garnered significant support, particularly in the Val di Susa region, where residents express concerns over environmental, social, and economic impacts of the project.

the context's unsuitability, led me to devote my energies elsewhere (as I will discuss in a moment).

The entry into ECPA of a younger lever has allowed me to find at a later stage, affinities and to be able to establish new collaborations, especially with Portuguese and Spanish colleagues, with whom we now have ongoing projects on immigration and acculturation issues. With Maria Fernandes-Jesus, currently a member of the board of ECPA, who moved from ISCTE-IUL in Lisbon to St. John University in York (England), we carried out a qualitative Italo-Portuguese study on the factors that lead people with a right-wing political orientation to adopt attitudes hostile to immigrants. Today, with Francesca Esposito, now ECPA president, who trained at ISPA in Lisbon with Maria João Vargas-Moniz and José Ornelas, and the Seville group of Manuel García Ramírez, Virginia Paloma, and Maria del Rocío Garrido, a cross-cultural study is up and running that aims to profile the population's attitudes to multicultural society.

1.2.5. The founding of the Community Psychology in Global Perspective journal

The founding of the journal Community Psychology in Global Perspective was an important step in terms of international networking. The idea was born out of a combination of a futile stimulus and an ambition: the stimulus came at the first round of the national scientific qualification in 2012, which required, among the titles, the editorship of a scientific journal. I had obtained it, so I did not need that title, but the thought crossed my mind that it would never occur to me that I would be able to direct a journal unless I found it myself.

The ambition was to create a scientific publication space for community psychology that would be truly international, indeed "global," and that would be open to qualitative, situated, research-intervention, and generally "eccentric" research. A space, in some ways, an alternative to the accredited journals in the field, available to our community of psychologists and psychologists. To start it up, I was able to count on the experience and relationship capital of colleagues such as Caterina Arcidiacono, Anne Brodsky, and Christopher Sonn (whom I met at an international conference and immediately perceived as akin, on a cultural and scientific as well as human level), co-founders and *associate editors* of the journal: together we built a board of people active on all five continents (and, among them, many of those I mentioned). Giving solidity and continuity to this project took time (the first issue came out in 2015), and the commitment it still requires shows no sign of abating, but I still think it was a good investment.

1.2.6. Horizon 2020 European projects
The latest, in order of time, piece of internationalization is participation in European Horizon 2020 projects. Horizon 2020 is a funding program of the European Commission to support

and promote research in the European space. The projects last two or three years and preliminarily require the construction of international consortia, i.e., collaborative agreements between research institutions from different countries that cooperate in the implementation of the project.

The first of these projects (RE.CRI.RE, 2015-18) led me to delve into the role of culture and the impact of cultural and symbolic variables on individual psychological processes and collective behaviors. The objective of the project was to map the symbolic universes prevalent in the population and the cultural milieu, meaning to identify fundamental orientations to understand their repercussions on how people relate to their surrounding environment, to institutions, and to those they perceive as similar or different across various areas (health, politics, immigration, sexual orientation, etc.), along with the policy implications. The research process was complex and conducted under tight timelines, but excellent coordination allowed us to achieve the set goals and enabled each partner to contribute collaboratively. The results of the project were presented in numerous scientific articles and three volumes published by Springer.

The second Horizon 2020 project (redirect), brings me back to the field of political psychology and participatory processes together with a group of political scientists. The two-year project aims to study current transformations in representative democracy at the national and supranational levels and the interconnection between different levels of territorial representation to understand whether the center of gravity is shifting from traditional forms of political intermediation to other forms of democratic representation and to propose strategies that can strengthen the democratic system at various levels.

1.2.7. Conclusions As this brief narrative suggests, the path taken to become a researcher in community psychology, with a good scientific output and several international relationships, has been shaped by both circumstances and individual choices. Certainly, for those who want to move toward an academic career, it is important to have a passion for study and research, but it is equally important to be able to read institutional and cultural contexts and to understand that the scientific enterprise is a collective rather than an individual one. It is therefore necessary to know how to stay within a community that is also made up of relationships, to grasp its resources, and not to be too limited by constraints.

1.3. Discovering, becoming, and being community psychologists
Evelyn De Simone (research fellow; 30 years old) In the shade of a pine forest, in the middle days of a scorching July, a student in her early twenties, in her second year of a bachelor's degree program in psychology, is studying the

first chapters of a community psychology textbook.

The scenario of my first encounter with community psychology undoubtedly lends itself to suggestive descriptions, through which I could hint at the profound change of perspective I experienced by trying my hand at the study of this discipline. To paraphrase the phrase attributed to John Lennon.

"Life is what happens to you while you're making other plans" ("Life is what happens to you while you're making other plans"), I could say that community psychology is what came (and recurred many times) in my life while I was busy making other plans. From a coherent academic path - three-year thesis and master's thesis in community psychology on participation, doctoral project on community resilience - personal and professional events have led me along directions that I thought were new, unconnected, sometimes fragmented, and which, instead, in hindsight are fitted into one big map.

1.3.1. From community psychology to communities I am a research fellow at the University of Salento working on a project about renewable energy communities. Additionally, for the past year, I have been the provincial coordinator for Fondazione Sylva, a non-profit organization focused on territorial regeneration through the reforestation of areas affected by *Xylella fastidiosa* and subsequently abandoned. Sustainability, energy transition, and reforestation were themes that seemed very distant from my aspirations when I was a student; what fascinated me about community psychology was its potential to address macro issues from a local perspective, using the strength of groups to promote individual and social well-being.

Meeting and gradually getting to know Professor Terri Mannarini in greater depth then made me fall in love with research and research intervention: I immediately appreciated the possibility of exploring territories, getting in touch with communities and studying them by cross-referencing quantitative data and qualitative data and, above all, the possibility of producing knowledge and at the same time producing change in territories.

The first time I felt like a community psychologist, even though I was only at the beginning of my studies, was while I was collecting data for my three-year thesis: I had to interview long-standing members of social movements, and I was in Messina to talk with the No Bridge Network. I was shy, fledgling, and had no militant background that could serve as my calling card with the participants, who, however, were very helpful and well-disposed toward me; the moment I sat down and started talking to them, asking questions about the time they devoted to the Network and all the potential factors that were hostile to participation, I felt perfectly at ease and realized that perhaps I had found what I "wanted to do when I grew up."

It was with the choice of the topic of the doctoral project that I began to think about the theories and methods of community psychology in a much more versatile and mature way: studying community resilience confronted me with powerful multidimensionality and multidisciplinarity; sometimes, I would stop and think, "All right, this is talked about more in engineering, how can I as a community psychologist add necessary elements of knowledge about this phenomenon? How, for example, can the structural elements of an area community become the subject of research and intervention in community psychology?"

This approach, very often unconscious, has guided my professional practice in recent years. After my Ph.D., for personal reasons, I moved away from research activity and started working in collaboration with different third-sector agencies on completely different activities: social and cultural planning, monitoring and evaluation, social reporting, and coordination of environmental agencies or projects. At the time when I was accepting certain assignments, I rarely thought I was doing it as a community psychologist, but now it is clear to me how, in all these collaborations, I was naturally exploring the different possibilities that my training offered me.

1.3.2. Social and cultural design, monitoring, and evaluation of projects Since 2017, I have worked with several cultural workers and third-sector entities, offering them support in the design and writing of social and cultural projects, as well as in the design and implementation of monitoring and evaluation plans. I have learned to understand the system of European structural funds and direct-disbursement funds, how to read calls, and how to activate strategic partnership networks.

Artistic residencies for the promotion of multiculturalism, dance workshops dedicated to women victims of violence, participatory urban regeneration projects, annual planning for a lug (urban youth laboratory) in a peripheral area of Puglia, visual and emotional education programs for the 6-11 age group: the range of themes and disciplines that characterize the projects I have built or evaluated is really wide, but some transversal aspects have always made me feel in my field. A good project is built with reliable partners, doing networking on the ground that far precedes the announcement of a call for proposals; a good project, moreover, only works if it is considered necessary and useful by the community and the target audience, and it is therefore always more functional to do participatory planning, involving the beneficiaries and the broader community in the primordial stages of defining objectives and designing activities; finally, a good project can be communicated and replicated if it is adequately monitored and evaluated using mixed methods, combining the need to enhance the specificities of the here-and-now, the context and the activities, with the need to obtain sufficiently generalizable data to assess the possibility of transferring the same project to new contexts.

1.3.3.   Social reporting In 2019, I responded to a regional call for people under 35 and founded Eutopia, an association that offers research, planning, and administration services. Eutopia was born mainly because we were studying the Third Sector Code and reasoning about social reporting: until then, it had been mainly the prerogative of for-profit entities (or large foundations) to communicate the social impact of their activities, internal *policies,* and investments; precisely with the introduction of the Third Sector Code, however, it was establishing itself as the main tool also for medium and small nonprofits to communicate with their stakeholders. We wondered, "If a large entity has the internal human resources or the financial resources to turn to large consulting agencies to draft the social report, how can small associations and cooperatives proceed to have a good social report every year?"

For Eutopia, I met with the cooperatives and associations that contacted us to avail themselves of our advice, collected the quantitative data needed for the report, conducted interviews at all levels of the structure to obtain information on the organizational climate, the indirect effects of the entities' activities on the lives of employees, and also met with representatives of the beneficiaries of the services offered by these entities to explore the strengths and weaknesses of the activities and, in the report, include not only the results achieved but also the goals for the following year, trying to hold together all the points of view that emerged.

I am sure that the way I write social reports – the extreme importance accorded not only to social impact but also to the collection of different competing perspectives within an entity or the projects implemented, in short, my whole approach to social reporting – would have been completely different if I had not observed contexts through the lens of community psychology.

1.3.4. Intersection of community psychology and environmental and energy transition projects And almost without realizing it, I approached the issues of energy transition and reforestation. In strategy documents, in directives, we often talk about community sometimes, community is given almost magical qualities, and although I am aware that a single community cannot reverse climate change, I am certain that new lifestyles and new processes of energy use might have a way of asserting themselves more easily in small local communities.

Information about climate change, the need to accelerate the energy transition, and the implementation of environmental protection actions are topics that are often discussed, but for some, they are so vast that they appear complex – a complexity that can almost lead to a sense of disempowerment for the individual. Instead, bringing these issues into local communities means not only studying strategies for local implementation but also triggering a process of information and engagement that is fueled by the relationships among community members. It is precisely in this field that I currently work, and like all ongoing processes, I have many more questions than answers.

The need to implement the use of energy from renewable sources and the difficulty for some citizens to have their plants are issues that the renewable energy community model seeks to address by proposing a solution through cooperation between entities and citizens in an association, a local distribution plant, sharing of useful space for the installation of production facilities; a local solution with impacts on people's daily lives and responding to a global problem. I am very interested in these kinds of experiences because I believe that even individuals who do not demonstrate a pro-environmental attitude can be involved in them and who, perhaps precisely because of the social influence of the community, may choose to join and consume clean energy.

A similar experience is what I carry on in my work with Sylva Foundation. One trillion trees: it was even said in the G20 final declaration, that is the number of trees to be planted by 2030 to slow climate change. But where do we start in the presence of such a large goal? From local stories: Puglia has lost millions of trees in the last decade due to *Xylella fastidiosa*. What has become of all the olive groves?

Productive areas have become productive again through replanting and conversion, but many Salento olive groves, especially after drying, have been abandoned.

Who to involve in reforestation projects? Is it right to run them as a top-down project, or can it also be done differently, with more citizen involvement? Still, in early 2023, I met with local governments, associations, businesses, and citizen committees and tried to imagine with them how the activity of planting new trees could help the area still protect its identity. It turned out that in many small communities, associations wanted to experiment with new environmentally sustainable production styles, agroforestry, community apiaries, etc. It was no longer my project, a project of Sylva, in each community, it became a project of and for the community.

1.3.5. From communities to community psychology It was the communities that made me rediscover myself as a community psychologist; it was during the first meetings to facilitate that I realized how much all my training was supporting me in proceeding in a way that no one had pointed out to me as the most effective for a given type of assignment. I still remember the first few times when I saw how my immersion in a context made projects more fluid, more desired, and more useful; at some point, I stopped thinking in advance about projects to be done in a place where I would simply meet the local actors and say, "Okay, let's go and listen to them," knowing that in the interaction itself, the real work would begin. I fully returned to community psychology, thanks also to a research grant at the University of Salento, a new curve on the path, a curve that is known and, at the same time, surprisingly still unknown.

1.4. Discovering: from community life to community psychology

Davide Boniforti (community social psychologist; 44 years old) To go back to the origin of a professional choice is an extraordinary journey through time, an intricate kaleidoscope of images and memories in which it is often difficult to find the origin of all changes. Therefore, I think several reasons contributed to the strength and clarity of my path, many of them scattered and growing over different seasons. Some came to life after my schooling as an industrial expert, which, for a variety of reasons, did not match my expectations and aspirations but still gave impetus to my desire to devote myself to the community. Among these aspirations was the pleasure of dwelling among the thoughts of various writers and philosophers that my literature teacher allowed me to get to know, gradually and involuntarily bringing me closer to the world of psychology and its wonderful opportunities: the pleasure of listening, the desire to understand the behaviors of others, but above all, the expenditures for collective action.

The passion for the community thus matured in everyday life in a town of 7,500 inhabitants a few kilometers from Milan, with a group of people mostly linked by friendship, with whom they shared a desire to change and to build a world that was increasingly on a human scale. Hence, I was born with the pleasure of taking an interest in those who were going through unpleasant conditions and moments in life. Deciding to be part of a missionary group and the subsequent confrontation with the various associations in the country were perhaps the first important steps in searching for possible forms of collaboration in me, however, clashing with the devitalizing difficulty of connecting points of view and histories, often very distant and filled with deep-rooted and long-standing prejudices. The frustration of not being able to understand each other, balanced by the youthful impetus of wanting to be heard, animated dreams and ideals that motivated me to spend myself in the field. My undergraduate years at the Catholic University in Milan pushed me to travel among the different fields of psychology. Among them, I discovered community psychology.

When I began attending **Professor Elena Marta's** lectures, I gradually became aware that what I was studying recalled and transformed that everyday life that I had spontaneously experienced over the years, through research and studies that restored further meaning to many processes:

network work, activation of people, a sense of community.

However, I graduated in Clinical Psychology as an opportunity became available to pursue a thesis on the transformation of the couple bond in migration processes **in Pakistani and Ghanaian cultures**. This was an important step, but more importantly, a confirmation of the desire to put interventions in my future agenda that were supportive of different cultures and attentive to people's life contexts. I would soon experience this more closely, combining my interest in migration with the passion for communities that had run through me since my youth.

1.4.1.   Meeting: from the periphery to professional awareness It was a chance encounter in the university corridors, a contact with a fellow student of mine who knew and shared with me these lively passions. She told me about a community intervention in a suburb of Milan. She left me the name of the contact person, an e-mail address, and a phone number. Looking for an institution where I could do my internship, I began to send several inquiries. At that moment, when I was about to embark without much conviction on a choice in the clinical field, I was contacted by the company **Metodi f**or an interview. From that day, I embarked on a path that, in a few years, led me to a new awareness.

The first day I arrived in Milan's San Siro neighborhood was not love at first sight. On the contrary, I felt invaded by an overwhelming uneasiness, overwhelmed above all by the complexity of problems that were difficult to understand and organize. Everything seemed excessively large and confusing. When I got off the subway and took the street that led me into the heart of the San Siro residential quadrangle, I was enraptured by its noise. A crossroads of cultures and backgrounds crisscrossed stores and streets, on the one hand, confirming my choice and, on the other, triggering in me the traditional fear of feeling out of one's comfort zone.

The **Neighborhood Contract**, which I began to learn about in those days, was an urban redevelopment project that the City of Milan was tackling in five of the city's suburbs. San Siro was one of them, with its historic courtyards, its vitality, but also the discomfort that plagued the public spaces, occupied by garbage, environmental neglect, and, above all, degradation, visible through the dilapidated conditions of many buildings that told of people's feelings of abandonment and likely difficult cohabitation. The Neighborhood Workshop, a physical space managed by the municipality activated to manage the redevelopment process, was an important and diverse crossroads where residents and various institutional stakeholders and local realities met to discuss the redevelopment work, but more importantly, to share problems and desires for change. It was a former store transformed to accommodate residents, some infuriated by the neighborhood's transformations or problems, others full of vitality and a desire to get involved. When no meetings were scheduled, the Workshop was a gymnasium of spontaneity. Families, the elderly, and young people would come in, even simply to greet the workers who, over time, they had come to know and consider as good interlocutors in their daily lives.

Important stories emerged, full of undertakings but also proposals. Among them, I remember a series of workshops held by the very inhabitants themselves, born out of interesting dialogues that were growing in the **Neighborhood Workshop**. We called it "**The Revival**" to empower their desire to transform a difficult context through simple actions, but it was important, especially for those who had fewer economic or relational opportunities

I thought I would stay in this project only for the canonical six months of postgraduate internship and then return to the world of clinical migration. Instead, I stayed for ten years, and during that time, my professional choice was strengthened. The Neighborhood Workshop years were my first "true love" for community psychology, allowing me to experience the flavor of relationships, to try to bet on the creativity of groups and to be able to bring out the beauty behind the immense and crippling problems. Faces, names, and surnames, stories that I could not help but listen to and walk through. I began to become more and more aware of what it means to bring out resources by listening to the concerns of the inhabitants, but at the same time, also the bewilderment of feeling powerless in the face of very ambitious problems and goals for change. And that is how the concept of empowerment and participatory planning took on life and concreteness.

One of the experiences I remember most vividly was the *Work in Progress* pathway, which, a few years after my internship, we initiated by involving associations, local authorities, and especially residents through the *photovoice tool*. "Living San Siro," the theme that started the photovoice collection, allowed us to gather shots from migrants, the elderly, and the young, focusing both on the bonds and mutual support among people in the neighborhood and on long-standing problems, such as environmental neglect and the abandonment of public spaces. The *photovoice* was an opportunity to broaden the reflection with other residents and interlocutors, especially the foreign women who were increasingly emerging as active protagonists in the neighborhood narrative. We thus gave birth to a periodic table to plan and boost initiatives and reflections around an evolving neighborhood. From the reading of fairy tales to the exchange of objects to the creation of a library to the various initiatives around the theme of caring for the neighborhood and people who carry psychosocial vulnerabilities. I do not hide the fatigue of that professional period. The complexity was often being able to cope with delicate diplomatic balances, institutional roles, political powers, and, above all, expectations, both of principals and residents. Many were times when

I went through that feeling of discouragement in not feeling up to dealing with that complexity I had heard about in my undergraduate years. Perhaps thanks in part to a bit of stubbornness, continuous study, and constant comparison with my patient colleagues, I better understood the meaning of professional action in my story. Meanwhile, the neighborhood in which I was working during those years began to change its skin. The square was redeveloped along with several buildings. The power of "beauty" unleashed a desire to get back into the game, for the practitioners, but especially for the inhabitants, by making the spaces come alive and bringing into play a sense of concreteness and possibility.

In the last phase of my operational mandate, we went through a period of greater collective demotivation, aggravated above all by a persistent media campaign that portrayed the neighborhood solely through its most problematic aspects. So, we decided to create an alternative narrative through the creation of a video, which we dubbed *The team that wins in San Siro,*[18] in which various citizens and organizations in the neighborhood recounted the interventions that had been taking place for years, including through the Neighborhood Workshop. It served as a communicative product for the city, but above all, to reactivate the enthusiasm and motivation of those who had been committed to their area for several years.

That passion for relationships and the beauty brought by people planning together to build something unprecedented helped me to overcome many struggles and to keep walking toward a new choice, that of joining the Italian Society of Community Psychology (SIPCO).

1.4.2. Generating: becoming psychologists of imagination I still remember the day I filled out the form downloaded from the website. In a way, that moment was a symbolic passage of great awareness for me. Choosing to make community psychology a profession thus opened new educational directions that intertwined several related fields, from anthropology to sociology, also turning to other disciplines, such as art, theater, and music. This lively combination gave rise to a creative whirlwind of thought and relationships, increasing more and more interest in exploring people and groups that could contribute to the approaches that had matured in those years. The experience of the Neighborhood Contract pushed me to new design opportunities through public and private calls on the themes of social cohesion and community welfare. From Milan, I thus began to move to other geographies and other contexts. The following years gave rise to numerous training and consulting opportunities in different Italian realities involving

---

18. The video can be found on YouTube: https://www.youtube.com/watch?v=c-gK0ec2gPBA.

organizations, public administrations, third-sector entities, and educational institutions.

And here we are in the present day. I think working as a community psychologist today is about being surprised by the narratives one encounters and setting up places where these narratives can be heard, confronted, and often transformed. One of the themes that has been fascinating me in recent years is the enhancement of what each person can retrieve from their past to bring it into the present of their community, generating spaces of imagination in which one can find answers to the most current concerns, but above all, feelings of hope that can re-motivate groups and communities by designing possible changes.

One of my greatest satisfactions is generating contexts of resonance across generations, backgrounds, and roles by setting up "future labs" through imaginative tools and methods, such as *Community Visioning* or the *Future Lab*. These spaces of creative co-construction, which use playful modes of intervention, are allowing me to savor anew the motivations for which I fell in love with community: the desire to regenerate forms of solidarity, welcome, and understanding.

In this creative motion, there is certainly no shortage of resistance, often held back by the difficulty of glimpsing margins for co-construction, by the cynicism of the news that covers the pages of newspapers, or by the comments that flood the main social networks. Others live in the painful narratives of their communities, which people are not always immediately willing to transform and therefore require to be heard to restore meaning to them.

As Korean philosopher Byung-Chul Han reminds us, a community is formed by listening to one another. A listening that for us psychologists often manifests itself in asking curious and interested questions, in allowing space for silences, in considering each person's "invisible and symbolic" environment as an essential component of life that can find a new configuration to the extent that we are able to set up situations in which we can connect through synergies and new relational geometries.

1.5. Community psychology, activism, and health promotion
Luana Valletta (PhD, contract lecturer; consultant in the Department of Public Health AUSL of Bologna, vice-president of the Emilia-Romagna Order of Psychologists and ENPAP CIG councilor, 3; 37 years old)

1.5.1. From first crushes to a solid relationship: travel, participation, moving, and integration "Don't wait for patients to lie on the couch but go into the contexts and become an agent of change." I was about 20 years old when, while touring Europe through Youth in Action programs, engaging politically in my country (Capua) and university politics, I was thunderstruck by social and community psychology.

Piero Amerio's book *Community Psychology* was an incredible journey for me. I was finally seeing a psychology that was articulated not only in "real" contexts but also in a well-defined historical, cultural, and value dimension, and not least strongly political and striving for change and social justice. The message then of not waiting for patients on the couch, of political positioning and responsibility of psychology, I think, marked much of my later career paths.

Books alone, however, are not enough. Numerous meetings and lecturers have been a unique source of inspiration. Although each meeting would deserve a special mention starting from the very first fall in love, I remember, for example, the passion, sense of humor, and deep openness and intellectual honesty of social psychology professor Roberto Fasanelli. Then I found Professor Claudia Chiarolanza a true example of consistency of how not only to teach community psychology well but how to apply it for real in every action as a teacher. I remember that she was the only one in the entire three-year period who, despite the high numbers, organized a visit to a community setting (a drug rehabilitation center) and asked us to identify and invite stakeholders from our area to class for moments of meeting and discussion about our communities.

During the last year of my bachelor's degree, I then had the opportunity to win a call from the university to attend the international community psychology conference (ECPA). It was a trip that allowed me to connect with so many different cultures, starting from the hostel I was staying in and ending with the *premeeting workshops*, the informal moments at the conference, and the amazing post-conference evenings. For example, I remember aperitifs and evenings spent with very nice professors whose academic and scientific weight I perhaps only discovered the next day, who, with great humility and simplicity, listened to you as an equal and from whom you were nourished with numerous insights and growth. Also, at that conference, I remember meeting another great teacher and role model of mine: Bruna Zani, a very motherly teacher, a career woman, and even with three children. Like when you fall in love and want to know more and more, so I started attending several *winter* and *summer schools* (e.g., on health inequalities, multidimensional analysis, participatory planning, etc.) and a substantial number of national and international conferences both as a learner and by presenting my research contributions or interventions.

I was often left alone but always found a very open and welcoming community of students and faculty. I remember an occasion when Professor Francescato, at a SIPCO conference, joined our table of students: a gesture I greatly appreciated because it broke those distances typical of the Italian academy by conveying all her enthusiasm and passion for this work.

During the last year of my bachelor's degree, I obtained a scholarship for a period of in-depth study and research in Lecce, Italy, supervised by Professor Terri Mannarini, an experience that became like a kind of first Erasmus and confirmed a strong interest in community psychology.

I had then found my professional home, and here began the most difficult and painful choice for me: drop everything and pursue this passion. To do my master's degree in Cesena in school and community psychology? Would it be the right choice? At the time, there was no dedicated master's program in Caserta and my parents were against me studying away from home, especially because of the economic difficulties of the period.

1.5.2. School and community psychology :Master in Cesena and scholarship in England During my master's degree years, while studying and working to support my studies, my passion and commitment to community psychology led me to get to know and appreciate the whole Bologna group, to keep traveling for conferences and training, finding myself organizing when elected to the SIPCO board. It was an important training ground. I still remember when, as one of the active minorities we were studying, together with other young students and researchers, we fought for a small revolution: no more distinctions between junior and senior members, everyone could be voted in and elected, even us young people.

In 2011, I graduated with Professor Zani with a thesis entitled *Pathways to Over-60 Civic Participation in the Cesena Territory* and won a scholarship for a research and internship period in Manchester at the Health Institute for Health and Social Change, under the supervision of Professor Judith Sixsmith. The English period was an invaluable training and life experience. How can we forget the encounter with critical psychology disability psychology but especially the lectures and exchanges with the fantastic Professor Carolyn Kagan and all the ferment of very political community psychology?

1.5.3. I graduated, now what? First steps, leaps, and bounds, double tracks, or sliding doors. When I returned from the English experience, I decided to give back to the community, together with the association Psychologists for the Territory, inside the Good Living Week in Forlì, the research, and reflections on what I had deepened on the topic of civic and political participation of citizens over 60. During those months, as I was finishing my postgraduate internship, I was also looking at different doctoral calls for study abroad, wondering whether to stay or leave and where I wanted to put down roots.

In September 2011, I started working with a fellowship at the public health service of the Emilia-Romagna region on a regional project concerning the health content of an active life, where I stayed for more than five years. Almost in parallel and deciding somewhat at the end, I applied for a PhD at the University of Bologna, entering as a PhD student without a fellowship.

Working in the region and embarking in parallel on the doctoral path reflected my dual interest in research and academia (due in part to my fascination with those conferences, the experiences abroad, and my interest in teaching/training) and the opportunity for more immediate, practical experiences, where from a project idea I could see the concrete realization of actions, interventions, and even small improvements in the contexts in which I worked.

It was not at all easy to cross and cultivate these two paths, which are so different, distant, and very high-performances. The work in the region kept me so busy. I would often come home after 7 p.m., learning and trying my hand at so many things with curiosity and passion but reducing the time devoted to my doctorate. Hoping to facilitate the sustainability of the two paths, I decided to research what I was following in the region, namely community projects under the regional "Gaining Health?" grant.

I make no secret that at several moments, I was tempted to discontinue my doctorate. That professional loneliness, not feeling part of an academic team, and the sacrifices, even in the personal and social spheres, that I was making to keep it all going were still very important and trying. In the end, however, I decided not to give up and to try hard. So, in 2016, I got my doctorate with Professor Elvira Cicognani with a thesis entitled *Building health with communities: analysis and evaluation of the quality of participation in health promotion interventions*. Then I remember the words of a medical colleague who told me: in Italy, the doctorate is not of much use, but you will see that it is always an extra title that can come in handy when the occasion arises. So, it was when the opportunity arose for me to teach at the University Studies Abroad Consortium (USAC, in Reggio Emilia), where I worked for more than four years, where a doctorate was needed as a minimum requirement to teach. Through the USAC experience, I was getting back what fascinated me about academia, which was teaching and working with young people. Plus, in the classroom, I was having a lot of fun building unconventional forms of interaction and teaching interactive and reflective lessons. The goals of each course also included fostering relationships among students, a sense of community and cooperation, often far removed from American models, and promoting health from themselves and their college community.

Among other teaching experiences, I cannot fail to mention leading my community psychology lab in 2017 for the master's degree in clinical psychology at the University of Bologna. Returning to one's university and on the other side of the chair (although I often stand in front of and among students) was a unique thrill. Teaching then continued with the innovative *service-learning* workshops that I continue today with Professors Cinzia Albanesi and Antonella Guarino.

As a trainer for local health authorities, I was called to address topics related to health promotion, community design, and work, as well as implementing the PRO.SA database (for which I was a reference in the region), organizational well-being, effective communication, and courses for walking leaders. But let's return to my work in the region on health promotion. During those years, as a community psychologist, I often followed and implemented all the projects related to community work, networking, and communication. In public services, many activities are carried out that are poorly known to citizens, so this communicative sensitivity often led me to follow and collaborate on the communication processes of various projects. What I remember most fondly is the creation of the health map, an interactive map where both citizens and healthcare professionals can find available opportunities in the area, perhaps close to home, including walking groups or smoking cessation groups. To this day, the map continues to be updated, and like a little child who has found their path and new parenthood, I hope it can grow, improve, and transform while always considering the communities that will use it.

The other major area grafted onto my doctoral path is community projects and the major strand of promoting motor activity. I have tried to give it my all, in tasks such as facilitating networking and constant comparison between different projects and institutions, structuring and negotiating moments of monitoring and evaluation, getting people to understand the importance of initiating real participatory processes, and innovating the way we do health promotion with groups and communities. Following them, then, under the guise of monitoring, from Rimini to Piacenza, seeing them grow from year to year was a very valuable opportunity to observe different possible designs for working with communities. Of course, communication of the projects and results (video documentary, live streaming, final event, booklet of experiences, and a scientific article in the "Journal of Community Psychology") has also been an important strand of my work. Certainly, innovating designs, unknowingly breaking some balances (I was doing things that other outsiders were charging handsomely for), all within a public service that is by nature slow, complicated, and resistant to change and evaluation, was a unique and unrepeatable training ground and opportunity for growth. I felt that at the level of values and as a professional responsibility, working in prevention was unparalleled, and often, good networking allowed for the sharing of the many resources, opportunities, and possibilities that on their own were in danger of being weaker or dispersed.

After my years in the Region, I started and continued in the field of health promotion at the direction of the public health department of Bologna. I currently follow a very stimulating project to develop participation processes for preadolescents in the Valsamoggia area. I lead smoking cessation groups, and I am on the steering committee of the regional prevention plan for the use of Bologna. Among the experiences that, as a community psychologist, I would like to mention is the creation of a laboratory to initiate paths of reflection and comparison on possible graft, connections, and new developments of the Regional Prevention Plan (2015-19), the Regional Social Health Plan (PSSR 2017-19) and the health houses (resolution of the Regional Council December 5, 2016, no. 2128).

1.5.4. Politics and representation As a community psychologist, political engagement has been an important part of my roots and professional fruits. As far as party politics are concerned, as a young girl I was a militant and candidate in the Greens; as a professional politician, on the other hand, I have been a student representative, founder of the university association *Menti in movimento*, elected to the SIPCO board for two terms, coordinator of AltraPsicologia Emilia-Romagna, founder of BuonAbitare, vice-president of the Order of Psychologists Emilia-Romagna (with delegation to communication), coordinator of the university and training commission and member of the health commission, and CIG ENPAP councilor.

On the other hand, there have been several experiences in local associationism (Giosef Caserta board, SIPS board - Italian Society for Health Promotion, Emilia-Romagna delegation). I would recommend to all young colleagues to gain experience in associationism, both for the *soft skills* that are acquired and, above all, for a whole network of relationships and work education that will come in handy. In all paths in associationism and politics, I have never been interested in power for power's sake, the vanity of certain visibility that some activities entail. I have always preferred and been comfortable in stooping down to work, listening to needs, weaving even in the background, figuring out what could be improved, and even sifting through multiple avenues in parallel.

Often without even realizing it, I have implemented what I learned from community psychology by trying to use power proactively and in a transformative way, especially for the professional community and the citizenry, which, regarding psychological needs today, sees a very unequal and not always appropriate supply. To date, professional politics engages much of my energy, but it is also a regenerative source, especially important when sacrifices begin to pay off, and small examples of change are seen.

For years, so many of us had been complaining about an outdated and unnavigable Order of Italian Psychologists website from cell phones, so as soon as I arrived in the Order, I set to work to finally make it usable, accessible, and more enjoyable, including consulting with colleagues on how to improve some sections. At last, I could take care of a site and social media that represent us and are useful to guide us in present or new services (e.g., implement a system of FAQs, infographic)

1.5.5. Grafts, roots, fruits, and future As I look back, I realize that there have been so many, even seemingly different, work, professional, and life experiences that have led me to where I am today. I am truly grateful to Professor Francescato for the insight (I hope appreciated by the students) to chronicle professional psychology and possible career paths, which I hope will serve as inspiration or food for thought on one's paths. Contributing to this book was an opportunity for reflection, looking at the whole path to connect some dots.

Today, I know that I would risk being turned off and terribly bored in doing one job and in one context. This circumnavigating constantly gives me new stimuli and riches that I can bring to other contexts, pollinating, grafting, or simply bringing new visions. Plus, when one island becomes more difficult (and in the audience, this can happen often), the other islands can help me find my balance of gratification and self-determination. Ultimately, always having the option to leave, to leave those islands when they become barren, or even just when different life choices are made, or new needs emerge, to this day, I consider it a unique freedom.

There are, however, veins, constants that I eventually find in each of my jobs or experiences in which I may have given me all only to turn elsewhere, and they are listening (to needs, desires, concrete proposals), activating inclusive, participatory processes (from a simple training to a project or consultancy on more complex cases), a strong drive for innovation and constant improvement, implementing new services (perhaps networked and codesigned) and communicating them, a strong curiosity and energy in not shirking even complex tasks and challenges with study, passion and incessant relational networking and activations of potentials (including my own). Every context, situation, and challenge for me has not only problems, criticalities, and difficulties but also resources to be sought, activated, or simply highlighted, enhanced, and strengthened.

And for the future? Perhaps I would like to work in teams of innovators, of people charged with energy and drive, and not only in professional politics but also in fields inherent in the generative concreteness of community psychology, and who knows that someday the opportunities that open will not turn toward these or other horizons and possibilities.

1.6. Becoming a community psychologist through community connections
Antonella Guarino (PhD student; 37 years old)

1.6.1. Early college education and encounter with community psychology In 2004, from a small town in the province of Taranto, I decided to migrate and enroll in the three-year degree program in Work and Organizational Psychology at Sapienza University of Rome. Halfway through my second year, I felt that the exclusive theoretical study did not stimulate me, and I could no longer find the meaning of what my studying was or the motivation. So, I decide to apply for an Erasmus scholarship in Lisbon. The experience proved to be very formative, particularly in terms of personal growth and reflection on my role and the contribution I could make as a 21-year-old student. Upon returning from Erasmus, I felt that I needed to find a practical place for these new reflections, which my studies could satisfy very little. The following year, I decided to apply to do community service abroad. This time, the destination was in perfect connection with the previous one from the point of view of both language and sociopolitical dynamics: I was going to Brazil, to the suburbs of Rio de Janeiro, for a year. This experience began with the word "community" already during the pre-departure training. The community was the physical place where the volunteers lived with the project coordinators, a community home, but it was also the neighborhood in which the local association operated. When I left, *comunidade* became a set of relationships, momentary and intense for me who was passing through (even if for a year) and lasting and strong for those who lived and experienced those places daily. I was finally living in a community (or *community*) and learning about its dynamics, conflicts, resources as well as responsibilities (probably much more evident since it was about helping relationships with people in poverty).

Upon returning from Brazil, I resumed my studies to finish them and came face to face with Professor Claudia Chiarolanza's community psychology (which will be one of those "weak" and bridging links for the rest of my education) that allowed me to integrate what I had experienced with what the books reported.

During this *first stage of life* as a psychologist-in-training, I had to overcome and unhinge many prejudices and stereotypes that were part of the imagination of the helping relationship and my future professionalism.

1.6.2. The importance of maintaining a look "beyond" and "in a different way" Having experienced community psychology in practice, I still did not know all the theories and constructs that the literature could bring to support my experiences and with which I could dialogue. After almost eight years of living in Rome, I decided to move to Bologna and enroll in a master's degree in School and Community Psychology. In 2012, Professor Bruna

Zani, during her first lecture, outlined the main features of the discipline and then invited us to consider doing a period of study abroad to obtain a double degree thanks to a partnership with the Universidad del Rosario in Bogota, Colombia. I remember that my eyes lit up, and having already noticed this opportunity from the online degree program presentation, I rushed to the professor to ask about how the experience would unfold. By now, going abroad and finding myself "beyond" what I already knew had become my mantra.

The departure would take place in August 2013, at the end of the first year, but in the meantime, I had the opportunity to dialogue and deepen the themes of Maritza Montero's (2007) critical community psychology and to keep calling to my mind the activities, processes, and relationships I had experienced in Brazil to integrate the practical experience, understand it, and reflect on it in depth. In addition, the themes of participation and citizenship attracted my interest, on the one hand, because I had experienced community service as a form of civic and political participation and an opportunity to contribute and show my commitment to a local and international community, feeling that my experience was part of this cause; on the other hand, because the themes of bottom-up participation and attention to the environment and sustainable lifestyles had been recalled during a presentation by the Transition Towns movement (i.e., "cities in transition," where communities large and small of citizens champion solutions to lessen the impact of pollution and other crises; cf. Zoli *et al*, 2018).

I pursued and successfully concluded my double degree experience thinking that, beyond the expendability of an international degree except in the context in which it was acquired, I had had a unique opportunity: to get to the heart of the dynamics of social inequality in the city of Bogotá, to learn about the social and political processes that determine such inequities, to understand how it is possible to intervene in such contexts taking into account the needs, the resources of each person. Returning to Bologna, I focused on my dissertation, resuming the interests that had aroused my curiosity during classes. My thesis project on the Transition Towns movement in Bologna, particularly in San Lazzaro, was kicking off, leading me to interview people who were collectively working to address a period of transition (already known by then!) through permaculture, care, and respect for nature and the environment. The biggest challenge was to take a project forward by bringing out the motivations and importance of dealing with it through the value lens of community psychology, considering the people, their communities of belonging, and the transformative contribution they can make.

I decide, therefore, to do the internship for six months inside the university, in the Laboratory of Community Psychology in Bologna, and for six months outside, in a CSM (mental health center) in the territory. It is in this year that I get closer to researching the experiences of the territory. I met the company of Arte & Salute, made up of actors (users and non-users of mental health services), technicians, and mental health professionals.

With Professor Zani, the process of evaluating this company began, which later led to extending the work to other companies in the regional project Theater and Mental Health (Francescato, Zani, 2017). In this way, research and my interest in theater found a meeting point and allowed me to be in contact with an "other" understood not only as a person suffering due to a mental health disorder but also as an alternative methodology and approach (theater) to rehabilitation that leverages people's interests, dreams and desires to be expressed in creative forms. It has been a great satisfaction for me to be able to bring my interests into research and understand how interventions can be developed that promote people's well-being. In this case, the shift from a purely individual and personal level (my own) to a collective level (the company) allowed me to understand how much my research (and professional) work could not be separated from a clear positioning towards people and context. The discovery of non-neutrality (thankfully!) before psychosocial facts was a great gift and responsibility for me.

1.6.3. University research and the present In 2015, having finished my postgraduate internship, I decided to embark on a far-from-easy path, that of a PhD. The modalities of research from the perspective of community psychology were, and are for me, extremely rooted in context and cannot be separated from a relationship (direct or indirect) with people and what is often called the "outside" with respect to the university, namely the community

Having passed the selections with a combination of wonder and satisfaction, I became part of the **research group on Psychosocial Factors, Forms of Participation, and Citizenship,** which investigates forms and processes of social, civic, and political participation. The research project funded by the Horizon 2020 CatchE-yoU program, Constructing AcTive CitizensHip with European Youth. Doing research within a European project was very challenging, given the different European partners with complementary scientific and disciplinary expertise who were addressing the issue of youth participation from multiple perspectives.

I had the opportunity to learn about different contexts and through a participatory action–research project (Albanesi *et al.*, 2007), to dialogue with Italian and Portuguese secondary school students. In fact, during the doctoral program, I participated in and evaluated an action–research intervention at school, worked with adolescents, and listened directly to their needs and desires. In sum, it was further evidence that the theoretical and methodological approaches of community psychology are essential to relate to and give voice to the other.

Also, during the same years, I had the opportunity to learn about and explore the topic of *service learning* through the Erasmus+ Europe Engage project. The theme of civic engagement aimed at university students and recognized in academia is related to the *service learning* approach, which combines experiential learning with engagement in community organizations. The importance of proposing such an approach in university classrooms and, thus, during the courses and workshops I have had the opportunity to organize lies in the promotion of responsibility and a sense of community among students who will be future professionals but remain citizens. Promoting responsibility and "opening the doors of the academy" to the community falls under the "third mission" but have always been core values of community psychology.

My years of research in academia continue, and I am currently working on youth participation in social justice along with the topic of mental health promotion in the community and at school. I believe these are issues that gather my commitment as a researcher but also as a citizen and that continued research on them can be extremely beneficial to communities. Continuing to combine research with community-based interventions, fostering collaboration, and co-construction of knowledge is at the heart of community psychology, which I will try to pursue in various contexts.

1.7. Intersecting action–research, political engagement, and social change: a story under construction through encounters, contexts, and borders Francesca Esposito (researcher and lecturer, community psychologist; 39 years old)

1.7.1. Introductory note Telling a story from the self is always a complex operation, as it involves pausing to think to develop a "thought about" (Carli, Paniccia, 2003). This ability to develop a meta-thought about lived experience is certainly a critical psychological skill for working with people, groups, and contexts. But sharing one's story with others whether in oral or written form, is also a political act, where that story is read in articulation – or interdependence, a key concept in community psychology – with the historical, social, economic, and political dimensions that frame that story. This reading and contextualization of one's own experience within a collective history are at the root of the feminist principle and method that "the personal is political," or the claim of the importance of understanding the political, social, and economic origins of one's personal experiences to organize and change society (the political value of every day). In community psychology, the theme of sharing stories – personal and professional – as a means of understanding the development of the discipline itself, in its complex heterogeneity, has been growing in popularity, gaining scientific and theoretical legitimacy (Kelly, Song, 2004; Rappaport, 2004; Ornelas, 2022).

1.7.2. London and the university years Born and raised in the port city of Livorno, where, from an early age, I had the opportunity to develop sensitivity and attention to the issue of social injustice and inequality, at the age of 19, in 2001, uncertain about my willingness to continue my studies, I decided to emigrate to London, not without the disappointment of my parents who wished me to continue my studies. In the British capital, at the time ruled by the Labour party (like the rest of the country), I had, probably for the first time in my life, the opportunity to experience firsthand the importance – and transformative power – of feeling part of a community. We were young people from all walks of life and backgrounds, eager to experience collective spaces of autonomy and sharing, spaces where desire and radical imagination led the way in our being together. And London, at the time a laboratory of interesting political, social, and artistic experiences, was the ideal context for our quest for experimentation.

Above all, this experience taught me the importance of creating contexts where people experience a sense of community and determine, from the bottom up, their own needs and desires, as well as the ways of being together with joy. Such an experience, moreover, made me realize that change, even when defined as "utopian," is always possible and that we can begin to build it starting with us and our daily lives. It is precisely this intense period in London that allowed me to create many meaningful relationships that serve as an antechamber to my arrival in Rome back in 2003 to undertake my studies in psychology

In the faculty of Via dei Marsi, located in the vibrant San Lorenzo neighborhood, I met the student movement, which I joined. Those were the years of critical reforms of the education and university system - the Moratti reform (2003-05) first and the Gelmini reform (2008-10) later - which the student collectives decided to denounce and oppose by all means, organizing protests, sit-ins, occupations, creative actions and moments of collective discussion and self-formation. During these years of ferment and intense activities, I had the opportunity, once again, to build community together with other students like me, many of whom became inseparable comrades.

I also had the opportunity to understand the crucial importance of critical thinking, thinking that can read personal experiences in constant articulation with the social, economic, and political matrix that frames them (in other words, I learned the urgency of ecological and contextualist thinking).

I joined the psych collective in which the issue of mental health and anti-psychiatry is explored, and I began my participation in the feminist collective La Mela di Eva, which would last almost until the time of my departure from

Rome in 2012. The years spent with fellow feminists, both in the collective and in the broader Roman feminist movement, were crucial for me. Years that contributed decisively to the formation of the lens through which I view and understand the world.

Throughout this continuous whirlwind of events, encounters, discoveries, experiences, and plots of relationships, I sought to continue my course of study in psychology. However, the experiences of movement made me increasingly critical of the teachings offered by my degree program, as well as of the kind of knowledge they promoted. A knowledge that I often felt was far removed from the very rich and intense experiences I was having in those years with so many comrades and companions in the classrooms of the university, in the streets, and the squares.

It was at this point, almost resigned to the inevitability of a split between political passion and professional construction, that I encountered community psychology. I still remember the feeling of joy, and probably also relief, when I began attending the lectures of Professor Donata Francescato, then head of the Chair of Community Psychology at the Faculty of Psychology I at Sapienza University of Rome. She was among the first, if not the first, to talk to us students not about Freud or the unconscious world but about gender inequalities in politics that needed to be transformed, about models of collaborative work with communities and organizations to produce local and bottom-up development; how to promote affective education in school contexts as a form of intervention to improve well-being and prevent dropout, often borne by the most disadvantaged boys and girls; how to support the creation and autonomous maintenance of self-help groups; and how to expand the social networks of people, especially the most marginalized (Francescato, Tomai, Ghirelli, 2002; Francescato, Putton, 2022). Meeting Francescato and the world of community psychology made me feel that I had found my dimension within psychology. That community with which I could finally reconcile political passion and commitment to social justice with building a professional identity in the field of psychology.

1.7.3. Working with feminist groups to counter gender-based violence When the time came to choose the first internship, I did not have much hesitation. I wanted to work with one of the feminist associations committed to combating gender-based violence and supporting women survivors in Rome.

The internship experience was very rich, and I became so passionate about it that I decided to continue my engagement in this area as an anti-violence worker. The anti-violence centers were and are for me - and for all of us who worked there in various capacities - laboratories of feminist political experimentation. Through individualized interviews, we accompanied the women who came to the centers on long and often painful paths of elaboration and emergence from violence. The function of the supportive interviews carried out by the women workers was - and continues to be - to offer women a space for reflection and narration of their individual histories, read no longer within a blaming or pathologizing framework, as often happens outside feminist spaces and shelters, but articulated within the patriarchal model that organizes family and gender relations in our society. This does not mean that in conversations with women, we only talk about patriarchy, as some sterile critics claim; on the contrary, an ecological vision of intervention is put into practice, articulating individual suffering in broader social, economic, and political dimensions.

Women workers also played the important role of facilitating women's access to natural resources in the community (e.g., social, health, and educational services in the area) and mediating this relationship where traditionally oppressive institutions for women were involved, such as the courts and the police (Federici, 2020). The experience of cohabitation with other women who have also passed through an experience of gender-based violence constitutes the other key element of the feminist politics of anti-violence centers, allowing them to create a sense of community and dynamics of mutual support and solidarity among women

It was precisely the internship and the following work experience as an anti-violence worker that led me, in 2010, to join the assembly of members of the newly formed BeFree - Social Cooperative against Trafficking, Violence and Discrimination in Rome: an experience that was certainly a marking experience for my future personal and professional trajectory.

Working with BeFree allowed me to broaden my view of gender-based violence intervention from the perspective of the contexts of intervention, the groups to be addressed, and finally, the strategies to be adopted. Indeed, the cooperative was operating simultaneously at several ecological levels. At the level of prevention, both primary and secondary, key contexts of intervention were schools (junior and senior secondary). Here, we carried out projects, often funded by Solidea – Institution of Women's Gender and Solidarity, of the Province of Rome, to reflect with students on affective behaviors, management of emotions in and out of the classroom, gender patterns, and relationships in society; causes and mechanisms of gender-based violence. In parallel with this significant area of work was the focus on training and raising awareness, both in the broader society – through ongoing participation in public events and media engagement – and among specialized professionals who, in various roles, interact with women experiencing gender-based violence, including doctors, nurses, social workers, educators, law enforcement, and journalists.

At the level, finally, of the intervention with women, BeFree used to run various anti-violence services and counters (today, these have also been joined by various refuge houses-anti-violence centers in the city of Rome and outside). It was within one of these services that the self-help experience *Le Fenici che volano verso Itaca* was born. I started the group under the supervision of Professor Tomai and then personally followed its course for several years, first as a support figure at the start of the group itself and then as an external supervisor. This experience and its analysis in terms of its impact on the lives of the participants were also the subject of my thesis work at the school of specialization in Health Psychology in Rome, and to this day, I continue to do training on self-mutual help in the BeFree cooperative's training courses for female workers.

A specific area of BeFree's engagement, which is always within the ecological and multilevel work that the cooperative carries out in the field of gender-based violence, is trafficking for exploitation in the sex industry.

1.7.4. Administrative detention It was 2009 when, in addition to running shelters and anti-violence services in various municipalities of Rome, BeFree opened a desk inside the Identification and Expulsion Center (CIE), as it was called at the time of Ponte Galeria, the largest administrative detention center in Italy. It was at this time that, for the first time in my life, I discovered the existence of such institutions, places so oppressive that I never thought they could exist in the 21st century

Going backward in my memories, the memory of the first time that, together with BeFree comrades, I crossed the threshold of this prison for foreign people guilty only of not having the "right" ID still resurfaces vividly. The very architecture of that place, designed to physically contain people and expel them from their home communities, was violent.

The feeling I most associate with the first period inside Ponte Galeria is a sense of shock, unease, and bewilderment. My comrades and I knew that many of the women detainees had experienced gender-based violence, including trafficking and sexual/labor exploitation. Despite this, these women were simply considered "illegal aliens," deserving only to be detained and subsequently deported back to their countries of origin (where, moreover, the violence of which they were victims had often begun and where their lives were therefore at risk). Some among these women, even more paradoxically, had been born and raised in Italy, but because they were not recognized as citizens, they were forcibly taken there, destined to be deported to countries of which they knew little or nothing, sometimes not even the language (this was the case for many women belonging to the Roma and Sinti communities).

The goal of our desk, like the others we ran in the area, was to meet with women prisoners, inform them about their rights and the gender-based violence laws in force in Italy, and provide them with psychosocial and legal support. Above all, we aimed to support these women in building their pathways to liberation from violence, both private and public/institutional. However, this turned out to be a more complex undertaking than we had initially thought.

Although the work we were doing-and which BeFree continued to do until the early 2020s when access to the center was suspended due to the COVID-19 pandemic-was hugely important, as it enabled many women to regain their freedom and return to living in their communities, the limited resources we had placed significant limits on our intervention. Building trust, which is the foundation of any feminist work with women survivors of gender-based violence, takes time and is not an easy process, especially in an environment as hostile and violent as a detention center. In the 3-4 hours per week that we were allowed to enter Ponte Galeria, we were only able to meet with a few women, and working with each of them took several weeks.

It was because of these challenges and the inherent limitations of our intervention in such a context, as well as because of the complex feelings that this experience aroused in me, that I decided to pursue my path with a doctorate in community psychology. My desire, or rather the urgency I felt at the time, was to make visible to all and sundry the inherent violence of such institutions and their futility.

1.7.5. Lisbon, Ph.D. and first steps into the world of research With these motivations and dreams, I left in 2012 for Lisbon, where years earlier, as part of an Erasmus experience, I had met Professor José Ornelas, founder of Portuguese community psychology and a staunch advocate of deinstitutionalization in mental health. Professor Ornelas, who has dedicated his life to fighting against psychiatric hospitals and all forms of institutional detention and segregation, immediately grasped the importance of the topic I was proposing and provided me with his unconditional support for the research project I had in mind.

The years spent in Lisbon and working with Professor Ornelas and his group taught me so much. To this experience, I owe much of who I have become today. During this time, I delved into so many theoretical and practical contributions of community psychology, especially the ecological and contextualist model of research and intervention on administrative detention and the various transformative projects in the mental health system centered on deinstitutionalization and the development of community-based services such as the Community Center that supports people with experience of mental illness in their recovery processes and the pursuit of social inclusion and citizenship through individualized housing, education, vocational training, employment, physical well-being, and mutual-help and advocacy activities.

Another innovative program, Housing First, has as its main goal to offer homeless people, often with substance abuse and mental health issues, direct access to individual housing and flexible, individualized support services. These programs have been validated at the European level (Ornelas, Sacchetto, Esposito, 2014).

In addition to active participation in these intervention programs, the years in Lisbon were crucial for me to take my first steps into the world of research. With the company and support of ISPA's community psychology research group, I worked intensively to conduct my study at the Ponte Galeria (Rome) administrative detention center and the UHSA (Unidade Habitacional de Santo António) administrative detention center, located on Portuguese soil. With my fellow spa students, as well as Professor Ornelas himself and Professor Maria João Vargas-Moniz, I participated in the first scientific conferences and published the first articles of my career (Esposito, Ornelas, Arcidiacono, 2015; Esposito, 2017).

The value of the research I undertook was recognized in 2021 by the SCRA (Society for Community Research and Action, Division 27 of the American Psychological Association) through the award of the Best Doctoral Thesis Award in Community Psychology. In the same year, the European Community Psychology Association (ECPA) also awarded me the first *ECPA Best Doctoral Thesis Award*, justifying this by the fact that my work demonstrated the urgency and need to close these centers and abolish detention.

In March 2019, having recently finished my Ph.D. in community psychology, I arrived in England to begin a postdoctoral fellowship at the University of Oxford's Centre for Criminology under the supervision of Professor Mary Bosworth, director of the Border Criminologies research center, after winning one of the British Academy's international fellowships. The theme of my research, which continues to this day, is "gender and detention," and the goal is to take an intersectional feminist analysis perspective to understand how constructions of gender, race, sexuality, and nationality (to mention a few) intertwine and shape women's experiences in these places.

The years at Oxford were very rich, and despite the COVID-19 pandemic that marked our lives by restricting them so much (especially in the first two years, 2020 and 2021), I had the opportunity to grow and learn a lot. I became more and more involved in the fight for the closure of detention centers and various places of segregation of migrant people, joining solidarity groups active against border violence.

As of September 2021, upon completion of my postdoctoral fellowship, I was hired in London, where I currently serve as a *lecturer* in the School of Social Sciences at the University of Westminster. There, I continue my research and have joined the Convict Criminology group coordinated by Andreas Aresti and Sacha Darke, who for years have been engaged in creating partnerships between universities and the prison system to provide education to incarcerated people who wish to resume or continue, their studies (Aresti, Darke, 2016; Dark, Aresti, Ellis-Rexhi, 2018)

The idea behind this group is to transform the university into a space accessible to those who have gone through experiences of repression and detention and where their experiences can be valued. Doing so also contributes to the goal of "decolonizing the academy," which has now become a daily slogan in British and U.S. universities. Our plan, however, is to expand the group more and more: in my small way, I would like to create an area dedicated to the study of detention and border violence, making sure that people who have this experience can, if they want, join our group and put their knowledge to value to build research projects that are ecological, collaborative, and capable of producing transformative change. In October 2022, I also got the news that I was selected among the five SCRA Fellow Awardees this year, first among Italians and perhaps even among Europeans, to receive the honor of such a mention. I hope that this *fellowship*, dedicated to supporting the outstanding work of young researchers (*early careers*) in community psychology, and the *mentorship* of Professor Brinton Lykes under the program will allow me to consolidate and expand my contributions and succeed in bringing a new perspective to psychology.

A perspective that will help young colleagues understand the current importance and urgency of engaging with the violence of border regimes, simultaneously with engagement with other "crises" of our time, and to develop what Francescato called a "planetary sense of community" to address "systemic problems such as climate change, racism, sexism, and socioeconomic and power disparities" (Francescato, 2020, p. 140).

2.   Community psychologists–entrepreneurs who created new services

2.1.   Progetto Tenda, a social cooperative in Turin Martina Cortese, 33; Elena Laureri, 53; Marco Maccarrone, 37; Vivienne Meli, 38; Sara Peters, 35; Giulia Santagata, 43 years old. Racamier Writes (1997), "I express myself in everyday terms by thinking in analytic terms." Paraphrasing this phrase, it would seem necessary to leave, at times, one's analyst's chair and remember that the private space of the setting is inscribed in the world. And in the world, phenomena that hardly come to tangle with professional practices. At the same time, we know how difficult it is for some people, often for entire social settings, to face psychological or psychotherapeutic settings.

It is not easy for us psychologists who are firmly committed to the third sector to define our role, representing this important hinge function. While we are involved in teams, in complex group dynamics, and real actions on behalf of beneficiaries, we are also called upon to constantly take a psychological gaze and adopt a style of thinking that is analytical, reflective, and capable of embracing complexity. Racamier's statement restores, with a certain intensity, a way of interpreting the work of the psychologist embedded in community and social settings, which is often developed in the seam between thought and action. A "speaking" action capable of restoring relational meaning to the practices of everyday life, healing, and transformation.

This reflection starts from our experience gained first as individuals and then developed as a collective, born internally at the social cooperative Progetto Tenda in Turin. In the areas of intervention in which we operate, in contact with severe vulnerability and marginality, we have frequently observed how, in the face of some individual and collective suffering, a psychotherapeutic intervention that we could define as "classic" would prove to be inadequate or not very usable.

We will try in this contribution to return part of the delicate and intricate weaving that we feel we undertake in our daily work: a dense dialogue and interweaving between individual and collectivity, between psychological gaze and pragmatic reality, between concrete actions and analytical thinking.

2.1.1. The contexts in which we operate Teamwork plays a crucial role within our organization. Groups of different professionals, from a multidisciplinary perspective, confront each other, assume a thinking arrangement, and become bearers of interventions that often have a transitional value between pragmatic reality and psychic reality. We start from here, from the services in which we are called to operate as individuals, which we describe with the gaze of the group:

- Reception services for people seeking asylum and holders of international protection;
- housing area, which addresses the housing needs of different segments of the population, starting with the Housing First methodology;
- Arcobirbaleno is a psycho-pedagogical parenting support center located on the northern outskirts of the city.

To better represent the different levels that intertwine and shape our daily work, we have chosen to associate each service with the narrative of a life story that we have encountered and that has been a source of learning for us.

2.1.2. Housing support for foreign women with international protection status

The Progetto Tenda cooperative has more than 20 years of experience in the field of hosting foreign women and men and has maintained a constant focus on changing migration processes over time. This has also led to the development of skills in accompanying the autonomy and territorial integration of people who are victims of human trafficking and unaccompanied foreign minors. The cooperative's gaze has always remained vigilant to grasp the new needs emerging from the social fabric, making itself available to rethink services based on the needs of users.

The services dedicated to women with whom we interface pertain to the integrated reception system SAI (*sistema accoglienza integrazione*)[19] and the L'Anello forte project[20] and are therefore aimed at women who are asylum seekers or already holders of international protection. The women's paths are managed at the request of the City of Turin, with which we maintain constant collaboration in monitoring the goals and outcomes of individual beneficiaries. Our reception services aim to build pathways to autonomy, which include the educational, labor, health, legal, and finally, the psychological side, which turns out to be fundamental for the fulfillment of the above. The stay of the beneficiaries within the projects is generally six months, during which we conduct cadenced follow-up interviews to assess the achievement of the goals shared in the initial phase of the project. At the end of this initial reception period, it is possible to request an extension from the central service if it has been identified that more time is needed to pursue the planned objectives or to establish new ones considering the needs that have emerged along the way.

To navigate within this dynamic, time-limited context, we opted to form multidisciplinary teams. This choice, on the one hand, enriches the readings of the migration phenomenon and allows the team continuous professional growth, and on the other hand, it is necessary to confront the users about the complexity of their paths.

Accompanying the subject's voluntary activation is the task of the team psychologist; it means restoring dignity and competence to the person and their experience, which would be lost if the practitioner substituted their action. In keeping with the *capability approach*, we believe that the moment the person can position himself independently, he can think of himself as a responsible agent of his change and pursue his own project goals. This allows their path to take on a personal meaning that is neither indicated by practitioners nor forced by the external context.

In addition to supporting the person and the team, the community psychologist maintains a comprehensive look at the processes, considering not only the practical goals of the mandate (Italian language learning, autonomy in document and health management, job search and stabilization) but also the resonances that achieving them has on the person's well-being.

19. https://www.retesai.it/.
20. Anti-trafficking network of Piedmont and Aosta Valley, a unique program of emersion, assistance and social integration in favor of foreigners and trafficked citizens.

The women with whom we interface daily are bearers of stories and journeys of distant worlds that require to be listened to and not neglected in accompanying them along their paths of integration. Migration produces a physical and emotional departure from the country of origin and a landing in a cultural, social, and relational context governed by different mechanisms and rules. It becomes a fundamental task for psychic integrity to "hold together" the dimension of the past with the demands of a present that is not only played out in the here and now but is loaded with expectations about the future: the journey as a chance to support the families left behind, and which sometimes represents a form of redemption from conditions of poverty and an opportunity for emancipation.

According to our experience in the field, the community psychologist supports this integration process by helping each user and the team maintain an overview of the person. To relate to this type of user, it becomes necessary to keep in mind the reasons why women leave and the important ruptures they bring with them. These ruptures take on the value of a loss (ties, social *status*, rhythms of life, language, group of belonging; Achotegui, 2000), which requires a long journey of acceptance and reconstruction of a new reference model.

The socially oriented psychological gaze allows us to look at these ruptures and their reconstruction, considering that psychological processes are always connected to social ones. Reception is, *first* and *foremost,* and the territory becomes an opportunity for users to rebuild their personal and collective well-being, both through relationships with other beneficiaries and by seizing tools and opportunities in the wider social fabric. For this movement to be therapeutic and not a source of discomfort and frustration, targeted work must be done to make the host community willing and competent.

Embracing the systemic perspective, it then becomes important to constantly move between microscopic and macroscopic planes, thinking about the interdependence of all these levels. Acknowledging the influence of what Bronfenbrenner calls the "exo-system" can trigger in the team the need to devote space and confrontation with the neighborhood, employers, children's teachers, and so on. And, as we widen the field more and more, we legitimize ourselves as practitioners to be spokespersons for political demands that can change the reception system itself, highlighting the priority of a culture of reception extended to the whole community, of which we are a part.

The story of Happy

Happy, a young woman from Nigeria, has come to Italy involved in human trafficking and forced into prostitution to pay off a very onerous debt contracted illegally in her home country with the organized crime syndicate that runs trafficking. In telling part of her story, Happy acquaints us with the vicissitudes of pain, violence, and loss of loved ones as early as childhood, a time when she is orphaned by her mother Happy is housed within our cooperative in an autonomous apartment where she shares common spaces with other women of different nationalities. She attends Italian language classes, follows the asylum application process in hopes of obtaining valid documents for her stay in Italy, and explores and cultivates her relationships in the territory.

The observation carried out in the daily household allowed us to encounter a phenomenon that, clinically, we could call "dissociative." In her relationship with caregivers and within the home, Happy expresses a childlike part of her that had distanced herself from the violence and pain of life. During the day at home, she wears baby girl pajamas, which fit her slender, petite body well, with teddy bears and pastel colors, and slippers shaped like stuffed animals. Her pastime is watching cartoons, one after another, and being moved or frightened by the characters' stories. Her favorite cartoons tell fantastic stories set in unreal places and experienced by imaginary creatures, fairies, talking animals, and witches. Happy likes to talk about her current school experiences related to learning Italian. She has all the air of a diligent student at a religious college of the past, seeking compliments and positive reinforcement: the tone of her voice is high-pitched, sweet, and with a sing along quality, like that of children when they are intent on attracting the attention and seeking the pampering of adults, showing them how good, praiseworthy, worthy of their affection.

This has been Happy, or rather the Happy, that the practitioners have been able to get to know so far. With them, she seems to be gradually building bonds of trust. But in other places and at other times of the day, or rather, say night, it is possible to meet another Happy, usually inaccessible, thanks to the strong dissociative system in place. What we have come to know of this second reality comes from the "accidental" encroachment of certain events that violently broke into the seemingly quiet daytime life she wants us to know. One morning, for example, we saw Happy show up with a swollen and bruised eye, a wound on her eyebrow, and other marks, similar to scratches, on her neck. She seemed surprised at our concern. She then told us how she had been robbed a week earlier, and the marks were evidence of that. She had not been robbed but assaulted and beaten. Happy, like other trafficked women, must deal with a complex relational system with people who are involved in exploitation and are more exposed to building difficult and problematic emotional bonds. Other incidents similar to this one have been repeated over time. These events that emerged accidentally, these breaks in the barriers between parallel lives, were accepted and partially used in the relationship with Happy as a possibility to tolerate her dual presence even though she couldn't access a dimension of critical dialogue about what was happening in separate contexts. In this affair, the dissociated parts of the self-represent the only way that Happy has identified to survive the numerous traumatic experiences that, in an ongoing way, have threatened her stability.

The reception system in which she is placed and the team that works with her can somehow support and assume a vicarious function of the self, showing the possibility of containment of incommunicable and parallel aspects. The caregivers, precisely through their relationship with Happy, which is not structured as a mere clinical relationship but appears to be embedded in a proposal for social integration, thus express the ability to resist and pause with her within her existential complexity.

In this example, it emerges how an individual clinical intervention focused on dialogue would perhaps have been extremely difficult for Happy, who, when questioned about her state of well-being or discomfort, used to respond with statements that ruled out any kind of problem. On the other hand, having the opportunity to bring to the team a reflection on Happy's intrapsychic functioning allowed both the recognition of her pain and restlessness and the preparation so that the network could be supportive to her in her moments of greatest fragility, giving meaning to the relapses and fragilities of her project hold, preventing them from being translated into terms of failure or exclusion.

In sum, a plural and integrated observational approach and the possibility of having a framework for social integration in a space of autonomy enabled Happy to cope with suffering and recognize his resources.

2.1.3. The housing area and methodology development Housing First The cooperative's housing area was established as a response to severe social marginality, putting housing needs first. All projects follow and take inspiration from the Housing First methodology (literally "housing first"). This model of intervention is seen as a pivotal action in the placement of homeless people in independent apartments in the fabric of the city, the first and essential step in the process of reintegration into the community, flanking and combining with interventions of accompaniment and support to the person, in a systemic perspective.

The Housing First approach has its roots in the United States, where Sam Tsemberis, considered its founder, initiated the *Pathways to Housing First* (pdf) program in 1992 in New York City, aimed at providing immediate access in independent apartments to chronically homeless people with mental health problems, supported on an ongoing basis by a team of social and health workers (Tsemberis, 2010). The goal was to shape a *housing* program that was affordable to people who were unable to meet the requirements placed on them by existing services. In the 1990s, it became evident that the tiered system (which offers various interventions done in different facilities that offer support, namely food, showers, bed for a night, etc.) had highly critical elements, especially for those who suffered from mental disorders and had substance addiction problems.

The type of service we offer today still differs substantially from the stepped approach in that it provides rapid access to a home in a community setting without requiring people to stop drinking to access the service, abstain from drug use, or join a mental health program in exchange for the home. At the same time, a flexible social-health accompaniment service is offered, tailored to the person's needs, for as long as needed. People are offered to be accompanied on a path to recover their well-being through immediate entry into an apartment that they will be able to live in freely, subject to weekly visits from the operators and an economic contribution of about 30 percent of their income (Cortese, Iazzolino, 2014).

An important part of Housing First is the promotion of participation in society on an equal basis with other people. The services offered aim to make sure that the person knows and can freely use the resources of the area to encourage a sense of belonging to the community. The Turin experience fits into this framework among the various projects initiated by members of the Housing First Italia network. The first Housing First service in Turin was born in 2014 from a collaboration between private social and public entities, with the Adults in Difficulty service of the city of Turin. Over the years, the service has grown and consolidated. Today, there are two Housing First projects, run by different agencies, with a total of 70 people taken in.

Following the Italian network guidelines, our team's choice has been to give privileged attention to chronically homeless people with mental health and addiction problems and to continue experimentation on different targets. Today, specific projects are active for family units in housing emergencies, young adults (*care leavers*), with specific expertise on unaccompanied foreign minors, people coming out of prison, and alternative measures to detention.

The story of Gian Maria

When we met Gian Maria, who was sleeping in an abandoned building after losing his job and his house, he told us that he was not interested in a house, he wanted a job: "... and how am I going to pay for a house if I don't have a job?" We returned a few days later with the keys to a mini apartment, brought nothing else, signed no covenants, and asked for no bail. He agreed, but for more than a year, it was war!

He was plagued by frightening dreams and fits of fear. He felt betrayed. Perennially on the alert, ready to signal his army to attack and avenge all those oppressed by power. His only thought was to retaliate for the wrong he had suffered, to unveil the great hoax we all undergo, the great hoax of a comfortable and submissive life. For Gian Maria, the equivalent of a cowardly act of submission to the enemy was re-doing his ID card, paying the late fee for registry registration, updating his health card, and filing for citizenship income. He just needed a job; then everything would work out. "They took it away from me. But the job is mine to give!" But Gian Maria also knew that he was in his late 60s, that he had two golden hands that had been stationary for a long time because his job no longer existed.

It took more than a year for us to be able to sign the educational pact that sealed our official entry into the Housing First project and the relationship between us. When we visited him, he was a torrent of words: childhood memories, past loves, work triumphs, the anger and pride of a stubborn man who falls, struggles, fights, and quits. His narrative was fragmented and unstoppable. It was hard to find margins and punctuation, and the chances to ask questions or comment were still rarer. We stood by, sharing the maelstrom

At the slow pace of patience, things began to change, and even his mood became more relaxed. One day, he showed us his health card; he had gone to renew it on his own because he needed the doctor: his sinusitis gave him no peace. Months later, we found out that he had gone to a mental health service in the city and had already had several interviews with a psychiatrist: "I'm going there because since the doctor sees me, I seem to be better!" Then came the unemployment certificate, the public transportation pass, and even the citizenship income. All practices are carried out independently.

For the past few months, a woman had entered Gian Maria's life. It was not an easy affair, with a woman with an equally emblematic and complex personality but also steadfast enough to agree not to date each other when one's malaise threatens to spill over too much into the other's life.

We proposed that he participate in the city's call for work sites. Challenged, he accepted. First in the ranking list! His reaction, unlike ours, was not one of joy. For the first time, Gian Maria showed his fragile and fearful side. From that day, he began to sleep badly and not see his partner because he knew he could not be supportive of her. From that day on, he began to call and ask for help, to express his difficulties and worries, and, while awkward, to accept support and understanding. He was afraid. His whole world could have undergone major changes. And indeed, the change began with further failure. Gian Maria, after struggling with severe panic attacks and unwillingly accepting pharmacological support, signed the contract and began to work. But he felt rusty and out of place. After less than two months, he resigned. It has now been more than three years since our meeting, and a relationship of trust has been established with Gian Maria. We offered him to do minor maintenance work for the cooperative's housing. He has regained a sense of ability and self-esteem. As he entered the homes, he also began to learn about other stories, proving to be a good observer and a good wingman in some educational interventions. Suddenly, something surprising happened: an old employer of his called him; he was expanding his workshop and would need his expertise. He was specifically looking for Gian Maria. We happened to witness the phone call and his emotion: "I told you I was expected at work."

It has been several months since Gian Maria returned to his job, and he has a different look on his face. This summer, he took his vacation in his home country, by the sea, reconnecting with his family. Smiling. We rarely see him lately. We eat pizza under the house from time to time. What we have learned is that power can be wielded if given back to the person and has surprising effects. Time for true self-determination cannot be set on paper, and people can be indulged and accompanied by change.

2.1.4. The reality of Arcobirbaleno The Arcobirbaleno family center was established about fifteen years ago through a collaboration between the Progetto Tenda cooperative and the educational services of the city of Turin as a center for children and parents (CBG). The center is in a suburban area of the city, densely populated by children and their families, often characterized by aspects of socioeconomic vulnerability and educational poverty. The idea underpinning the center's work is that *all* families need material, emotional, and psychological support from institutions and the community to grow and, above all, to grow well. For this reason, space and time have been offered for fifteen years to children, girls, and their caregivers within which they can socialize, confront each other, and be supported educationally, psychologically, and socially. In addition to the early childhood educators and other figures who deal in a privileged way with pedagogical and educational issues, several psychologists with different training and professional experience are part of the work team.

The main activities carried out as forms of psychological support are:

- Group pathways to support the perinatal period for migrant women;
- Group meetings for parents who have children with disabilities;
- Meetings dedicated to the exchange and health promotion of children and adults, with the involvement of peers;
- Educational and psychological support for offender fathers.

Other interventions include the reading and mediation work that a community psychologist does, particularly with foreign mothers, so that their *knowing how to do* (and *know how to be*) with their own caregiving and educational modalities can be motivated and mediated with services such as school or pediatrician, promoting mutual understanding and recognition, overcoming "cultural misunderstandings."

What characterizes these tools is that they foster therapeutic action in nontherapeutic contexts where an individual clinical proposal would struggle to take root. The group, at first circumscribed to the participants, is gradually transformed into a community that becomes educating and becomes the container within which some people can begin a process of reworking and changing their modes of functioning. The relationship becomes a multifaceted tool through various forms and declinations: sometimes it is physical closeness, sometimes it is doing together, we can hold each other's child, support each other, learn, and teach in turn.

The Arcobirbaleno center is a hybrid, informal place that can be traced back to the group of so-called *supplementary services* with regards to the offerings of the institutional educational system; families in the area spontaneously access the activities or are referred by the collaboration with public social health services (e.g., social service, child neuropsychiatry, juvenile court). In recent years, the center has also become a reference for the cooperative's in-house shelter services: a protected place in which to meet the most vulnerable families and women in the community.

Sandra's story

Sandra is a woman of about 36, a mother of three children, expecting her fourth, and of Nigerian nationality. At her first meeting with the center's coordinator, she hardly speaks, walks close to the walls, and always keeps her gaze downward; she agrees to participate in our workshops probably because she was "advised" to do so by an institutional figure. Among the activities in which she takes part is the group for pregnant women with children up to one-year-old; the meetings are led by a psychologist and the Nigerian linguistic-cultural mediator. Even within the group, initially, Sandra does not speak, but after a few meetings, she begins to talk about herself and share thoughts, doubts, and questions about childcare in Italy and Nigeria. In this case, the psychologist's role is to bring out the emotions, both individual and collective, that accompany the experience of motherhood in a foreign country and to facilitate the mediation work that each migrant person undertakes in terms of the host context. Some participants in the group are only united by their Nigerian nationality and the experience of motherhood, the rest of the group is very heterogeneous: those who have only arrived in Italy a few months ago and those who have been in Italy for more than ten years; those who have a single child born in Italy and those who live a double experience of parenthood, here and in the country of origin; those who are single mothers and those who have a partner beside them. The multiple characteristics allow each woman to mirror parts of herself in the others, to observe from the outside the same experience or phase of life they have gone through, reframing and understanding them from a distance, recognizing the learnings and changes that may have resulted.

As time passed, Sandra was able to recount how scared she was during her first visit to the center, her fear that someone might "take her children away."

On another occasion, during a one-on-one interview to talk about enrolling her recently-born son in the daycare center, she was able to recount that she had been a victim of trafficking. She had never spoken about it. Sandra was thus accompanied to the cooperative's in-house service for emerging possible victims of trafficking, where she embarked on a journey with other colleagues. She has continued to attend the mothers' group and has become a peer figure, a bridge for other mothers expressing fear of encountering social services and institutions; the ability to tell their experience, after understanding it, becomes a useful tool in the hands of the Other. Therein lies the power of community.

2.1.5. The I.magine collective The birth of our collective draws impetus precisely from the health emergency of the COVID-19 pandemic, in contrast to the feeling of suspension we have experienced in recent years. We had begun to question ourselves, a few months before the state of the pandemic was declared, about the possibility of forming a transversal group of psychologists in a cooperative, a context where we could pour out and compare the experiences gained in the different services where we operated. In taking the first steps to put ourselves more into play and discussion, we chose to be accompanied by an external trainer, to whom our gratitude and appreciation go for his guidance along the road we had started and taken: Marco Iazzolino.

I.magine emerged through the windows of video conferencing software, enabling us to connect despite the distances during a time marked by closures, isolation, and the disbelief of what we were all collectively facing. The desire to envision a better tomorrow, inspired by John Lennon's song, led to the group's name. Here, the "I" becomes central – the individual placing themselves at the heart of a transformative design process. Reimagining becomes our watchword; the spirit that animates us, dynamic and creative, allows us to question the constant changes, ruptures, and transformations that our time places before us, both on a personal level and within work contexts.

2.1.6. Antifragility as a lens through which to observe processes of psychosocial change We found an interesting interpretive key starting with the concept of antifragility. Lebanese philosopher Nassim Nicholas Taleb (2014) introduced the concept of antifragility to reflect on the different strategies adopted to cope with critical events that are unpredictable, have a strong collective impact, and cannot be immediately explained through linear causality (Taleb, 2008). Antifragile is a type of response opposite to fragility, as it takes advantage of everything that pertains to the "extended family of disorder," such as uncertainty, chaos, volatility, and the unknown. The author urges a new consideration of disorder and error, shifting the gaze to complex and interdependent systems and reflecting on the difference between adaptation and change. Central to this thinking has been the shift in focus from the person to the context: it is not the person alone who must become antifragile, but it is the services and the community that must try to be antifragile, creating the conditions for each person to self-determine. For us, "creating the conditions" for change to happen was a good summary of this process, which we tried to apply both internally by contributing to the development of the services in which we were already operating and externally by initiating training and consulting that allowed us to interface with organizations that, like us, were driven by curiosity and a desire for experimentation.

Ours is still an ongoing process, about which it is difficult to conclude. What has been important for us as individuals and as a group is the opportunity to put into practice knowledge and a view of the world. And to see its potential on multiple levels: in working with people and in contexts at serious risk of vulnerability and marginalization, in formal and informal organizations, and our personal and professional lives.

2.2.   Building a community, an "im"-possible enterprise? The story of an academic spin-off Patrizia Meringolo, 75; Moira Chiodini, 51; Cristina Cecchini, 38; Elisa Guida, 38

2.2.1. Our story: Why create a community psychology spin-off? In 2011, a legislative measure[21] was passed that allowed the establishment of innovative start-ups originated by academic institutions. In other words, start-ups - guaranteed by a faculty member belonging to the university - could "commercialize" products originating from research. Initially, this possibility was mainly used by science-technology departments to put results from patents on the market. Realizing, however, that we community psychologists and psychologists also had much to offer the area, we worked to establish our spin-off.

Our *enterprise* (in every sense of the word) had several purposes:

* To spread a community-based approach not only within the scientific community but also among potential local users;

* To preserve and build upon the know-how and experiences gained through various research initiatives designed to address emerging social needs and

---

21. University of Florence Spin-off Regulations, https://www.unifi.it/upload/ sub/ statute_normative/spinoff_regulation.pdf.

- To create employment opportunities for young people engaged in this field and aligned with our way of working.

It was not at all easy, first because it was not immediate to make people understand the meaning of the *community* perspective, and secondly, because the establishment of a spin-off forced us to "reverse" the usual planning of a research project. We had to start with the product we intended to offer, its peculiarities, the needs it was intended to address, and the potential area of demand.

What we knew were the requests from local or third-sector entities and the interventions we had verified as most promising. Positive experiences in European planning certainly helped us to structure our ideas in a useful way. Above all, the expertise in the group was valuable: people with both a university background (essential for the spin-off proposal) and an entrepreneurial *mentoring* background (essential for structuring and developing a credible *business plan*).

Finally, in 2013, LabCom was officially approved. Research and Action for Psychosocial Wellbeing,[22] the first psychology spin-off in Tuscan universities and the first in Italy with a community-based approach. Almost ten years have passed, and our experience has been increasingly refined. Current active members are Moira Chiodini (president), Cristina Cecchini and Camillo Donati (board of directors), Elisa Guidi, Laura Remaschi, Patrizia Meringolo and Carlo Volpi (scientific committee).

2.2.2. What do we do? We specialize in training, consulting, research (especially action–research and participatory action–research), evaluation and impact assessment, and other services aimed at overcoming critical issues and promoting well-being in social, organizational, and educational contexts. We use qualitative (our specialty) and quantitative approaches, *mixed methods*, and creative approaches, valuing the specificities of each. Central to our mission is the investment in research and development to devise new models and tools. Our university background allows us to have a national and international network of expertise to propose innovative interventions based on the latest research. Just as central is the need to combine scientific rigor with the creation of personalized interventions tailored to the specific needs of "clients," who are not just those who buy a product but become co-constructors of participatory activities.

2.2.3. Some of our models and tools The Community Impact (IC) Model (Meringolo, Volpi, Chiodini, 2019) is an original impact evaluation model that aims to understand changes in a complex social system, supporting the

22. https://www.lab-com.it/.

actions that cause them and constructing new and more effective narratives to represent it.

The IC is characterized by evaluation methodologies that can combine different approaches and tools. Through the acquisition of "quali-quantitative" data, it is, therefore, possible to assess the "material" and "immaterial" outcomes of interventions and projects. In addition, the evaluation process contributes to the strengthening of existing networks by facilitating the participation and expression of the views of all stakeholders. The IC makes it possible to identify and measure effects at both the micro (target group) and macro (community) levels.

The model consists of six actions representing the stages of a circular process of evaluation, implementation, and development:

1. Create responsible leadership and support group accountability;

2. Transfer knowledge and create innovation;

3. Turn *bad* data into useful data for evaluation;

4. Building an effective narrative;

5. Create added value for interventions;

6. Promoting partnerships.

Impact assessment is combined with participatory monitoring tools (e.g., consensus conference, world café, open space technology, swot analysis) that enable active involvement of all stakeholders and target groups to increase levels of awareness, *decision-making* capabilities, and the empowerment process.

Another model of our creation is Community Actions for Resilience and Empowerment (CARE; Chiodini, Meringolo, Cecchini, 2020), which is based on a *community-based approach* aimed at promoting resilience in individuals, groups, and communities. It takes the form of a training and intervention program tested in the *Toscana da ragazzi* project. *Resilience and Lifestyles*, which involved more than 1,000 teachers in about 60 Tuscan secondary schools. The framework is the systems approach and action–research and consists of four interrelated phases: research, training, action plan, and evaluation (based on the IC model).

The CARE model focuses on three priorities associated with resilience:

1. Increase *close* relationships and *caring*;

2. Encourage perceived social support;

3. Activate creative problem-solving.

The model seeks to foster elements of resilience such as information sharing and communication, creating a positive and safe environment, and finding new outcomes/meanings to move from resilience to empowerment.

The participatory consensus conference (Cecchini, Donati, 2020) is a tool through which a group, usually consisting of experts in different fields, is guided in reaching an agreement on specific issues. Participants in a consensus conference can be different categories of stakeholders, chosen according to the objectives of the issue of interest. Through a series of structured steps, in which participants can discuss specific issues from the sharing of working materials, the consensus conference promotes involvement and agreement to establish best practices and guidelines on issues characterized by high complexity.

Evocative Cards (Gavrilovici, Dronic, Remaschi, 2020) are an intervention method that integrates a psychosocial approach with narrative and artistic techniques, developed from working in communities, listening to stories, and sharing moments of difficulty and resources for coping. This tool encourages the use of individual and group creative thinking. It can be used in the initial moments of a journey to work on motivations and expectations and to facilitate the expression of emotions. By activating creativity, it enables innovative problem-solving solutions to emerge.

2.2.4. Examples of projects implemented by LabCom We list below some projects implemented at LabCom.

1. *Empowerment.* We mention below two of the interventions on empowerment, both of which invovled women.

- Women and Democracy. Feminist Spaces, Transformative Practices and Political Participation of Women with Disabilities in Palestine: the project, coordinated by Cospe,[23] aimed to support women's rights in Palestine, through the promotion of social and relational changes to foster women's participation in the public sphere; the facilitation of *feminist* spaces to foster networking related to women's rights; and a dialogue between change-makers and the local community. LabCom, in collaboration with Cospe, carried out training activities in the West Bank and Gaza area (conducted online due to the pandemic situation), supporting local researchers in promoting women's empowerment, with a special focus on women with disabilities.

- Women In Transition – with:[24] funded by the Waldensian Table and coordinated by the Society of Reason, aimed to promote pilot experiences aimed at individual and collective empowerment among women in detention. LabCom conducted the impact evaluation, verifying the positivity of the workshop activities carried out in prison. It noted how having undertaken empowerment actions within a prison institution,

23. https://www.cospe.org/.
24. https://www.societadellaragione.it/progetti/wit-women-in-transition/.

385

certainly a difficult and unusual setting, had demonstrated the feasibility of resource-based rather than deficit-based interventions even in a depersonalizing and afflicting institution.

2. *Social inclusion and promotion*

- Tenuta futura:[25] were developed, in partnership with Proteo Fare Sapere Tuscany,[26] service learning paths in secondary schools in Tuscany, centered on the enhancement of the territory and from a legality education perspective, starting with the use of the Suvignano "estate" confiscated from organized crime. Service learning involves both male and female students and teachers to ensure continuity and sustainability of the experience. LabCom conducted the impact evaluation according to the ci model.

- Creativity-Language-Education-Sociality (CLES) and "Promoting Skills in Youth to Overcome Obstacles through Inclusive Pathways" (pop): the projects were promoted by Fondazione Cassa di Risparmio di Firenze and Fondazione Il Cuore si scioglie, under the patronage of the City of Florence. Both were aimed at unaccompanied foreign minors (MSNA) hosted by a residential community in the Florence area. LabCom oversaw the impact evaluation: in CLES, this made it possible to identify and support outcomes such as the strengthening of partnership and responsible leadership and the development of skills by operators to be able to foster the inclusion and autonomy of MSNA. The impact evaluation of the subsequent pop project found outcomes related to the increase in perceived social support and relationship skills in minors, the development of neighborly relationships, and the strengthening of territorial networks.

- Inclusive Enterprises (COOB consortium - Tuscany Region, "Support for Inclusive Enterprises"): a significant group of projects involved in supporting a consortium of type B cooperatives (COOB),[27] which provides - as is the norm - for the employment of disadvantaged individuals. LabCom's work developed over time, initially defining the main dimensions of the social inclusion program to increase participants' awareness, improve decision-making processes, and define the strategic plan. This was followed by a discussion and selection of best practices through the participatory consensus conference. Then, applying the IC model, LabCom evaluated the impact of the projects, looking at strengthening partnerships among stakeholders to co-construct a quality system based on sharing the value of work placements, enhancing the

25. http://www.tenutafutura.it/2020/.

26. http://www.proteotoscana.it/home/

27. https://www.coob.it/.

positive outcomes of interventions, acquiring tools to optimize data collection and analysis; and finally building actions to develop local cooperation policies.

- Probation: is a project coordinated by the Society of Reason in collaboration with the Offices of External Criminal Execution to develop an innovative pathway to replace prison sentences with access to community service. LabCom conducted the impact evaluation of the experience, involving operators and users through qualitative-quantitative tools and creating a replicable *model* in other associations.

3. *The "Youth Projects."* In the years 2022 and 2023, LabCom conducted two surveys in two municipalities in the Florentine area-Pontassieve (*Voce ai giovani, Pontassieve ascolta*) and *San Casciano in Val di Pesa* (*Giù la maschera: fai sentire la tua voce!*) to explore emerging lifestyles and needs in young people following the coronavirus health emergency. The results from the surveys were used for the activation of specific community pathways and actions in response to the identified needs.

4. Participations in European projects. We recall two of them: the first - Prevention of Violent Radicalization and Violent Actions in Intergroup Relations (test) - was of great interest in elaborating the state-of-the-art prevention of violent radicalization (Cecchini, Guidi, Meringolo, 2019; Meringolo, 2020; Meringolo, Cecchini, Donati, 2022). The second, *Youth Rites of Passage in Europe* (YOUROPE), allowed us to explore, including using creative methods, the current meaning of rites of passage to adulthood.

2.2.5. **Strengths and weaknesses** Let us briefly list what are, in our perception, the strengths and weaknesses of the LabCom experience. Strengths:

- Group cohesion and the passion that each of us put into the work;

- The multi-competencies of members and associates, with no distinction between *clinical* community psychology and community *social* psychology, but (proudly) centered on a *community-based* approach;

- The corporate structure, because being a social cooperative (and not, for example, an LLC), becomes a *presentational message* to those requesting a service;

- Overcoming the individualistic approach of academic research traditionally entrusted to a single faculty member, without adequate enhancement of the working group;

- Community psychology is a bridge between the clinical and the political (Amerio, 2000), where politics is understood as a strategy for collectively addressing problems;

- Taking care of our national and international network, seeking to be present in cultural debate and social contexts.

- *Critical issues*:
- Still, the difficulty of getting people to understand who we are and what we do: community psychology, in Italy and especially outside academia, is not sufficiently known;
- The economic crisis, which has undermined the resources of our potential "clients" for interventions in the social, forcing us - and, sometimes, stimulating us - to "invent" *light* projects, albeit of equal scientific and application consistency;
- The social and sociopolitical crisis, so that investments for cohesion and solidarity are not always adequately appreciated (and financed), just when they would be of fundamental importance. Borrowing an expression from Francescato and Zani (2010): we are increasingly in the situation of *more needed, less wanted.*

2.2.6. Suggestions and proposals for the future Creating a spin-off was, and is, an opportunity: we are always convinced that it is worth using. In the future, it would be interesting to create networks of spin-offs that are similar or at least share the same interests. Or even – as we are already experimenting locally – collaborating with those who offer *different products* but can benefit from action–research interventions or participatory impact assessment. A crucial aspect concerns dissemination and communication: making known what we do is important not only to create new networks and new opportunities for comparison but also to give scientific value to experiences to make science and *public engagement* proceed jointly.

2.3. All-inclusive, or making community through music Claudia Lorenzi, (psychologist, musician; 46 years old). I live in Bordighera, in the far west of Liguria. I graduated in Psychology in Turin in 2000 with a thesis on community psychology with Professor Piero Amerio; the main theme was the value of volunteering and the mutual-help group. When I chose my two-year path, I thought that the "clinical and community" direction might be suitable for me because I have always believed in the resources that people bring within themselves and in the strength of shared work. Since I started working, I have developed the ability to network and activate collaborations, albeit limited and small in scope, highlighting the risky resources that my area has.

Tuttinclusi is a project born ten years ago from a collaboration I formed with the music band Borghetto San Nicolò Città di Bordighera. I studied music from a very young age and completed a specialization in music therapy:

since then, I have combined my two acquired skills in psychology and music therapy to design and implement pathways for people with disabilities that are strongly imbued with the culture of participation and self-determination. Frequenting the musical environment in my area has allowed me to get to know musicians and music teachers, as in the case of the bandmaster, Professor Luca Anghinoni, and the head of the band-oriented music school that was born within the band itself, Professor Diandra Di Franco.

At the time, I was following a boy with Fragile X syndrome, I had known him for five years: by then, the individual path had exhausted its potential, and I understood that it was necessary to think about inclusion within a group, being able to rely on a marked musical intelligence and excellent motor coordination. I thought of the band because this reality already has the inclusive characteristics necessary to accommodate the variety and uniqueness of musicians, the most disparate and diverse, so I felt that my patient could also be part of that group.

2.3.1. Principles of project design and implementation Tuttinclusi was born as an innovative project within the band context aimed at structuring a group of pre-adolescent and adolescent students made up of both neurotypical and disabled girls and boys. The underlying idea of this group is that music can be enjoyed and played by anyone who has the skills and motivation to do so and that it can serve as a pretext for group formation, discovering one's motivation and predisposition, confronting social rules, experimenting with one's limits to overcome them when possible, and engaging with authority, diversity in all its forms, and the uniqueness of each individual.

There are three highlights of this project, and they are characterized by an evolutionary process not only of the individual, and therefore of psychological, social, and technical growth, but also of the band group and community that brings in gradually new resources for the inclusion of new elements, and the process of sensitization and activation of resources in the area (the civil community, social and health services):

- The evolution from individual music therapy pathway to music teaching pathway for children and youth with disabilities;
- The evolution from inclusion in the ensemble music group to the band group for all students in the school resulted in the activation of personal, group, and social resources that allow for change and mutual adaptation to the new reality;
- The growth of awareness on the part of social and health workers, as well as civil society, of the potential of a reality with strong cultural roots in the area.

Participation in the project places the individual in the position of measuring himself or herself against a small, protected community in which everyone, respecting their abilities, capacities, and uniqueness, contributes to smooth functioning. Activities, both musical and non-musical, are offered from a supportive and non-supportive perspective to stimulate the individual's resources and personal coping strategies of the learners and, in parallel, the adaptive and reorganization skills of the group and each section into which the training is divided on the musical level.

2.3.2. Musical and group goals and objectives Tuttinclusi has goals in the psychological and musical spheres, pursued by the whole group: we are convinced that only by paying attention to group dynamics can we increase the group's sense of belonging, the management of any conflicts, and the growth not only of knowledge but also of everyone's well-being.

The aims from the Intra and interpsychic point of view of the project are:

- Promote suitable activities that favor contexts that encourage the activation of individual and group resources;
- Promote positive interdependence;
- Promote individual and group responsibility;
- Promoting mental well-being;
- To stimulate and increase critical thinking;
- To increase the use of an artistic medium for self-expression. The purposes from the musical point of view are:
- Promote musical practice within a peer group;
- Know and recognize instruments and musical roles within the group;
- Learning basic music practice techniques of specific instruments.

The inclusion of disabled children, mild, moderate, and severe, in the school's ensemble music group, is carefully evaluated by the team formed by me as a music therapist psychologist and the two referents of the ensemble music group, Professors Di Franco and Anghinoni. In most cases, placement is proposed to the family after a period of music therapy of at least one year. The student is placed when he or she has achieved sufficient interpersonal skills, can stay in the room with a peer group without being frightened by it, and is not bothered by noise.

Participation in the group also presupposes that all students attend individual lessons, which are useful in making a comprehensive assessment of the student from a physical point of view (playing a musical instrument is primarily a physical action, and each instrument requires a different degree of physical effort), psychological (the part each instrument plays indicates the role that instrument plays in the piece, and the musician must be able to support it not only technically, but also psychologically, whether it is a first part of an accompanying part) and social (knowing the characteristics of the members of each section allows for conscious and sustainable insertions to be made both for the new entrant and for the micro group that welcomes him or her).

2.3.3. Collaborations with local stakeholders The growth of the children's skills and our experience has allowed us to look around and create collaborations with public and third-sector entities.

In fact, for the past four years, there has been an active collaboration with the City of Ventimiglia music band, active in the Intemelia area, for the inclusion of two boys. For three years, there has been an active collaboration with social services, which not only contribute financially by providing grants to families to attend courses but have also identified the project as a useful proposal for the development of transversal skills for children and young people with disabilities to be included in their life project.

Similarly, health services from both the referral district and the adjacent district make children aware that Tuttinclusi can be the node of a network that supports families and patients and can support and boost the therapies provided by the public service itself, particularly psychomotricity and speech therapy. From 2015 to 2020, we had a stable collaboration with the High School of Human Sciences, San Remo plexus, which allowed some students to gain additional experience in their schooling by taking an active role in the project and increasing their skills and knowledge in the field of disability psychology. Right from the start, we welcomed trainees from the three-year music therapy school APIM (Italian professional association of music therapists) in Genoa to carry out practical training.

We have, in addition, had the opportunity to speak about our project at several conferences both in Liguria and outside the region: these have been opportunities to compare with other musical, educational, and band realities that have allowed us to rethink some aspects of our project, but also to confirm how innovative, functional and interesting it is.

Before the pandemic, we had the opportunity to do concerts in collaboration with associations active in our area that understood and appreciated the innovative aspect of the project. For us, it was an opportunity to play in places admirable for their beauty, giving prominence and value to the work and the path that the students have done; it was also an opportunity to get out of school and introduce ourselves to the population, contributing to the dissemination of a traditional culture that is in dialogue with innovation.

2.3.4. The role of families in supporting the project Families play a very important role because they are the second driving force. We are convinced that only through the creation of a solid collaboration between school and family and a network among families can the project have continuity. A lot of time and energy is devoted to them presenting the initiatives and

explaining the choices in a word to make them aware of and participate in the life of the school. The school head makes herself available to listen individually to questions, doubts, and concerns, find shared solutions, bring out the resources that each family brings, and enhance them within the group. Before the pandemic, we used to organize shared outings and lunches that had a high community value; attending performances in the area was an opportunity to meet, help each other, and enjoy the show all together.

2.3.5. Critical issues and prospects The project itself has achieved a good balance between attention to social, psychological, and technical-musical aspects. A satisfactory balance has also been achieved within the band group, where the school students are now the majority but have been included with constant work to respect the positions of the veterans, who are not always open to innovations in a traditional context of PAR excellence. Good relationship with social and health services. Society in our area is beginning to learn about this reality and to appreciate the group's musical talents and indispensable social value.

2.3.6. Conclusions The Tuttinclusi project stands as an innovative experience in the band world because it emphasizes community-making to welcome children and young people who have very different levels of functioning, and it is based on an analysis of the strengths and abilities that each person brings to the group, starting from a common interest: music. Community-building, already *in use* in the band experience, is sustained and augmented through attention to the welcoming group even before the most fragile musicians, to allow them to enter a place where they are thought of and recognized in their musical role, which for us becomes a metaphor for the social role each person plays.

This experience stands as a node in a network formed by the voluntary sector, social and health services, and families. We have built relationships of collaboration and trust with health services, which recognize Tuttinclusi as an alternative or supportive pathway to the therapies they offer; with social services, which can propose the pathway within the life project of children and young people; and with families, who find a place that welcomes them even before a place of care and rehabilitation.

Finally, Tuttinclusi also networked with schools: in fact, secondary school children were welcomed in the years before the pandemic and enjoyed field experience, thus increasing their knowledge, know-how, and adding value to their social studies course.

In conclusion, this is an ongoing and transforming project that, starting from the bottom, seeks to create the conditions to increase the sense of sharing and belonging for young and very young people in an area that offers them few resources.

References

Abel E. (2000), Psychosocial Treatment for Battered Women: A Review of Empirical Research, in "Research on Social Work Practice", 10, 1, pp. 55-77.

Achotegui J. (2000), Los duelos de la migración: una perspectiva psicopatológica y psicosocial, in E. Perdiguero, J. M. Comelles (eds.), Medicina y cultura, Editorial Bellaterra, Barcelona, pp. 88-100.

Adams G. et al. (2015), Decolonizing Psychological Science: Introduction to the Special Thematic Section, in "Journal of Social and Political Psychology", 3, 1, pp. 213-38. agosti a. (2006), Gruppo di lavoro e lavoro di gruppo. Aspetti pedagogici e didattici, FrancoAngeli, Milano.

Akhurst J. (2020), A South African Perspective on Community Psychology Practice Competencies, in "Journal of Community Psychology", 48, 6, pp. 2108-23. albanesi c., cicognani e., zani b. (2007), Sense of Community, Civic Engagement and Social Well-Being in Italian Adolescents, in "Journal of Community & Applied Social Psychology", 17, 5, pp. 387-406.

Albee G.W. (1996), Revolutions and Counterrevolutions in Prevention, in "American Psychologist", 51, 11, pp. 1130-3.

Allen G.J. et al. (2017), Community Psychology and the Schools: A Behaviorally Oriented Multilevel Preventive Approach, Routledge, London.

Allen J., mohatt g. v. (2014), Introduction to Ecological Description of a Community Intervention: Building Prevention through Collaborative Field Based Research, in "American Journal of Community Psychology", 54, 1-2, pp. 83-90.

Amerio P. (1996), Scenari sociali e norme morali, in Id. (a cura di), Forme di solidarietà e linguaggi della politica, Bollati Boringhieri, Torino.

id. (a cura di) (2000), Psicologia di comunità, Il Mulino, Bologna.

id. (2003), Quale psicologia per quale comunità, in B. Gelli (a cura di), Comunità, rete e arcipelago, Carocci, Roma (2 a ed.), pp. 59-67.

id. (2004), Problemi umani in comunità di massa. Una psicologia tra clinica e politica, Einaudi, Torino

id. (2009), L'action research tra psicologia sociale e politica, in "Ricerche di Psicologia", 3-4, pp. 23-50.

Amerio P., Croce M. (2000), Le reti sociali, in Amerio (2000).

Ansmann l. et al. (2012), Patients' Perceived Support from Physicians and the Role of Hospital Characteristics, in "International Journal for Quality in Health Care", 24, 5, pp. 501-8.

Antonucci T. C. et al. (2001), Widowhood and Illness: A Comparison of Social Network Characteristics in France, Germany, Japan, and the United States, in "Psychology and Aging", 16, 4, pp. 655-65.

Arcidiacono C. (2017a), Psicologia di comunità per le città. Rigenerazione urbana a Porta Capuana, Liguori, Napoli.

id. (2017b), The Community Psychologist as a Reflective Plumber, in "Global Journal of Community Psychology Practice", 8, 1, https://www.gjcpp.org/pdfs/1Arcidiacono_ FINAL.pdf.

Arcidiacono C., Di Napoli i. (a cura di) (2012), Sono caduta dalle scale... I luoghi e gli attori della violenza di genere, FrancoAngeli, Milano.

id. (2021), Violent Dad in Child Shoes: A Moment Before. vidacs Serious Game in a Multi-Dimensional Action Research Promoting Awareness about Gender-Based Violence Perpetrators, Fedoa-Federico ii University Press, Napoli.

Arcidiacono C., Gelli B., Putton A. (a cura di) (1996), Empowerment sociale. Il futuro della solidarietà : modelli di psicologia di comunità, FrancoAngeli, Milano.

Arcidiacono C., Lavanco G., Novara C. (2021), La prospettiva psicodinamica, Arcidiacono et al. (2021).

Arcidiacono C., Procentese F. (2010), Participatory Research into Community Psychology within a Local Context, in "Global Journal of Community Psychology Practice", 1, 2, https://journals.ku.edu/gjcpp/article/view/20919.

Arcidiacono C., Tuccillo F. (2017), Movie come strumento di condivisione e rappresentazione, in Arcidiacono (2017), pp. 82-4.

Arcidiacono C., Tuozzi T., Procentese F. (2015), Community Profiling in Participatory Action Research, in L. A. Jason, D. S. Glenwick (eds.), Handbook of Methodological Approaches to Community-Based Research: Qualitative, Quantitative, and Mixed Methods, Oxford University Press, Oxford-New York, pp. 355-64. arcidiacono c. et al. (a cura di) (2021), Psicologia di comunità, 2 voll., Franco Angeli, Milano.

Aresi G. et al. (2017), Practicing Community Psychology through Mixed Methods Participatory Research Designs, in "World Futures", 73, 7, pp. 473-90.

Aresti A., Darke S. (2016), Practicing Convict Criminology: Lessons Learned from British Academic Activism, in "Critical Criminology", 24, pp. 533-47.

Arloti M., Barberis M., Kazepov Y. (2008), Indicatori sociali e indici: qualche istruzione per l'uso, in G. Agnetti et al., Territori per la salute mentale. Manuale per la valutazione delle politiche di inclusione sociale, FrancoAngeli, Milano, pp. 124-44. Attneave C. l. (1980), Networking Families in Crisis, in "Contemporary Psychology", 25, 2, pp. 162-78.

Attree P. et al. (2011), The Experience of Community Engagement for Individuals: A Rapid Review of Evidence, in "Health & Social Care in the Community", 19, 3, pp. 250-60.

Atz A. H., Bender e. i. (1976), Self-Help Groups in Western Society: History and Prospects, in "The Journal of Applied Behavioral Science", 12, 3, pp. 265-82.

Aubry T. et al. (2013), Social Role Valorization in Community Mental Health Housing: Does It Contribute to the Community Integration and Life Satisfaction of People with Psychiatric Disabilities?, in "Journal of Community Psychology", 41, 2, pp. 218-35.

Autiero M. et al. (2020), Combatting Intimate Partner Violence: Representations of Social and Healthcare Personnel Working with Gender-Based Violence Interventions, in "International Journal of Environmental Research and Public Health", 17, 15, p. 5543.

Avallone F. (1999), Conoscere le organizzazioni. Strumenti di ricerca e di intervento, Guerini, Milano

Avallone F., Borgogni L. (2007), Convivenza ed efficacia nelle organizzazioni, in "Rassegna di psicologia", Quaderno speciale, 1, pp. 5-7.

Avallone, F., Farnese, M. L.. (2005), Culture organizzative. Modelli e strumenti di ricerca e intervento, Guerini, Milano.

Avissar N. (2016), Psychotherapy, Society, and Politics: From Theory to Practice, Palgrave Macmillan, New York.

Back M. D., Schmukle S. C., Egloff B. (2008), Becoming Friends by Chance, in "Psychological Science", 19, 5, pp. 439-40.

Badolato G., Di iullo M. G. (1979), Gruppi terapeutici e gruppi di formazione, Bulzoni, Roma.

Baggott R., Forster R. (2008), Health Consumer and Patients' Organizations in Europe: Towards a Comparative Analysis, in "Health Expectations", 11, 1, pp. 85-94.

Baggott R., Jones K. L. (2011), Prevention better than Cure? Health Consumer and Patients' Organisations and Public Health, in "Social Science & Medicine", 73, 4, pp. 530-4.

Bagnara S., Misiti R. (1978), Psicologia ambientale, Il Mulino, Bologna.

Ban Y. et al. (2021), The Effect of Fear of Progression on Quality of Life among Breast Cancer Patients: The Mediating Role of Social Support, in "Health and Quality of Life Outcomes", 19, 1, pp. 1-9.

Bandura A. (1969), Principles of Behavior Modification, Holt, Rinehart & Winston, New York.

id. (1978), Social Learning Theory of Aggression, in "Journal of Communication", 28, 3, pp. 12-29.

Barak A., Boniel-Nissim M., Suler J. (2008), Fostering Empowerment in Online Support Groups, in "Computers in Human Behavior", 24, 5, pp. 1867-83.

Barker R. G. (1968), Ecological Psychology: Concepts and Methods for Studying the Environment of Human, Stanford University Press, Stanford (ca).

id. (1978), Habitats, Environments and Human Behavior, Jossey-Bass, San Francisco (ca)

Barker C., Pistrang N. (2005), Quality Criteria under Methodological Pluralism: Implications for Conducting and Evaluating Research, in "American Journal of Community Psychology", 35, 3, pp. 201-12.

Barnes J. A. (1954), Class and Communities in a Norwegian Island Parish, in "Human Relations", 7, 1, pp. 39-58.

Barrera M. (1980), A Method for the Assessment of Social Support Networks in Community Survey Research, in "Connections", 3, 3, pp. 8-13.

Barrera M., Anlay S. L. (1983), The Structure of Social Support: A Conceptual and Empirical Analysis, in "Journal of Community Psychology", 11, 2, pp. 133-43.

Barrera M., Fleming C. F., Khan F. S. (2004), The Role of Emotional Social Support in the Psychological Adjustment of Siblings of Children with Cancer, in "Child: Care, Health and Development", 30, 2, pp. 103-11.

Barrera m., Sandler I. N., Ramsay T. B. (1981), Preliminary Development of a Scale of Social Support: Studies on College Students, in "American Journal of Community Psychology", 9, 4, pp. 435-47.

Bartholomew B., Sama S. (2020), Conscious Coaching: The Art and Science of Building, CreateSpace Independent Publishing Platform (Kindle Edition).

Bauer S. et al. (2003), Use of Text Messaging in the Aftercare of Patients with Bulimia Nervosa, in "European Eating Disorders Review: The Professional Journal of the Eating Disorders Association", 11, 3, pp. 279-90.

Benedetti M., Mebane M. E., Onacea D. (2010), Promozione del dialogo interculturale in un quartiere multietnico attraverso una ricerca intervento sui profili di comunità, in "Psicologia di comunità", 11, 1, pp. 87-97.

Bennett C. C. (1965), Community Psychology: Impressions of the Boston Conference on Education of Psychologists for Community Mental Health, in "American Psychologist", 20, 10, pp. 832-5.

Benson J. K. (1988), Il reticolo interorganizzativo come una economia politica, in S. Zan (a cura di), Logiche di azione organizzativa, Il Mulino, Bologna, pp. 187-216.

berger p. l., luckmann t. (1967), The Social Construction of Reality: A Treatise in the Sociology of Knowledge, Penguin. London.

Berkman L.F., Glass T. (2000), Social Integration, Social Networks, Social Support, and Health, in "Social Epidemiology", 1, 6, pp. 137-73.

Berkman L. F., Syme S. L. (1979), Social Networks, Host Resistance and Mortality: A Nine-Year Follow-Up Study of Alameda County Residents, in "American Journal of Epidemiology", 109, 2, pp. 186-204.

Berkman L. F. et al. (2000), From Social Integration to Health: Durkheim in the New Millennium, in "Social Science & Medicine", 51, 6, pp. 843-57.

Bermudez G. (2019), Community Psychoanalysis: A Contribution to an Emerging Paradigm, in "Psychoanalytic Inquiry", 39, 5, pp. 297-304.

Bernardi L. (2003), Channels of Social Influence on Reproduction, in "Population Research and Policy Review", 22, 5, pp. 427-555.

Biasi B. Bonaiuto P. (1992), Il disegno come strumento per attivare e comunicare emozioni in psicoterapia, in "Attualità in psicologia", 7, pp. 101-20.

Biegel D. E. (1984), Help Seeking and Receiving in Urban Ethnic Neighborhoods: Strategies for Empowerment, in "Preventions and Human Services", 3, 2-3, pp. 119-43.

Biggio G. (2008), La valutazione tra sistemi di gestione e pratiche empiriche, in L. Borgogni (a cura di), Valutazione e motivazione delle risorse umane nelle organizzazioni, FrancoAngeli, Milano.

Bisogni F., Dolcetti F., Pirrotta S. (2021), Emotional Textual Analysis as a Semiotic Action-Research Method to Work with Emotions within Organisations, in "Organisational and Social Dynamics", 21, 1, pp. 152-70.

Block P. (1987), The Empowered Manager: Positive Political Skills at Work, Jossey Bass, San Francisco (ca)-London.

Bobbio N. (1990), L'età dei diritti, Einaudi, Torino.

Bogat G. A., Jason L. (2000), Toward an Integration of Behaviorism and Community Psychology, in Rappaport, Seidman (2000), pp. 101-14.

Bonaiuto M. (2017), La psicologia ambientale in Italia: evoluzione storica e prospettive di sviluppo, in "Giornale italiano di psicologia", 44, 1, pp. 9-50.

id. (2023), Psicologia sociale della transizione energetica, in D. G. Myers et al., Psicologia sociale, McGraw-Hill, Milano (4 a ed.).

Bonaiuto M. et al. (2016), Place Attachment and Natural Hazard Risk: Research Review and Agenda, in "Journal of Environmental Psychology", 48, pp. 33-53.

Bonaiuto P. et al. (1996), Visual Defense or Facilitation Processes Favored by Alarming or Playful Colours, in C. M. Dickinson, I. J. Murray, D. Carden (eds.), John Dalton's Colour Vision Legacy, Taylor & Francis, London, pp. 723-31.

Bonazzi G. (2002), Come studiare le organizzazioni, Il Mulino, Bologna.

Bond M. A., Serrano-García i., Keys c. b. (2017), Community Psychology for the 21 st Century, in M. A. Bond et al. (eds.), apa Handbook of Community Psychology, vol. 1: Theoretical Foundations, Core Concepts, and Emerging Challenges, American Psychological Association, Washington dc, pp. 3-20.

Bonnes M., Secchiaroli G. (1979), Il centro di Milano: spazio e significato nella rappresentazione cognitiva di una grande città, in "Applicazioni psicologiche", 1, pp. 233-55.

id. (1992), Psicologia ambientale. Introduzione alla psicologia sociale dell'ambiente, La Nuova Italia Scientifica, Roma.

Bonnes M. et al. (2006), Le aree naturali protette per la promozione di consapevolezza, sensibilità e impegno ambientali, in L'educazione ambientale nelle aree protette del Lazio, Edizioni arp – Agenzia regionale parchi, Roma, pp. 19-52.

Boog B. W. (2003), The Emancipatory Character of Action Research, Its History and the Present State of the Art, in "Journal of Community & Applied Social Psychology", 13, 6, pp. 426-38.

Borg M. B. jr. (2004), Community Intervention as Clinical Case Study, in "Clinical Case Studies", 3, 3, pp. 250-70.

id. (2010), Community Psychoanalysis: Developing a Model of Psychoanalytically-Informed Community Crisis Intervention, in N. Lange, M. Wagner (eds.), Community Psychology: New Directions, Nova Science Publishers, Hauppauge (ny), pp. 1-66.

Borg M. B. jr., Lynch M. J. (2005), A Collaborative and Consultative Approach toward Understanding and Change in an Immigrant Catholic Parish: A Case Study, in "Consulting Psychology Journal: Practice and Research", 57, 2, pp. 142-52.

Borgogni L. (1994), Valutazione e motivazione delle risorse umane nelle organizzazioni, FrancoAngeli, Milano.

Borgogni L. et al. (2005), ocs – Organizational Checkup System. Come prevenire il burnout e costruire l'engagement, Giunti, Firenze.

Borman L. D. (1975), Exploration in Self-Help and Mutual Aid, Northwestern University, Evanston (il).

Bost K. K. et al. (2002), Structural and Supportive Changes in Couples' Family and Friendship Networks across the Transition to Parenthood, in "Journal of Marriage and Family", 64, 2, pp. 517-31.

Bostock J. (1998), Developing Coherence in Community and Clinical Psychology: The Integration of Idealism and Pragmatism, in "Journal of Community and Applied Social Psychology", 8, 5, pp. 363-71.

Bourg D., Whiteside K. (2010), Vers une démocratie écologique. Le citoyen, le savant et le politique, Seuil, Paris.

Boyd N., Nowell B. (2014), Psychological Sense of Community: A New Construct for the Field of Management, in "Journal of Management Inquiry", 23, 2, pp. 107-22.

Boyd N. et al. (2018), Sense of Community, Sense of Community Responsibility, and Public Service Motivation as Predictors of Employee Well-Being and Engagement in Public Service Organizations, in "The American Review of Public Administration", 48, 2, pp. 428-43.

Braun K. L. et al. (2014), Research on Indigenous Elders: From Positivistic to Decolonizing Methodologies, in "The Gerontologist", 54, 1, pp. 117-26.

Brodsky A. E., O'campo p. j., Aronson r. e. (1999), psoc in Community Context: Multi-Level Correlates of a Measure of Psychological Sense of Community in Low-Income, Urban Neighborhoods, in "Journal of Community Psychology", 27, 6, pp. 659-79.

Brodsky A. E., Scheibler J. E. (2011), Quando l'empowerment non è abbastanza. Un argomento a favore della resilienza multilivello in situazioni caratterizzate de estreme disuguaglianze di potere, in "Psicologia di comunità", 2, pp. 55-64.

Brody C. et al. (2017), Can Self-Help Group Programs Improve Women's Empowerment? A Systematic Review, in "Journal of Development Effectiveness", 9, 1, pp. 15-40.

Bronfenbrenner U. (1986), Ecologia dello sviluppo umano, Il Mulino, Bologna.

Brown L. D., Tang X., Hollman R. L. (2014), The Structure of Social Exchange in Self-Help Support Groups: Development of a Measure, in "American Journal of Community Psychology", 53, 1, pp. 83-95.

Brucker P. S., Mckenry p. c. (2004), Support from Health Care Providers and the Psychological Adjustment of Individuals Experiencing Infertility, in "Journal of Obstetric, Gynecologic, & Neonatal Nursing", 33, 5, pp. 597-603.

Bruner J. (1988), La mente a più dimensioni, Laterza, Roma-Bari. bruscaglioni m. (1982), Il comportamento organizzativo, in M. Bruscaglioni, E. Spaltro (a cura di), La psicologia organizzativa, FrancoAngeli, Milano.

id. (1992), Il self-empowerment come anello di collegamento tra formazione e cambiamento, in "Quaderno proposte risfor", 1, pp. 31-52.

id. (1994a), La società liberata, FrancoAngeli, Milano.

id. (1994b), Orizzonte empowerment, in "risfor", numero speciale.

Id. (2005), Per una formazione vitalizzante. Strumenti professionali, FrancoAngeli, Milano.

id. (2007), Persona empowerment. Poter aprire nuove possibilità nel lavoro e nella vita, FrancoAngeli, Milano.

Bryan K. S., Klein D. K., Elias M. J. (2007), Applying Organizational Theories to Action Research in Community Settings: A Case Study in Urban Schools, in "Journal of Community Psychology", 35, 3, pp. 383-98

Brytting T., Tollestad C. (200 2000), Managerial Thinking on Value-Based Management, in "International Journal of Value-Based Management", 13, 1, pp. 55-77.

Bucci F. et al. (2021), Bringing Mental Health back into the Dynamics of Social Coexistence: Emotional Textual Analysis, in M. Borcsa, C. Willig (eds.), Qualitative Research Methods in Mental Health: Innovative and Collaborative Approaches, Springer, Cham, pp. 193-218.

Bucholz E. M. et al. (2014), Effect of Low Perceived Social Support on Health Outcomes in Young Patients with Acute Myocardial Infarction: Results from the Variation in Recovery. Role of Gender on Outcomes of Young ami Patients (virgo) Study, in "Journal of the American Heart Association", 3, 5, doi: 10.1161/JAHA.114.001252.

Bunning K. et al. (2020), Empowering Self-Help Groups for Caregivers of Children with Disabilities in Kilifi, Kenya: Impacts and Their Underlying Mechanisms, in "PloS One", 15, 3, https://doi.org/10.1371/journal.pone.0229851.

Burrell G., Morgan G. (1979), Sociological Paradigms and Organizational Analysis: Elements of the Sociology of Corporate Lite, Heinemann, London-Exeter.

Burti L. et al. (2005), Does Additional Care Provided by a Consumer Self-Help Group Improve Psychiatric Outcome? A Study in an Italian Community-Based Psychiatric Service, in "Community Mental Health Journal", 41, 6, pp. 705-20.

Butera F. (1990), Il castello e la rete. Impresa, organizzazioni e professioni nell'Europa degli anni '90, FrancoAngeli, Milano.

id. (1995), Bachi, crisalidi e farfalle. L'evoluzione dei Parchi scientifici e tecnologici verso reti organizzative autoregolate, FrancoAngeli, Milano.

id. (2008), Organizzazione del lavoro della conoscenza e formazione, in "for. Rivista per la formazione", 76, pp. 7-10.

id. (2021), Affrontare la complessità. Per governare la transizione ecologica, Ambiente,   Milano.

Bucci F. et al. (2021), Bringing Mental Health back into the Dynamics of Social Coexistence: Emotional Textual Analysis, in M. Borcsa, C. Willig (eds.), Qualitative Research Methods in Mental Health: Innovative and Collaborative Approaches, Springer, Cham, pp. 193-218.

Brown S. (1985), Treating the Alcoholic: A Developmental Model of Recovery, Wiley, New York.

Brucker p. s., Mckenry p. c. (2004), Support from Health Care Providers and the Psychological Adjustment of Individuals Experiencing Infertility, in "Journal of Obstetric, Gynecologic, & Neonatal Nursing", 33, 5, pp. 597-603.

Bruner J. (1988), La mente a più dimensioni, Laterza, Roma-Bari.

Bruscaglioni M. (1982), Il comportamento organizzativo, in M. Bruscaglioni, E. Spaltro (a cura di), La psicologia organizzativa, FrancoAngeli, Milano.

Bruscagioni M. (1992), Il self-empowerment come anello di collegamento tra formazione e cambiamento, in "Quaderno proposte risfor", 1, pp. 31-52.

id. (1994a), La società liberata, FrancoAngeli, Milano.

id. (1994b), Orizzonte empowerment, in "risfor", numero speciale.

id. (2005), Per una formazione vitalizzante. Strumenti professionali, FrancoAngeli, Milano.

id. (2007), Persona empowerment. Poter aprire nuove possibilità nel lavoro e nella vita, FrancoAngeli, Milano.

Bryan S. S., klein D. K., Elias M. J. (2007), Applying Organizational Theories to Action Research in Community Settings: A Case Study in Urban Schools, in "Journal of Community Psychology", 35, 3, pp. 383-98.

Brytting T, Trollestad C. (2000), Managerial Thinking on Value-Based Management, in "International Journal of Value-Based Management", 13, 1, pp. 55-77.

Bucci F. et al. (2021), Bringing Mental Health back into the Dynamics of Social Coexistence: Emotional Textual Analysis, in M. Borcsa, C. Willig (eds.), Qualitative Research Methods in Mental Health: Innovative and Collaborative Approaches, Springer, Cham, pp. 193-218.

Bucholz E. M. et al. (2014), Effect of Low Perceived Social Support on Health Outcomes in Young Patients with Acute Myocardial Infarction: Results from the Variation in Recovery. Role of Gender on Outcomes of Young Patients (virgo) Study, in "Journal of the American Heart Association", 3, 5, doi: 10.1161/JAHA.114.001252

Bunning K. et al. (2020), Empowering Self-Help Groups for Caregivers of Children with Disabilities in Kilifi, Kenya: Impacts and Their Underlying Mechanisms, in "PloS One", 15, 3, https://doi.org/10.1371/journal.pone.0229851.

Burrell G., Morgan G. (1979), Sociological Paradigms and Organizational Analysis: Elements of the Sociology of Corporate Lite, Heinemann, London-Exeter.

Burti L. et al. (2005), Does Additional Care Provided by a Consumer Self-Help Group Improve Psychiatric Outcome? A Study in an Italian Community-Based Psychiatric Service, in "Community Mental Health Journal", 41, 6, pp. 705-20.

Butera F. (1990), Il castello e la rete. Impresa, organizzazioni e professioni nell'Europa degli anni '90, FrancoAngeli, Milano.

id. (1995), Bachi, crisalidi e farfalle. L'evoluzione dei Parchi scientifici e tecnologici verso reti organizzative autoregolate, FrancoAngeli, Milano.

id. (2008), Organizzazione del lavoro della conoscenza e formazione, in "for. Rivista per la formazione", 76, pp. 7-10.

id. (2021), Affrontare la complessità. Per governare la transizione ecologica, Ambiente, Milano.

Caia G., Ventimiglia F., Maass A. (2010), Container vs. Dacha: The Psychological Effects of Temporary Housing Characteristics on Earthquake Survivors, in "Journal of Environmental Psychology", 30, 1, pp. 60-6.

Calpbinici p., Terzioglu f., koc g. (2021), The Relationship of Perceived Social Support, Personality Traits and Self-Esteem of the Pregnant Women with the Fear of Childbirth, in "Health Care for Women International", November, pp. 1-15.

Campbell d. t. (1969), Reforms as Experiments, in "American Psychologist", 24, 4, pp. 409-29.

Campbell d. t., Stanley j. c. (1963), Experimental and Quasi-Experimental Design for Research and Teaching, in N. L. Gage (ed.), Handbook of Research on Teaching, Rand McNally, Skokie (il).

id. (1966), Experimental and Quasi-Experimental Designs for Research, Rand McNally, Skokie (il).

Campbell J. C. (2002), Health Consequences of Intimate Partner Violence, in "The Lancet", 359, 9314, pp. 1331-6.

Caplan g. (1964), Principles of Preventive Psychiatry, Basic Books, New York.

Caplan r. d. (1979), Social Support, Person-Environment Fit, and Coping, in L. A. Furman, J. P. Gordus (eds.), Mental Health and the Economy, W. E. Upjohn Institute for Employment Research, Kalamazoo, pp. 89-131.

Caprara G. V., Accursio G. (1994), Psicologia della personalità, Il Mulino, Bologna.

Caputo A. (2013), Cultural Models Shaping Stalking from a Content Analysis of Italian Newspapers, in "ejop – Europe's Journal of Psychology", 9, 3, pp. 443-60.

id. (2015), The Local Culture as a Means to Explore the Processes of Social Coexistence: A Case Study on a Neighborhood in the City of Rome, in "Community Psychology in Global Perspective", 1, 2, pp. 22-39.

Caputo A., Tomai M. (2020), A Systematic Review of Psychodynamic Theories in Community Psychology: Discovering the Unconscious in Community Work, in "Journal of Community Psychology", 48, 6, pp. 2069-85.

Caputo A. et al. (2020), Towards a Community Clinical Psychology? Insights from a Systematic Review of Peer-Reviewed Literature, in "Community Psychology in Global Perspective", 6, 2-1, pp. 128-43.

Carbone A. et al. (2021), Close Family Bonds and Community Distrust: The Complex Emotional Experience of a Young Generation from Southern Italy, in "Journal of Youth Studies", 25, 8, pp. 1052-71.

Cargo M., Mercer S. L. (2008), The Value and Challenges of Participatory Research: Strengthening Its Practice, in "Annual Review of Public Health", 29, pp. 325-50.

Carli R. (2020), The Setting and the Interpretation in Psychoanalysis as a Clinical Practice, in "Rivista di psicologia clinica", 15, 2, pp. 5-27.

Carli R., giovagnoli F. (2011), A Cultural Approach to Clinical Psychology: Psychoanalysis and Analysis of the Demand, in S. Salvatore, T. Zittoun (eds.), Cultural Psychology and Psychoanalysis: Pathways to Synthesis, Info Age Publishing, Charlotte (nc), pp. 117-50.

Carli R., Paniccia R. M. (1981), Psicosociologia delle organizzazioni e delle istituzioni, Il Mulino, Bologna.

id. (2003), Analisi della domanda. Teoria e intervento in psicologia clinica, Il Mulino, Bologna.

Carli R., Paniccia R. M., Giovagnoli F. (2010), L'organizzazione e la dinamica inconscia, in "Rassegna italiana di sociologia", 51, 2, pp. 183-204.

Carli R. et al. (2016), Emotional Textual Analysis, in L. A. Jason, D. S. Glenwick (eds.), Handbook of Methodological Approaches to Community-Based Research: Qualitative, Quantitative, and Mixed Methods, Oxford University Press, Oxford-New York, pp. 111-20.

Carli R. et al. (2016), Emotional Textual Analysis, in L. A. Jason, D. S. Glenwick (eds.), Handbook of Methodological Approaches to Community-Based Research: Qualitative, Quantitative, and Mixed Methods, Oxford University Press, Oxford-New York, pp. 111-20.

Carolissen R. L., Duckett P. S. (2018), Teaching toward Decoloniality in Community Psychology and Allied Disciplines: Editorial Introduction, in "Community Psychology", 62, 3-4, pp. 241-9.

Carrere J. et al. (2020), The Effects of Cohousing Model on People's Health and Wellbeing: A Scoping Review, in "Public Health Reviews", 41, 22, https://doi.org/10.1186/ s40985-020-00138-1.

Casale O., Piva P. (2005), Strumenti per il welfare locale. Lavorare con piacere. Equilibrio tra vita e azienda, Futura, Roma.

Case A. D. (2017), Reflexivity in Counterspaces Fieldwork, in "American Journal of Community Psychology", 60, 3-4, pp. 398-405.

Caselli R. (1993), Le clearing houses, ovvero i Centri di sostegno per i gruppi di self-help, in "Volontariato oggi. Studi, ricerche e collegamento fra le associazioni ed i gruppi", 10, novembre, pp. 2-4.

Cassese S. (2020), Il buongoverno. L'età dei doveri, Mondadori, Milano.

Castelein S. et al. (2008), The Effectiveness of Peer Support Groups in Psychosis: A Randomized Controlled Trial, in "Acta Psychiatrica Scandinavica", 118, 1, pp. 64-72.

Cecchi m. (1993), Gruppi di auto-mutuo-aiuto: caratteristiche, funzioni, obiettivi, in "Il seme e l'albero", 1, pp. 9-15.

Cecchini C., Donati C. (2020), Involving Local Communities: Participatory Meetings with Stakeholders, in P. Meringolo (ed.), Preventing Violent Radicalisation in Europe: Multidisciplinary Perspectives, Springer, Cham, pp. 135-55.

Cecchini C., Guidi E., Meringolo P. (2019), Conoscenze e competenze psicologiche per la prevenzione della radicalizzazione violenta: l'esperienza italiana del Progetto prova, in "Psicologia di comunità", 2, pp. 56-74.

Chakravarty S., Jha A. N. (2012), Health Care and Women's Empowerment: The Role of Self Help Groups, in "Health, Culture, Society", 2, 1, pp. 115-28.

Chandra A. et al. (2016), Drivers of Health as a Shared Value: Mindset, Expectations, Sense of Community, and Civic Engagement, in "Health Affairs", 35, 11, pp. 1959-63.

Chaveepojnkamjorn W. et al. (2009), A Randomized Controlled Trial to Improve the Quality of Life of Type 2 Diabetic Patients Using a Self-Help Group Program, in "The Southeast Asian Journal of Tropical Medicine and Public Health", 40, 1, pp. 169-76.

Cherniss C. (2002), 2001 Division 27 Presidential Address: Emotional Intelligence and the Good Community, in "American Journal of Community Psychology", 30, 1, pp. 1-11.

Chiodini M, Meringolo P., Cecchini C. (2020), Care: A Community-Based Resilience Training Programme, in C. Cefai, R. Spiteri (eds), Resilience in Schools: Research and Practice, Centre for Resilience & Socio-Emotional Health, University of Malta, Msida, pp. 117-41.

Chioneso N. A., Brookins c. c. (2013), Coming to Get and Needing to Keep: Participation within a Membership Association for Black Scholars, in "Journal of Black Psychology", 41, 1, pp. 49-74.

Christens B. D. (2012), Toward Relational Empowerment, in "American Journal of Community Psychology", 50, 1-2, pp. 114-28.

Chung J. E. (2013), Social Interaction in Online Support Groups: Preference for Online Social Interaction over Offline Social Interaction, in "Computers in Human Behavior", 29, 4, pp. 1408-14.

Cicognani E., Albanesi C., Zani B. (2012), Il senso di comunità, in B. Zani (a cura di), Psicologia di comunità. Prospettive, idee, metodi, Carrocci, Roma, pp. 163-93.

Cicognani.E et al. (2008), Social Participation, Sense of Community and Social Well-Being: A Study on American, Italian and Iranian University Students, in "Social Indicators Research", 89, pp. 97-112.

Cicognani E. et al. (2015), Sense of Community and Empowerment among Young People: Understanding Pathways from Civic Participation to Social Well-Being, in "voluntas. International Journal of Voluntary and Nonprofit Organizations", 26, 1, pp. 24-44.

Cicognani E. et al. (2020), Quality of Collaboration within Health Promotion Partnerships: Impact on Sense of Community, Empowerment, and Perceived Projects' Outcomes, in "Journal of Community Psychology", 48, 2, pp. 323-36.

Cipolla C. (1987), Gli indicatori socio-sanitari, in P. P. Donati (a cura di), Manuale di sociologia sanitaria, La Nuova Italia Scientifica, Roma, pp. 183-200.

Cleri M. (1999), Dall'analisi della domanda alla differenziazione della risposta, in E. Giusti, D. Francescato (a cura di), Empowerment e clinica. Integrazione di tecniche per l'autopotenziamento in psicologia clinica di comunità e psicoterapia umanistica integrata, Kappa, Roma.

id. (2021), Vita, abilità, lavoro. Riconoscere, esprimere ed educare la presenza umana nell'esperienza professionale, Armando, Roma.

id. (2022), Il metodo integrato di educazione socio-affettiva, in Francescato, Putton (2022), pp. 189-95.

Clinard M. B. (1970), Slums and Community Development: Experiments in Self-Help, Free Press, New York.

Cobo-rendón R. et al. (2020), Perceived Social Support and Its Effects on Changes in the Affective and Eudaimonic Well-Being of Chilean University Students, in "Frontiers in Psychology", 11, https://doi.org/10.3389/fpsyg.2020.590513.

Cohen A. et al. (2012), Sitting with Others: Mental Health Self-Help Groups in Northern Ghana, in "International Journal of Mental Health Systems", 6, 1, pp. 1-8.

Cohen B. (1990), The Elderly Mistique: Impediment to Advocacy and Empowerment, in "Generations", 14, pp.13-6.

Cohen C., Sokolovsky j. (1978), Schizophrenia and Social Networks: Ex-Patients in the Inner City, in "Schizophrenia Bulletin", 4, 4, pp. 546-60.

Cohen J. (1988), Statistical Power Analyses for the Behavioral Sciences, Lawrence Erlbaum Associates, Hillsdale (nj) (2 nd ed.).

Cohen L., Manion L. (1994), Research Methods in Education, Routledge, London (4 th ed.).

Cohen S., Hoberman H. M. (1983), Positive Events and Social Supports as Buffers of Life Change Stress, in "Journal of Applied Social Psychology", 13, 2, pp. 99-125.

Cohen S., Wills T. A. (1985), Stress, Social Support, and the Buffering Hypothesis, in "Psychological Bulletin", 98, 2, pp. 310-57.

Colletti F. (1991), Formazione manageriale e formazione specialistica: il settore dei servizi, in D. Forti (a cura di), Orizzonte formazione. L'apprendere nelle organizzazioni degli anni '90, FrancoAngeli, Milano.

Collins N. L., Ford m. b., Feeney b. c. (2010), An Attachment-Theory Perspective on Social Support in Close Relationships, in L. M. Horowitz, S. Strack (eds.), Handbook of Interpersonal Psychology: Theory, Research, Assessment, and Therapeutic Interventions, Wiley, Oxford, pp. 209-31.

Commission on social determinants of health (2008), Closing the Gap in a Generation: Health Equity through Action on the Social Determinants of Health, in "Global Public Health", 6, 1, pp. 102-5.

Commoner B. (1971), The Closing Circle: Nature, Man, and Technology, Knopf, New York.

Contessa G., Sberna M. (a cura di) (1981), Per una psicologia di comunità, clued, Milano.

Converse J. jr., Foa U. G. (1993), Some Principles of Equity in Interpersonal Exchanges, in U. G. Foa et al. (eds.), Resource Theory: Explorations and Applications, Academic Press, Cambridge (ma), pp. 31-9.

Cook T. D., Campbell C. T. (1979), Advanced Level Quasi-Experimentation: Design and Analysis Issues for Field Settings, Rand McNally, Skokie (il).

Cooper D. (1972), La morte della famiglia, Einaudi, Torino.

Cornwall A., Brock K. (2005), Beyond Buzzwords: "Poverty Reduction", "Participation" and "Empowerment" in Development Policy, United Nations Research Institute for Social Development, Geneva.

Corrao S. (2000), Il focus group, FrancoAngeli, Milano.

Cortese C., Iazzolino m. (2014), Innovare per includere: le sfide dell'approccio Housing first/Innovation for social inclusion: Challenges of Housing First, Paper for the espanet Conference Sfide alla cittadinanza e trasformazione dei corsi di vita: precarietà, invecchiamento e migrazioni (Università degli Studi di Torino, Torino, 18-20 settembre), in https://www.researchgate.net/publication/281748404_Innovare_per_includere_le_sfide_dell'approccio_Housing_First_in_Italia.

Costa M. E., Duarte c. (2000), Violência familiar, Ambar, Porto.

Costello A. et al. (2009), Managing the Health Effects of Climate Change: Lancet and University College London Institute for Global Health Commission, in "The Lancet", 373, 9676, pp. 1693-733.

Cotinaud O. (1975), Groupe et analyse institutionelle, Éditions du Centurion, Paris. coulombe s., krzesni d. a. (2019), Associations between Sense of Community and Wellbeing: A Comprehensive Variable and Person-Centered Exploration, in "Journal of Community Psychology", 47, 5, pp. 1246-68.

Cowen E. L. (1973), Social and Community Interventions, in "Annual Review of Psychology", 24, pp. 423-72.

Croce M. (1995), Il lavoro di rete tra tecnica e partecipazione, in Id., L'intervento di rete. Concetti e linee d'azione, Edizioni Gruppo Abele, Torino, pp. 3-9.

Croce M., Merlo R. (1989), La rete sociale, in "Animazione sociale", 16, pp. 5-14. id. (1991), Reti che ammalano, reti che curano. Esperienze ed ipotesi di utilizzo clinico della rete sociale, in "Dei delitti e delle pene", 3, pp. 47-68.

Croce M., Oliva F. (1995), Considerazioni sul rapporto tra professionisti e self-help, in "Animazione sociale", 12, pp. 53-7.

Cuneo G. (1992), Riprogettare l'impresa per competere in un mondo senza confini, Il Sole 24 Ore Libri, Milano.

Cunningham B. (1976), Action Research: Toward a Procedural Model, in "Human Relations", 29, 3, pp. 215-38.

Cuthill M., Fien J. (2005), Capacity Building: Facilitating Citizen Participation in Local Governance, in "Australian Journal of Public Administration", 64, 4, pp. 63-80.

Cutrona C. E., Russell d. w. (1987), The Provisions of Social Relationships and Adaptation to Stress, in D. Perlman, W. Jones (eds.), Advanced Personal Relationships, jai Press, Greenwich (ct), pp. 37-67.

Cyril S. et al. (2015), Exploring the Role of Community Engagement in Improving the Health of Disadvantaged Populations: A Systematic Review, in "Global Health Action", 8, 1, doi: 10.3402/gha.v8.29842.

Dallago L. (2006), Che cos'è l'empowerment, Carrocci, Roma.

D'amato A., Majer V. (2005), m-doq10: Majer-D'Amato Organizational Questionnaire 10 manuale, Organizzazioni Speciali, Firenze.

Darke S., Aresti A., Ellis-Rexhi N. (2018), Supporting Prisoners into Academia, in V. Friso, L. Decembrotto (a cura di), Università e carcere. Il diritto allo studio tra vincoli e progettualità, Guerini, Milano, pp. 217-37

Darke S. et al. (2020), Prisoner University Partnerships at Westminster, in S. S. Shecaira, L. G. B. Ferrarini, J. D. M. Almeida (eds.), Criminologia. Estudos em Homenagem ao Alvino Augusto de Sá, D'Plácido, Belo Horizonte, pp. 475-98.

D'eaubonne F. (2022), Feminism or Death: How the Women's Movement Can Save the Planet, Verso Books, London.

De Dominicis S., Schultz P. W., Bonaiuto M. (2017), Protecting the Environment for Self-Interested Reasons: Altruism Is not the Only Pathway to Sustainability, in "Frontiers in Psychology", 8, https://doi.org/10.3389/fpsyg.2017.01065.

De Grada E. (1969), Elementi di psicologia di gruppo, Bulzoni, Roma.

De Piccoli N., Fedi A., Cicognani E. (2021), La prospettiva psicosociale, in Arcidiacono et al. (2021), vol. 1, pp. 28-47.

De Vries M. F. K. (2001), Creating Authentic Organizations: Well-Functioning Individuals in Vibrant Companies, in "Human Relations", 54, 1, pp. 101-11.

Di Iullo M. G. (1999), Le fasi di vita dei gruppi, in A. Putton (a cura di), Empowerment e scuola, Carocci, Roma, pp. 95-104.

Di Maria F. (2005), Psicologia per la politica. Metodi e pratiche, FrancoAngeli, Milano.

Di Maria F., Lavanco G. (a cura di) (1993), Nel nome del gruppo. Gruppoanalisi e società, FrancoAngeli, Milano.

Di Maria F., Falgares G. (2021), Elementi di psicologia dei gruppi. Modelli teorici e ambiti applicativi, McGraw-Hill, Milano.

Dimatteo M. R. (2004), Social Support and Patient Adherence to Medical Treatment: A Meta-Analysis, in "Health Psychology", 23, 2, p. 207.

Di Napoli I., Dolce P., Arcidiacono C. (2019), Community Trust: A Social Indicator Related to Community Engagement, in "Social Indicators Research", 145, 2, pp. 551-79.

Di Napoli I. et al. (2019a), Ending Intimate Partner Violence (ipv) and Locating Men at Stake: An Ecological Approach, in "International Journal of Environmental Research and Public Health", 16, 9, https://doi.org/10.3390/ijerph16091652.

Di Napoli I. et al. (2019b), Trust, Hope, and Identity in Disadvantaged Urban Areas: The Role of Civic Engagement in the Sanità District (Naples), in "Journal of Community Psychology in Global Perspective", 5, 2, pp. 46-62.

Dittmer D. L, Riemer M. (2012), Fostering Critical Thinking about Climate Change: Applying Community Psychology to an Environmental Education Project with Youth, in "Global Journal of Community Psychology Practice", 4, 1, https://www.gjcpp.org/pdfs/2012-010-final-20121230.pdf.

Dixxson-Declève S. et al. (2022), Earth for All: A Survival Guide for Humanity, New Society Publishers, Kindle Edition.

Dohrenwend B. et al. (1992), Socioeconomic Status and Psychiatric Disorders: The Causation-Selection Issue, in "Science", 255, 5047, pp. 946-52.

Droste C. (2015), German Cohousing: An Opportunity for Municipalities to Foster Socially Inclusive Urban Development?, in "Urban Research & Practice", 8, 1, pp. 79-92.

Druiff P. (2001), Psychodynamic Therapy with Low-Income Women: The "Talking Cure" as a Desirable and Alternative Intervention, Doctoral dissertation, Stellenbosch University, https://pdfssemanticscholar.org/f33e/5024359a73b461dca652ca45a85fe5bf641b.pdf.

Dudgeon P. et al. (2017), Facilitating Empowerment and Self-Determination through Participatory Action Research: Findings from the National Empowerment Project, in "International Journal of Qualitative Methods", 16, 1, https://doi.org/10.1177/1609406917699515.

Dunkel-Schetter C., Sandman C. A., Wadhwa p. (2000), Maternal Social Support Predicts Birth Weight and Fetal Growth in Human Pregnancy, in "Psychosomatic Medicine", 62, 5, pp. 715-25.

Dunn R. (2022), A Natural History of the Future: What the Laws of Biology Tell Us about the Destiny of the Human Species, Basic Books, New York.

Durrett C. (2022), Cohousing Communities: Designing for High-Functioning Neighborhoods, Wiley, Hoboken (nj).

Dworski-Riggs D., Langhout R. D. (2010), Elucidating the Power in Empowerment and the Participation in Participatory Action Research: A Story about Research

Team and Elementary School Change, in "American Journal of Community Psychology", 45, 3, pp. 215-30.

Eaton J., Radtke B. (2010), Cbm Community Mental Health Implementation Guidelines, in https://www.yumpu.com/en/document/read/31312155/cbmcommunity-mental-health-cmh-implementation-guidelines-.

Eea – european environment agency (2019), European Union Emission Inventory Report 1990 – 201 under the unece Convention on Long-Range Transboundary Air Pollution (lrtap),

Eea technical report n. 9, https://www.eea.europa.eu/ data-and-maps/indicators/main-anthropogenic-air-pollutant-emissions/eea-2011.

Ehmayer C., Reinfeldt S., Gtotter S. (2000), Agenda 21 as a Concept for Sustainable Development, Paper presented at iii Panel of Experts (Wien, May 11-13), non pubblicato.

Elias M. J., Dalton J. H., Godin S. (eds.) (1987), A Survey of Graduate Education in Community Psychology, in "The Community Psychologist", 20, 3, pp. 10-33.

Enriquez E (1980), Interrogation au paranoia, enjeu de l'intervention, in "Sociologie et société", xi, 2, pp. 79-104.

Epplein M. et al. (2011), Quality of Life after Breast Cancer Diagnosis and Survival, in "Journal of Clinical Oncology", 29, 4, pp. 406-12.

Esposito F. (2012), Gruppi di auto-aiuto con donne vittime di un'esperienza di violenza nell'ambito delle relazioni intime, in "Funzione Gamma", https://www.funzionegamma.it/wp-content/uploads/2022/11/auto-aiuto-donne29i.pdf.

id. (2017), Practicing Ethnography in Migration-Related Detention Centers: A Reflexive Account, in "Journal of Prevention & Intervention in the Community", 45, 1, pp. 57-69.

Esposito F., Ornelas J., Arcidiacono C. (2015), Migration-Related Detention: A Focus on the Italian Context, in A. Cunha, M. Silva, R. Frederico (eds.), The Borders of Schengen, Peter Lang, Bruxelles, pp. 177-91.

Esposito F. et al. (2018), From Rehabilitation to Recovery: A Selfhelp Experience to Regain Quality of Life after Violence, in "The Journal of Special Education and Rehabilitation", 19, 3-4, pp. 85-104.

Eysenbach G. et al. (2004), Health Related Virtual Communities and Electronic Support Groups: Systematic Review of the Effects of Online Peer to Peer Interactions, in "bmj", 328, 7449, https://doi.org/10.1136/bmj.328.7449.1166.

Fairweather G. W. (1967), Methods for Experimental Social Innovation, Wiley, New York.

Falk R. (1994), The Making of Global Citizenship, in B. van Steenbergen (ed.), The Condition of Citizenship, sage, Thousand Oaks (ca), pp. 127-40.

Farrell S. J., Aubry t., Coulombe d. (2004), Neighborhoods and Neighbors: Do they Contribute to Personal Well-Being?, in "Journal of Community Psychology", 32, 1, pp. 9-25.

Fawcett S. B., Mathews r. m., Fletcher r. k. (1980), Some Promising Dimensions for Behavioral Community Technology, in "Journal of Applied Behavior Analysis", 13, 3, pp. 505-18.

Fawzy F. I. et al. (1993), Effects of an Early Structured Psychiatric Intervention, Coping, and Affective State on Recurrence and Survival 6 Years Later, in "Archives of General Psychiatry", 50, pp. 681-9.

Federici S. (2020), Revolution at Point Zero: Housework, Reproduction, and Feminist Struggle, pm Press, Oakland.

Feldman P. J. et al. (2000), Maternal Social Support Predicts Birth Weight and Fetal Growth in Human Pregnancy, in "Psychosomatic Medicine", 62, 5, pp. 715-25.

Feola G. (2020), Capitalism in Sustainability Transitions Research: Time for a Critical Turn?, in "Environmental Innovation and Societal Transitions", 35, pp. 241-50.

Fernández J. S. (2018), Toward an Ethical Reflective Practice of a Theory in the Flesh: Embodied Subjectivities in a Youth Participatory Action Research Mural Project, in "American Journal of Community Psychology", 62, 1-2, pp. 221-32.

Fernández J. S. et al. (2021), Roots and Routes toward Decoloniality within and outside Psychology Praxis, in "Review of General Psychology", 25, 4, pp. 354-68.

Fernández-Feito A. et al. (2015), Face-to-Face Information and Emotional Support from Trained Nurses Reduce Pain during Screening Mammography: Results from a Randomized Controlled Trial, in "Pain Management Nursing", 16, 6, pp. 862-70.

Ferrario F. (1993), Il lavoro di rete nel servizio sociale. Gli operatori fra solidarietà e istituzioni, Carocci, Roma.

Figueres C. (2020), Paris Taught Me how to Do What Is Necessary to Combat Climate Change, in "Nature", 577, 7791, pp. 470-1.

Fisher K. (1993), Leading Self-Directed Work Teams: A Guide to Developing New Team Leadership Skills, McGraw-Hill, New York.

Flasch P., Murray C. E., Crowe A. (2017), Overcoming Abuse: A Phenomenological Investigation of the Journey to Recovery from Past Intimate Partner Violence, in "Journal of Interpersonal Violence", 32, 22, pp. 3373-401.

Flood R. L., Jackson M. C. (1988), Cybernetics and Organization Theory: A Critical Review, in "Cybernetics and Systems", 19, pp. 13-33.

id. (1991), Creative Problem Solving: Total Systems Interventions, Wiley, Chichester.

Focardi F. et al. (a cura di) (2006), Indagine conoscitiva nazionale dei Gruppi di auto aiuto, in https://www.cesvot.it/sites/default/files/type_documentazione/allegati/indagine_conoscitiva_naz_gruppi_auto_aiuto.pdf

Folgheraiter F. (1990), Operatori sociali e lavoro di rete. Saggi sul mestiere di altruista nelle società complesse, Erickson, Trento.

id. (1997), La community care nella prospettiva del lavoro di rete, C. M. Mozzanica, R. Granata, C. Castelli (a cura di), (Dis)agio giovanile negli itinerari di community care, FrancoAngeli, Milano.

id. (2004), Teoria e metodologia del servizio sociale. La prospettiva di rete, Franco Angeli, Milano.

Fondazione istituto Andrea Devoto (1999), Indagine conoscitiva sulle associazioni di auto aiuto e di tutela della salute, condotta negli anni 1998-1999, cesvot, Firenze.

Fosti G. (a cura di) (2013), Rilanciare il welfare locale. Ipotesi e strumenti: una prospettiva di management delle reti, Egea, Milano.

Fraisse P., Piaget J., Reuchlin M. (1990), Psicologia sperimentale. Storia e metodo, Einaudi, Torino.

Francescato D. (1975), Psicologia ambientale: schemi e immagini di una città. Un contributo di ricerca, Bulzoni, Roma.

id. (1977a), Psicologia di comunità, Feltrinelli, Milano.

id. (1977b), Psicologia di comunità: un nuovo ruolo per lo psicologo, in "Giornale italiano di psicologia", 1, pp. 11-63.

id. (1982), Tecniche di analisi organizzativa multidimensionale, dispense Ecopoiesis, Scuola di specializzazione in Psicologia di comunità, Roma, non pubblicato

id. (1983), Psicologi di comunità nei servizi materno-infantili delle usl: dal rapporto con il cliente all'intervento nel sociale, in Francescato, Contesini, Dini (1983).

id. (1992a), A Multidimensional Perspective of Organizational Change, in "Systems Practice", 5, pp. 129-46.

id. (1992b), Quando l'amore finisce, Il Mulino, Bologna.

id. (1994), Figli sereni di amori smarriti. Ragazzi e adulti dopo la separazione, Mondadori, Milano.

id. (2000), Community Psychology Intervention Strategies as Tools to Enhance Participation in Projects Promoting Sustainable Development and Quality of Life, in "Gemeinde", 50.

id. (2010), Amarsi da grandi, Mondadori, Milano.

id. (2019), "With Greta Let's Save the Planet": A Multi-Faced Rebellion to Get Action on Climate Change, in "La camera blu. Rivista di studi di genere", 20, pp. 117-29.

id. (2020), Why We Need to Build a Planetary Sense of Community, in "Community Psychology in Global Perspective", 6, 2/2, pp. 140-64.

id. (2022), Formare più psicologi di comunità per affrontare problemi locali e globali e promuovere futuri migliori, in "Giornale italiano di psicologia", xlix, 4, pp. 713-26.

Francescato D., Contesini A., Dini S. (a cura di) (1983), Psicologia di comunità. Esperienze a confronto, Il Pensiero Scientifico, Roma.

Francescato D., Francescato G. (1974), Famiglie aperte, la comune. Analisi socio-psicologica delle comuni nordamericane, con una nota sulle comuni italiane, Feltrinelli, Milano.

Francescato D., Ghirelli G. (1988), Fondamenti di Psicologia di comunità, La Nuova Italia Scientifica, Roma.

Francescato D., Leone l., Traversi M. (1993), Oltre la psicoterapia. Percorsi innovativi di psicologia di comunità, La Nuova Italia Scientifica, Roma.

Francescato D., Mebane M. E. (2015), Learning Innovative Methodologies to Foster Personal, Organizational and Community Empowerment through Online and Face-to-Face Community Psychology Courses, in "Universitas Psychologica", 14, 4, pp. 1209-20.

Francescato D., Aber M. S. (2015), Learning from Organizational Theory to Build Organizational Empowerment, in "Journal of Community Psychology", 43, 6, pp. 717-38.

Francescato D. Mebane M.E:. (2018), Globalizzazione e innovazione tecnologiche: nuove sfide per gli psicologi di comunità del xxi secolo, in "Rivista di psicologia di comunità", 1, pp. 9-19.

Francescato D., Mebane M. E., Vecchione m. (2017), Gender Differences in Personal Values of National and Local Italian Politicians, Activists and Voters, in "International Journal of Psychology", 52, 5, pp. 406-

Francescato D., Putton A. (1995), Star meglio insieme. Oltre l'individualismo: imparare a crescere e collaborare con gli altri, Mondadori, Milano.

id. (2022), Star bene insieme a scuola. Strategie per la promozione del benessere relazionale e il welfare di comunità, Carocci, Roma.

Francescato D., Putton A., Cudini S. (1986), Star bene insieme a scuola: strategie per un'educazione socio-affettiva dalla materna alla scuola media inferiore, La Nuova Italia Scientifica, Roma.

Francescato D., Putton A., Toraldo M. (1997), L'educazione socioaffettiva per genitori, in B. Zani, M. L. Pombeni (a cura di), L'adolescenza, bisogni soggettivi e risorse sociali, Il Ponte Vecchio, Cesena.

Francescato S., Rosa V., Pellegrini R. (1997), Empowerment and Change among Women Working in Public and Private Sector Organizations, in "Quaderni di psicologia del lavoro", 5, pp. 89-97.

Francescato D., Tomai M. (2001), Community Psychology: Should there Be a European Perspective? in "Journal of Community & Applied Social Psychology", 11, 5, pp. 371-80.

id. (2002), I profili di comunità nell'era della globalizzazione, in M. Prezza, M. Santinello (a cura di), Conoscere la comunità, Il Mulino, Bologna, pp. 39-65.

Francescato D., Tomai M., Foddis A. (2002), I fattori di efficacia nei gruppi di auto-aiuto, in "Psicologia della salute", 2, pp. 145-56.

Francescato D., Tomai M., Ghirelli G. (2002), Fondamenti di psicologia di comunità. Principi, strumenti, ambiti di applicazione, Carocci, Roma.

Francescato D., Tomai M., Mebane M.E. (2004), Psicologia di comunità per la scuola, l'orientamento e la formazione. Esperienze faccia a faccia e online, Il Mulino, Bologna.

Francescato D., Tomai M., Solimeno a. (2008), Lavorare e decidere meglio in organizzazioni empowering ed empowered. L'Analisi Organizzativa Multidimensionale e la formazione empowering come strumenti di intervento nei contesti di lavoro, FrancoAngeli, Milano.

Francescato D, Zani B. (2013), Community Psychology Practice Competencies in Undergraduate and Graduate Programs in Italy, in "Global Journal of Community Psychology Practice", 4, 4, https://www.gjcpp.org/pdfs/2013-001CCSI-20131018.pdf.

Francescato D .,Zani B- (2017), Strengthening Community Psychology in Europe through Increasing Professional Competencies for the New Territorial Community Psychologists, in "Global Journal of Community Psychology Practice", 8, 1, https://www.gjcpp.org/pdfs/

Francescato D. et al. (2000), Chi ha voglia di studiare? Psicologia di comunità e orientamento nelle scuole medie superiori del Sud d'Italia, in Santinello, Crespi, Vieno (2000).

Francescato D. et al (2007), Empo: una scala di misurazione dell'empowerment personale e politico, in "Giornale italiano di psicologia", 34, 2, pp. 465-90.

Francescato D. et al. (2009), Increasing Students' Perceived Sociopolitical Empowerment through Online and Face-to-face Community Psychology Seminars, in "Journal of Community Psychology", 37, 7, pp. 874-94.

Francescato D. et al. (2012), Promoting Social Capital, Empowerment and Counter-Stereotypical Behavior in Male and Female Students in Online CSCL Communities, in H. Cuadra-Montiel (ed.), Globalization: Education and Management Agendas, Process Manager Martina Durovic, Rijeka, pp. 75-108.

Francescato D. et al. (2017), Dispositional Characteristics, Relational Well-Being and Perceived Life Satisfaction and Empowerment of Elders, in "Aging & Mental Health", 21, 10, pp. 1052-7.

Folgheraiter F. (1990), Operatori sociali e lavoro di rete. Saggi sul mestiere di altruista nelle società complesse, Erickson, Trento.

id. (1997), La community care nella prospettiva del lavoro di rete, C. M. Mozzanica, R. Granata, C. Castelli (a cura di), (Dis)agio giovanile negli itinerari di community care, FrancoAngeli, Milano.

id. (2004), Teoria e metodologia del servizio sociale. La prospettiva di rete, Franco Angeli, Milano.

Fondazione istituto Andrea Devoto (1999), Indagine conoscitiva sulle associazioni di auto aiuto e di tutela della salute, condotta negli anni 1998-1999, cesvot, Firenze.

Fosti G. (a cura di) (2013), Rilanciare il welfare locale. Ipotesi e strumenti: una prospettiva di management delle reti, Egea, Milano.

Fraisse P., Piaget J., Reuchlin M. (1990), Psicologia sperimentale. Storia e metodo, Einaudi, Torino.

Fridays for future italia (2020), La crisi climatica, in https://www.fridaysfor futureitalia.it/crisi-climatica.

Friedan B. (1963), The Sexual Sell, in http://www.consumerculture.amdigital. co.uk/FurtherResources/Essays/TheSexualSell.

Fryer D. (1995), Benefit Agency? Labour Market Disadvantage, Deprivation and Mental Health, in "The Psychologist", 8, 6, pp. 265-72.

Gallino L. (1987), L'attore sociale. Biologia, cultura e intelligenza artificiale, Einaudi, Torino.

Gardner H. (1994a), Intelligenze multiple, Anabasi, Milano.

id. (1994b), L'educazione delle intelligenze multiple, Anabasi, Milano.

Gartner A., Riessman f. (eds.) (1984), The Self-Help Revolution, Human Sciences Press, New York.

Gatti F., Procentese F. (2020), Open Neighborhoods, Sense of Community, and Instagram Use: Disentangling Modern Local Community Experience through a Multilevel Path Analysis with a Multiple Informant Approach, in "tpm. Testing, Psychometrics, Methodology in Applied Psychology", 27, 3, pp. 313-29.

id. (2021), Experiencing Urban Spaces and Social Meanings through Social Media: Unravelling the Relationships between Instagram City-Related Use, Sense of Place, and Sense of Community, in "Journal of Environmental Psychology", 78, doi: 10.1016/ j.jenvp.2021.101691.

id. (2022), Ubiquitous local community experiences: unravelling the social added value of neighborhood-related social media, in "Psicologia di comunità", 2, pp. 56-79.

Gatti F., Procentese F., Schouten A. P. (2022), People-Nearby Applications Use and Local Community Experiences: Disentangling Their Interplay through a Multilevel, Multiple Informant Approach, in "Media Psychology", 26, 3, pp. 278-305.

Gavrilovici O., Dronic A., Remaschi L. (2020), Innovative Methods for the Interventions in Preventing Violent Radicalisation, in P. Meringolo (ed.), Preventing Violent Radicalisation in Europe: Multidisciplinary Perspectives, Springer (e-book), pp. 59-84.

Gelli B., Mannarini T. (1999), Il mentoring, Carocci, Roma.

id. (2014), La mente politica: leadership, potere, carisma, in "Psicologia di comunità", 2, pp. 11-22.

Gergen K. J. (1999), An Invitation to Social Construction, sage, Thousand Oaks (ca).

Ghaderi A., Scott B. (2003), Pure and Guided Self-Help for Full and Sub-Threshold Bulimia Nervosa and Binge Eating Disorder, in "The British Journal of Clinical Psychology", 42, pp. 257-69.

Ghosh D. et al. (2017), Social Network Strategies to Address hiv Prevention and Treatment Continuum of Care among At-Risk and hiv-Infected Substance Users: A Systematic Scoping Review, in "aids and Behavior", 21, 4, pp. 1183-207.

Gibson K., Sandenbergh R., Swartz L. (2001), Becoming a Community Clinical Psychologist: Integration of Community and Clinical Practices in Psychologists' Training, in "South African Journal of Psychology", 31, 1, pp. 29-35.

Gibson K., Swartz L. (2008), Putting the "Heart" back into Community Psychology: Some South African Examples, in "Psychodynamic Practice", 14, 1, pp. 59-75.

Giddens A. (1991), Modernity and Self-Identity, Polity Press, Cambridge.

"Division of Community Psychology N letter", 15, 4, pp. 4-5.

Glenwick D. (1982), Community Psychology in the 80s: A Discipline for All Seasons or One whose Time Has Passed

Goldenberg I. I. (1971), Build Me a Mountain: Youth, Poverty, and the Creation of New Settings, The mit Press, Cambridge (ma).

Gone J. P. (2011), Is Psychological Science A-Cultural?, in "Cultural Diversity and Ethnic Minority Psychology", 17, 3, pp. 234-42.

Goodenow C. (1993), The Psychological Sense of School Membership among Adolescents: Scale Development and Educational Correlates, in "Psychology in the Schools", 30, 1, pp. 79-90.

Goodman I. A. et al. (2004), Training Counseling Psychologists as Social Justice Agents: Feminist and Multicultural Principles in Action, in "The Counseling Psychologist", 32, 6, pp. 793-836.

Gordon M. (1987), Organizzare e condurre un meeting, Armenia, Milano.

Gordon T. (1974), Teacher Effectiveness Training, Peter Weiden, New York

id. (1979), "Parents efficaces". La méthode Gordon expérimentée et vécue, Belfond, Paris.

Gore S. (1978), The Effect of Social Support in Moderating the Health Consequences of Unemployment, in "Journal of Health and Social Behavior", 19, 2, pp. 157-65

Gottlieb B. H. (1981), The Developmental Application of a Classification Scheme of Informal Helping Behaviors, in "Canadian Journal of Science", 10, 2, pp. 105-15.

Granovetter M. (1973), The Strength of Weak Ties, in "American Journal of Sociology", 78, 6, pp. 1360-80.

id. (1974), Getting a Job: A Study of Contacts and Careers, Harvard University Press, Cambridge (ma).

Grasso M., Cordella B., Pennella A. R. (2016), L'intervento in psicologia clinica, Carocci, Roma (nuova ed.).

Graven L. J., Grant J. S. (2014), Social Support and Self-Care Behaviors in Individuals with Heart Failure: An Integrative Review, in "International Journal of Nursing Studies", 51, 2, pp. 320-33.

Gray R. E. et al. (1997), Interviews with Men with Prostate Cancer about Their Self-Help Group Experience, in "Journal of Palliative Care", 13, 1, pp. 15-21.

Green L. W. (2001), From Research to "Best Practices" in Other Settings and Populations, in "American Journal of Health Behavior", 25, 3, pp. 165-78.

Greenwood D. I., Levin M. (1998), Introduction to Action Research: Social Research for Social Change, sage, London-Thousand Oaks (ca).

Greenwood R. M. et al. (2022), Structure and Agency in Capabilities-Enhancing Homeless Services: Housing First, Housing Quality and Consumer Choice, in "Journal of Community & Applied Social Psychology", 32, 2, pp. 315-31.

Gregory R. J. (2001), The Spirit and Substance of Community Psychology: Reflections, in "Journal of Community Psychology", 29, 4, pp. 473-85.

Gugerty M. K., Biscaye p., Anderson c. l. (2019), Delivering Development? Evidence on Self-Help Groups as Development Intermediaries in South Asia and Africa, in "Development Policy Review: The Journal of the Overseas Development Institute", 37, 1, pp. 129-51.

Guimarães P. R. jr. (2020), The Structure of Ecological Networks across Levels of Organization, in "Annual Review of Ecology, Evolution and Systematics", 5, 1, pp. 433-60.

Gulcur L. et al. (2007), Community Integration of Adults with Psychiatric Disabilities and Histories of Homelessness, in "Community Mental Health Journal", 43, 3, pp. 211-28.

Haldane V. et al. (2019), Community Participation in Health Services Development, Implementation, and Evaluation: A Systematic Review of Empowerment, Health Community, and Process Outcomes, in "PloS One", 14, 5, doi:10.1371/journal. pone. 0216112.

Haley J. (1969), An Editor's Farewell, in "Family Process", 8, 2, pp. 149-58.

Hamman W. R. (2004), The Complexity of Team Training: What We Have Learned from Aviation and Its Applications to Medicine, in "Quality and Safety in Health Care", 13, Suppl. 1, pp. i72-9.

Hanley J. et al. (2018), The Social Networks, Social Support and Social Capital of Syrian Refugees Privately Sponsored to Settle in Montréal: Indications for Employment and Housing during Their early Experiences of Integration, in "Canadian Ethnic Studies", 50, 2, pp. 123-48.

Hanson P. (1981), Learning Through Groups: A Trainer's Basic Guide, University Associates, San Diego (ca).

Harari Y. N. (2019), 21 lezioni per il xxi secolo, Bompiani, Milano.

Harper G. W. et al. (2004), Diverse Phases of Collaboration: Working Together to Improve Community-Based HIV Interventions for Adolescents, in "American Journal of Community Psychology", 33, 3-4, pp. 193-204.

Hartman S. (1987), Therapeutic Self-help Group: A Process of Empowerment for Women in Abusive Relationships, in C. M. Brody (ed.), Women's Therapy Groups: Paradigms of Feminist Treatment, Springer, New York, pp. 67-81.

Hartmann W. E., St. Arnault D. M., Gone J. P. (2018), A Return to the Clinic for Community Psychology: Lessons from a Clinical Ethnography in Urban American

Indian Behavioral Health, in "American Journal of Community Psychology", 61, 1-2, pp. 62-75.

Hayes S. C. et al. (2006), Acceptance and Commitment Therapy: Model, Processes and Outcomes, in "Behaviour Research and Therapy", 44, 1, doi: 10.1016/j.brat. 2005.06.006.

Heller A, Vajda M. (1970), Családforma és kommunizmus ["Forme de famille et communisme"], in "Kortárs", 2, pp. 1655-60.

Heller K.,Swindler R. W. (1983), Social Networks, Perceived Social Support, and Coping with Stress, in R. D. Felner et al. (eds.), Preventive Psychology: Theory, Research and Practice in Community Intervention, Pergamon Press, New York, pp. 87-103.

Heller K. et al. (1984), Psychology and Community Change, The Dorsey Press, Homewood (il).

Hess R. (1984), Thoughts on Empowerment, in "Prevention in Human Services", 3, 2-3, pp. 227-30.

Hetherington E. et al. (2015), Preterm Birth and Social Support During Pregnancy: A Systematic Review and Meta-Analysis, in "Paediatric and Perinatal Epidemiology", 29, 6, pp. 523-35

Hickel J. (2019), Is It Possible to Achieve a Good Life for All within Planetary Boundaries?, in "Third World Quarterly", 40, 1, pp. 18-35.

Higgins J. W. (1999), Citizenship and Empowerment: A Remedy for Citizen Participation in Health Reform, in "Community Development Journal", 34, 4, pp. 287-307. hill-briggs f. et al. (2021), Social Determinants of Health and Diabetes: A Scientific Review, in "Diabetes Care", 44, 1, pp. 258-79.

Hinshelwood R. D., Skogstad W. (eds.) (2000), Observing Organisations: Anxiety, Defence and Culture in Health Care, Routledge, London.

Hipkins J. et al. (2004), Social Support, Anxiety and Depression after Chemotherapy for Ovarian Cancer: A Prospective Study, in "British Journal of Health Psychology", 9, 4, pp. 569-81.

Hirigoyen M. F. (2000), Molestie morali. La violenza perversa nella famiglia e nel lavoro, Einaudi, Torino.

Hirsch b. j. (1980), Natural Support Systems and Coping with Major Life Changes, in "American Journal of Community Psychology", 8, 2, pp. 159-72.

Hobfoll S. E. (1990), Person-Environment Interaction: The Question of Conceptual Validity, in P. Tolan et al. (eds.), Researching Community Psychology: Issue of Theory and Methods, American Psychological Association, Washington dc, pp. 164-7.

Hoekstra-Weebers J. E. et al. (2001), Psychological Adaptation and Social Support of Parents of Pediatric Cancer Patients: A Prospective Longitudinal Study, in "Journal of Pediatric Psychology", 26, 4, pp. 225-35.

Holloway M. (1961), Heavens on Earth: Utopian Communities in America, Turnstile, New York.

Holt-Lunstad J. Rubles C, Barra D. A. (2017), Advancing Social Connection as a Public Health Priority in the United States, in "American Psychologist", 72, 6

Holt-lunstad J., Smith T. B., Layton J. B. (2010), Social Relationships and Mortality Risk: A Meta-Analytic Review, in "PLoS Medicine", 7, 7, https://doi. org/10.1371/journal.pmed.1000316.

Holt-lLunstad J. et al. (2015), Loneliness and Social Isolation as Risk Factors for Mortality: A Meta-Analytic Review, in "Perspectives on Psychological Science", 10, 2, pp. 227-37.

House J. S. (1981), Work Stress and Social Support, Addison-Wesley, Reading (ma).

House J. S., Kahn r. (1985), Measures and Concepts of Social Support, in S. Cohen, S. Syme (eds.), Social Support and Health, Academic Press, Orlando (fl), pp. 83-108.

Housman J., Dorman S. (2005), The Alameda County Study: A Systematic, Chronological Review, in "American Journal of Health Education", 36, 5, pp. 302-8.

Høyer M. et al. (2011), Health-Related Quality of Life among Women with Breast Cancer: A Population-Based Study, in "Acta oncologica", 50, 7, pp. 1015-26.

Hughey J. et al. (2008), Empowerment and Sense of Community: Clarifying Their Relationship in Community Organizations, in "Health Education & Behavior", 35, 5, pp. 651-63.

Humphreys K., Blodgett J. C., Wagner T. H. (2014), Estimating the Efficacy of Alcoholics Anonymous without Self-Selection Bias: An Instrumental Variables ReAnalysis of Randomized Clinical Trials, in "Alcoholism: Clinical and Experimental Research", 38, 11, pp. 2688-94.

Humphreys K., Moos R. (2001), Can Encouraging Substance Abuse Patients to Participate in Self-Help Groups Reduce Demand for Health Care? A Quasi-Experimental Study, in "Alcoholism: Clinical and Experimental Research", 25, 5, pp. 711-6.

Humphreys K., Rappaport J. (1993), From the Community Mental Health Movement to the War on Drugs: A Study in the Definition of Social Problems, in "American Psychologist", 48, 8, pp. 892-901.

Hunter M. R., Gillespie B. W., Chen S. Y. (2019), Urban Nature Experiences Reduce Stress in the Context of Daily Life Based on Salivary Biomarkers, in "Frontiers in Psychology", 10, 722, doi: 10.3389/fpsyg.2019.00722.

Hurley A. L., Sullivan P., McCarthy J. (2007), The Construction of Self in Online Support Groups for Victims of Domestic Violence, in "British Journal of Social Psychology", 46, 4, pp. 859-74.

Huutoniemi K. (2015), Interdisciplinarity as Academic Accountability: Prospects for Quality Control across Disciplinary Boundaries, in "Social Epistemology: A Journal of Knowledge, Culture and Policy", 30, 2, pp. 163-85.

Hyman J. B. (2002), Exploring Social Capital and Civic Engagement to Create a Framework for Community Building, in "Applied Developmental Science", 6, , pp. 196-202.

Ikeda A. et al. (2013), Social Support and Cancer Incidence and Mortality: The jphc Study Cohort ii, in "Cancer Causes Control", 24, 5, pp. 847-60.

Ilgen D. R. et al. (2005), Teams in Organizations: From Input-Process-Output Models to imoi Models,, in "Annual Review of Psychology", 56, pp. 517-43.

Ilioudi S. et al. (2012), Health-Related Virtual Communities and Social Networking Services, in A. A. Lazakidou (ed.), Virtual Communities, Social Networks and Collaboration, Springer, New York, pp. 1-13.

Imf (2019), imf Annual Report 2019: Our Connected World, in https://www.imf.org/external/pubs/ft/ar/2019/eng/assets/pdf/imf-annual-report-2019.pdf.

Insel P. M., Moos R. H. (1974), Psychological Environments: Expanding the Scope of Human Ecology, in "American Psychologist", 29, 3, pp. 179-88.

Iscoe I. (1974), Community Psychology and the Competent Community, in "American Psychologist", 29, 8, pp. 607-13.

Istat (2022), Popolazioni e famiglie, in Annuario statistico italiano 2022, https://www. istat.it/it/files/2020/12/C03.pdf.

id. (2023), Rapporto annuale 2023. La situazione del paese, in https://www.istat.it/storage/rapporto-annuale/2023/Rapporto-Annuale-2023.pdf.

Ittelson w. h. et al. (1974), An Introduction to Environmental Psychology, Holt, Rinehart and Winston, New York.

Jackson M. C., Keys P. (1984), Towards a System of System Methodology, in "Journal of Operational Research Society", 35, pp. 473-86.

id. (eds.), New Directions in Management Science, Aldershot-Gower, HantsBrookfield.

Jaques E. (1966), Lavoro, creatività e giustizia sociale, Boringhieri, Torino.

Jason L. A., Ase D. M. (2016), Community-Clinical Psychology, in J. C. Norcross et al. (eds.), APA Handbook of Clinical Psychology: Roots and Branches, American Psychological Association, Washington dc, pp. 201-22.

Jason L. A. et al. (2019), Introduction to the Field of Community Psychology, in Id. (eds.), Introduction to Community Psychology: Becoming an Agent of Change, https:// press.rebus.community/introductiontocommunitypsychology/, pp. 3-22.

Jenkins R. A. (2016), Clinical Community Psychology: Reflections on the Decades Following Swampscott, in "American Journal of Community Psychology", 58, 3-4, pp. 269-75.

Johnson A. et al. (2017), Emotional Socialization and Child Conduct Problems: A Comprensive Review and Meta-Analysis, in "Clinical Psychology Review", 54, pp. 65-80.

Johnson D. M., Zlotnick C. (2009), Hope for Battered Women in Domestic Violence Shelters, in "Professional Psychology: Research and Practice", 40, 3, pp. 234-41.

Johnson D. W., Johnson F. (1975), Joining Together Group Theory and Group Skills, Prentice-Hall, Englewood Cliffs.

Jones E., Gerard H. B. (1967), Foundations of Social Psychology, Wiley, New York. jordan a. h., lovett b. j., sweeton J. L. (2012), The Social Psychology of Black White Interracial Interactions: Implications for Culturally Competent Clinical Practice, in "Journal of Multicultural Counseling and Development", 40, 3, pp. 132-43.

Kagan C. et al. (2011), Critical Community Psychology, Blackwell/John Wiley, Chichester.

Kang H. Y., Yoo Y. S. (2007), Effects of a Bereavement Intervention Program in Middle-Aged Widows in Korea, in "Archives of Psychiatric Nursing", 21, 3, pp. 132-40.

Kanter E., Halter M. (1973), The Housewifing of Women, Domesticating Men: Equality between the Sexes in Urban Communes, Paper presented at the Annual Convention of the American Psychological Association (Montréal, August 1) in J. Heiss (ed.), Marriage and Family Interaction, Rand McNally, Chicago (il) (2 nd ed.) 1976.

Kaplan G. A. et al. (1987), Mortality among the Elderly in the Alameda County Study: Behavioral and Demographic Risk Factors, in "American Journal of Public Health", 77, 3, pp. 307-12.

Katz A. H. (1981), Self-Help and Mutual Aid: An Emerging Social Movement, in "Annual Review of Sociology", 7, pp. 129-55.

Katz A. H., Bender E. (1976), The Strength in Use: Self-Help Groups in the Modern World, Franklin Watts, New York.

Kavafis C. (1992), Itaca, in Id., Settantacinque poesie, Einaudi, Torino, pp. 63-5.

Kawa M. H. (2017), Influence of Perceived Social Support and Meaning in Life on Fighting Spirit: A Study of Cancer Patients, in "International Journal of Advanced Education and Research", 2, pp. 86-93.

Keiffer C. H. (1984), Citizen Empowerment: A Developmental Perspective, in "Prevention in Human Services", 3, 2-3, pp. 9-36.

Kelly J. G. (1987), The Ecology of Empowerment: A Study of Two Advocacy Organizations, Nathalie P. Voorhees Center for Neighborhood and Community Improvement, University of Illinois at Chicago.

id. (1990), Changing Contexts and the Field of Community Psychology, in "American Journal of Community Psychology", 18, 6, pp. 769-92.

Kelly J. G., Song A. (eds.) (2004), Six Community Psychologists Tell Their Stories: History, Contexts, and Narrative, Haworth Press, Philadelphia (pa).

Kerr M. E., Bowen M. (1990), La valutazione della famiglia. Un approccio terapeutico basato sulla teoria boweniana, Astrolabio, Roma.

Kieffer J. (1982), The Development of Empowerment: The Development of Participatory Competence among Individuals in Citizen Organizations, in "Division 27 Newsletter", 16, 1, pp. 13-5.

King R. A., Shelley c. a. (2008), Community Feeling and Social Interest: Adlerian Parallels, Synergy and Differences with the Field of Community Psychology, in "Journal of Community and Applied Social Psychology", 18, 2, pp. 96-107.

Kloos B., Johnson R. L. (2017), Prospects for Synergies and Symbiosis: Relationships between Community Psychology and Other Subdisciplines of Psychology, in M. A. Bond et al. (eds.), APA Handbook of Community Psychology, vol. 1: Theoretical Foundations, Core Concepts, and Emerging Challenges, American Psychological Association, Washington dc, pp. 169-88.

Knowles M. S. (1996), La formazione degli adulti come autobiografia. Il percorso di un educatore tra esperienza e idee, Raffaello Cortina Editore, Milano.

Kofahl C. et al. (2014), Self-Help Friendliness: A German Approach for Strengthening the Cooperation between Self-Help Groups and Health Care Professionals, in "Social Science & Medicine", 123, pp. 217-25.

Koh E., Twemlow S. (2016), Towards a Psychoanalytic Concept of Community (ii):Relevant Psychoanalytic Principles, in "International Journal of Applied Psychoanalytic Studies", 13, 2, pp. 124-141.

Koh Yah G., Castillo León T. (2014), Trabajo Colaborativo con Mujeres. Una Experiencia en Clínica Comunitaria, in "Psicoperspectivas", 13, 2, pp. 121-32.

Konrath S. H., O'Brien E. H., Hsing C. (2011), Changes in Dispositional Empathy in American College Students Over Time: A Meta-Analysis, in "Personality and Social Psychology Review", 15, 2, pp. 180-98.

Korchin S. J. (1977), Psicologia clinica moderna, Borla, Roma.

Kotler P., Wingard D. L. (1989), The Effect of Occupational, Marital and Parental Roles on Mortality: The Alameda County Study, in "American Journal of Public Health", 79, 5, pp. 607-12.

Krivacska J. J. (1990), Designing Child Sexual Abuse Prevention Programs: Current Approaches and Proposal for the Prevention, Reduction and Identification of Sexual Misure, Charles C. Thomas Publishing, Springfield (il).

Kumar N. et al. (2019), Social Network, Mobility, and Political Participation: The Potential for Women's Self-Help Groups to Improve Access and Use of Public Entitlement Schemes in India, in "World Development", 114, pp. 28-41.

Kunnes R. (1972), Radicalism and Community Mental Health, in H. G. Gottesfeld (ed.), The Critical Issues in Community Mental Health, Behavioral Publications, New York, pp. 303-12.

Kuppelomäki M. (2003), Emotional Support for Dying Patients: The Nurses' Perspective, in "European Journal of Oncology Nursing", 7, 2, pp. 120-9.

Lake D., Wendland J. (2018), Practical, Epistemological, and Ethical Challenges of Participatory Action Research: A Cross-Disciplinary Review of the Literature, in "Journal of Higher Education Outreach and Engagement", 22, 3, pp. 11-42.

Lane S., Sawaia B. (1991), Community Social Psychology in Brazil, in "Applied Psychology: An International Review", 40, pp. 119-42.

Lang F. R. (2004), Social Motivation across the Life Span, in F. R. Lang, K. L. Fingerman (eds.), Growing Together: Personal Relationships across the Lifespan, Cambridge University Press, Cambridge, pp. 341-67.

Langford C. P. H. et al. (1997), Social Support: A Conceptual Analysis, in "Journal of Advanced Nursing", 25, 1, pp. 95-100.

Langher V. et al. (2019a), Symbols and Meanings in a Suburban's Local Culture: A Case Study, in "Rassegna di psicologia", 36, 1, pp. 21-38.

Langher V. et al. (2019b), Trauma e riparazione in una comunità colpita dal sisma: una lettura clinica dinamica per l'intervento post-emergenziale, in "Psicologia della salute", 3, pp. 98-121.

Lapassade G. (1975), Socioanalyse et potentiel humain, Gauthier Villars, Paris.

Lara junior N., Ribeiro C. T. (2009), Intervenções psicossociais em comunidades. Contribuições da psicanálise, in "Psicologia & Sociedade", 21, 1, pp. 91-9.

Lasch C. (1991), The True and Only Heaven: Progress and Its Critics, W. W. Norton & Co., New York-London.

id. (2019), The Culture of Narcissism, in R. Wilkinson (ed.), American Social Character, Westview Press, Boulder (co), pp. 241-67.

id. (2021), La cultura del narcisismo. L'individuo in fuga dal sociale in un'età di disillusioni collettive, Neri Pozza, Vicenza.

Latkin C. A., Sherman s., knowlton a. (2003), hiv Prevention among Drug Users: Outcome of a Network-Oriented Peer Outreach Intervention, in "Health Psychology", 22, 4, pp. 332-9.

Lauriola M., Tomai M. (2019), Biopsychosocial Correlates of Adjustment to Cancer during Chemotherapy: The Key Role of Health-Related Quality of Life, in "The Scientific World Journal", 2019, doi: 10.1155/2019/9750940.

Lavanco G., Novara C. (2013), La psicologia dinamica di comunità, in G. Mannino (a cura di), Anima, cultura, psiche. Relazioni generative, FrancoAngeli, Milano, pp. 63-85.

Laverack G. (2006), Improving Health Outcomes through Community Empowerment: A Review of the Literature, in "Journal of Health, Population and Nutrition", 24, 1, pp. 113-20.

Laverack G., Wallerstein N. (2001), Measuring Community Empowerment: A Fresh Look at Organizational Domains, in "Health Promotion International", 16, 2, pp. 179-85.

Leahy-Warren P. et al. (2020), The Experiences of Mothers with Preterm Infants within the First-Year post Discharge from nicu: Social Support, Attachment and Level of Depressive Symptoms, in "bmc Pregnancy Childbirth", 20, 1, doi: 10.1186/s12884020-02956-2.

Leavy R. (1983), Social Support and Psychological Disorder: A Review, in "Journal of Community Psychology", 11, pp. 3-21.

Lee M. S., Ryu Y. M., Hwang E. Y. (2014), The Experience of Self-Help Group Activities among Women with Breast Cancer in Korea, in "Korean Journal of Adult Nursing", 26, 4, pp. 466-78.

Leitão M. N. C. (2014), Women Survivors of Intimate Partner Violence: The Difficult Transition to Independence, in "Revista da escola de enfermagem da usp", 48, https://doi.org/10.1590/S0080-623420140000600002.

Lencioni P., Sransky C. (2002), The Five Dysfunctions of a Team: A Leadership Fable, Jossey-Bass, Hoboken (nj).

Leon A., Montenegro M. (1998), Return of Emotion in Psychosocial Community Research, in "Journal of Community Psychology", 26, 3, pp. 219-27

Leonard M., Graham S., Bonacum D. (2004), The Human Factor: The Critical Importance of Effective Teamwork and Communication in Providing Safe Care, in "Quality and Safety in Health Care", 13, suppl. 1, pp. i85-90.

Leone L. (1993), Reti, reticoli e politiche pubbliche, in Francescato, Leone, Traversi (1993), pp. 97-138.

Leone L., Prezza M. (2005), Costruire e valutare i progetti nel sociale. Manuale operativo per chi lavora su progetti in campo sanitario, sociale, educativo e culturale, FrancoAngeli, Milano.

Levin L. S., Idler E. L. (1981), The Hidden Health Care System: Mediating Structures and Medicine, Ballinger Publishing Company, Cambridge (ma).

Levine M., Perkins D. V. (1987), Principles of Community Psychology: Perspectives and Applications, Oxford University Press, New York

Levy L. H. (1979), Processes and Activities in Groups, in M. Lieberman, L. Borman (eds.), Self-Help Groups for Coping with Crisis, Jossey-Bass, San Francisco (ca), pp. 234-71.

Lewin K. (1946), Action Research and Minority Problems, in "Journal of Social Issues", 2, 4, pp. 34-46.

id. (1947), Frontiers in Group Dynamics: Concept, Method and Reality in Social Science, Social Equilibria and Social Change, in "Human Relations", 1, 1, pp. 143-53.

id. (1948), Resolving Social Conflicts: Selected Papers on Group Dynamics, Harper & Brothers, New York.

id. (1972), Teoria e sperimentazione in psicologia sociale, Il Mulino, Bologna. L

Lewin K., Lippitt R., White R. K. (1939), Patterns of Aggressive Behavior in Experimentally Created "Social Climates", in "The Journal of Social Psychology", 10, 2, pp. 269-99.

Lewis C., Pearce J., Bisson J. I. (2012), Efficacy, Cost-Effectiveness and Acceptability of Self-Help Interventions for Anxiety Disorders: Systematic Review, in "The British Journal of Psychiatry", 200, 1, pp. 15-21.

Liang B., Tummala-Narra P., West J. (2011), Revisiting Community Work from a Psychodynamic Perspective, in "Professional Psychology: Research and Practice", 42, 5, pp. 398-404.

Lilla M. (2017), The Once and Future Liberal: After Identity Politics, Harper Collins, New York.

Linney J. A., Reppucci n. d. (1982), Research Design and Methods in Community Psychology, in P. C. Kendall, J. N. Butcher (eds.), Handbook of Research Methods in Clinical Psychology, Wiley, New York, pp. 535-66.

Longden E., Read J., Dillon J. (2018), Assessing the Impact and Effectiveness of Hearing Voices Network Self-Help Groups, in "Community Mental Health Journal", 54, 2, pp. 184-8.

Lu Y. (2012), Household Migration, Social Support, and Psychosocial Health: The Perspective from Migrant-Sending Areas, in "Social Science & Medicine", 74, 2, pp. 135-42.

Lutfey K., Freese J. (2005), Toward Some Fundamentals of Fundamental Causality: Socioeconomic Status and Health in the Routine Clinic Visit for Diabetes, in "American Journal of Sociology", 110, 5, pp. 1326-72.

Lynch K. (1960), The Image of the City, The MIT Press, Cambridge (ma).

Macgeorge E. L., Feng B., Burleson B. (2011), Supportive Communication, in M. L. Knapp, J. A. Daly (eds.), The sage Handbook of Interpersonal Communication, sage, Thousand Oaks (ca), pp. 317-54.

Mackenbach J. P. et al. (2008), Socioeconomic Inequalities in Health in 22 European Countries, in "New England Journal of Medicine", 358, 23, pp. 2468-81.

Maddux J. E., Tangney J. P. (eds.) (2011), Social Psychological Foundations of Clinical Psychology, Guilford Press, New York.

Maguire L. (1987), Creating Ties and Maintaining Support: Networking and Self Help, in H. J. Altman (ed.), Alzheimer's Disease: Problems, Prospects, and Perspectives, Springer, Boston (ma), pp. 279-84.

id. (1994), Il lavoro sociale di rete, Erickson, Trento.

Maiter S. et al. (2008), Reciprocity: An Ethic for Community-Based Participatory Action Research, in "Action Research", 6, 3, pp. 305-25.

Maldonado-Torres N. (2007), On the Coloniality of Being: Contributions to the Development of a Concept, in "Cultural Studies", 21, 2-3, pp. 240-70.

id. (2016a), Outline of Ten Theses on Coloniality and Decoloniality, in https://fondation-frantzfanon.com/wp-content/uploads/2018/10/maldonado-torres_outline_of_ten_theses-10.23.16.pdf.

id. (2016b), Transdisciplinaridade e decolonialidade, in "Sociedade e Estado", 31, 1, pp. 75-97.

Malkoç A., Mutlu A. K. (2019), Mediating the Effect of Cognitive Flexibility in the Relationship between Psychological Well-Being and Self-Confidence: A Study on Turkish University Students, in "International Journal of Higher Education", 8, 6, pp. 278-87.

Malley J., Fernández J. L. (2010), Measuring Quality in Social Care Services: Theory and Practice, in "Annals of Public and Cooperative Economics", 81, 4, pp. 559-82.
mancoske r., standifer d., cauley c. (1994), The Effectiveness of Brief Counseling Services for Battered Women, in "Research on Social Work Practice", 4, 1, pp. 53-63.

Mannarini T., Fedi A. Trippetti S. (2010), Public Involvement: How to Encourage Citizen Participation, in "Journal of Community & Applied Social Psychology", 20, 4, pp. 262-74

Mannarini T., Rochira A., Talò C. (2014), Negative Psychological Sense of Community: Development of a Measure and Theoretical Implications, in "Journal of Community Psychology", 42, 6, pp. 673-88.

Mannarini T., Salvatore S. (2019), Making Sense of Ourselves and Others: A Contribution to the Community-Diversity Debate, in "Community Psychology in Global Perspective", 5, 1, pp. 26-37.

Mannarini T. et al. (2012), The Role of Affects in Culture-Based Interventions: Implications for Practice, in "Psychology", 3, 8, pp. 569-77.

Mannarini t. et al. (2021), Basic Human Values and Sense of Community as Resource and Responsibility, in "Journal of Community & Applied Social Psychology", 31, 2, pp. 123-41.

Manning C., Clayton S. (2018), Threats to Mental Health and Wellbeing Associated with Climate Change, in Id. (eds.), Psychology and Climate Change, Academic Press, Cambridge (ma), pp. 217-44.

Mannino F. V., Maclennan B. W., Shore M. F. (eds.) (1975), The Practice of Mental Health Consultation, Gardner Press, New York.

Manstead A. S. R. (2018), The Psychology of Social Class: How Socioeconomic Status Impacts Thought, Feelings, and Behaviour, in "British Journal of Social Psychology", 57, 2, pp. 267-91.

Marcomin F., Cima L. (2018), L'ecofemminismo in Italia. Le radici di una rivoluzione necessaria, Il Poligrafo, Padova.

Marecek J., Kravetz D. (1998), Power and Agency in Feminist Therapy, in I. B. Seu, C. Heenan (eds.), Feminism and Psychotherapy, sage, London, pp. 13-29. marks m. a.,

Mathieu J. E., Zaccaro S. J. (2001), A temporally Based Framework and Taxonomy of Team Processes, in "Academy of Management Review", 26, 3, pp. 356-76.

Marmot M. G., Syme S. L. (1976), Acculturation and Coronary Heart Disease, in "American Journal of Epidemiology", 104, 3, pp. 225-46.

Marrow A. J. (1977), Kurt Lewin fra teoria e pratica, La Nuova Italia, Firenze.

Marsella A. J., Snyder k. (1981), Stress, Social Supports and Schizophrenic Disorders: Toward an Interactional Model, in "Schizophrenia Bulletin", 7, 1, pp. 152-63.

Marta E., Pozzi M. (2004), Generatività e volontariato: quale connessione?, in C. Arcidiacono (a cura di), Volontariato e legami collettivi, FrancoAngeli, Milano, pp. 188-212.

Marta E., Scabini E. (2007), Famiglia e comunità: promuovere e rigenerare legami, reti, generatività sociale, in "Psicologia di comunità", 1, 3, pp. 9-20.

Marta E. et al. (2016), When Living and Working Well Together in Organizations Changes into Good Social Coexistence: The Talent Club Case, in "World Futures", 72, 5-6, pp. 266-8

Marta E. et al. (2017), Quando la generatività si fa concreta nel sociale. L'impegno civico e l'azione sociale, in Centro di ateneo studi e ricerche sulla famiglia, La generatività nei legami familiari e sociali, Vita e Pensiero, Milano, pp. 137-53.

Martin P., Lounsbury D., Davidson W. (2004), AICP as a Vehicle for Improving Community Life: An Historic-Analytic Review of the Journal's Contents, in "American Journal of Community Psychology", 34, pp. 163-73.

Martín-Baró I., Aron A., Corne S. (1994), Writings for a Liberation Psychology, Harvard University Press, Cambridge (ma).

Martinez K. K., Wong s. e. (2009), Using Prompts to Increase Attendance at Groups for Survivors of Domestic Violence, in "Research on Social Work Practice", 19, 4, pp. 460-3.

Martínez R. S. et al. (2011), Changes in Perceived Social Support and Socioemotional Adjustment across the Elementary to Junior High School Transition, in "Journal of Youth and Adolescence", 40, 5, pp. 519-30.

Martini E.R. (1996), Ricerca partecipata e sviluppo di comunità, in C. Arcidiacono,B. Gelli, A. Putton (a cura di), Empowerment sociale. Il futuro della solidarietà: modelli di psicologia di comunità, FrancoAngeli, Milano.

Martini E. R., Sequi R. (1988), Il lavoro nella comunità. Manuale per la formazione e l'aggiornamento dell'operatore sociale, Carocci, Roma.

id. (1995), La comunità locale. Approcci teorici e criteri di intervento, Carocci, Roma. martini e. r., torti a. (2003), Fare lavoro di comunità. Riferimenti teorici e strumenti operativi, Carocci, Roma.

Maton K. I. (1990), Towards the Use of Qualitative Methodology in Community Psychology Research, in P. Tolan et al. (eds.), Researching Community Psychology: Issue of Theory and Methods, American Psychological Association, Washington dc, pp. 153-6.

Matzat J. (2013). Self-Help Meets Science: Patient Participation in Guideline Development, in "Zeitschrift für Evidenz, Fortbildung und Qualität im Gesundheitswesen", 107, 4, pp. 314-9.

Maxwell J. C. (2002), Teamwork Makes the Dream Work, J. Countryman, Nashville (tn).

id. (2013), The 17 Indisputable Laws of Teamwork: Embrace Them and Empower your Team, HarperCollins, New York.

Maynard M.T. et al. (2015), Team Cohesion: A Theoretical Consideration of Its Reciprocal Relationships within the Team Adaptation Nomological Network, in E. Salas,W. B. Vessey, A. X. Estrada (eds.), Team Cohesion: Advances in Psychological Theory, Methods and Practice, Emerald, Bingley, pp. 83-111.

Mazerolle M. J., Singh G. (2002), Social Support and the Reduction of Discouragement after Job Displacement, in "The Journal of Socio-Economics", 31, 4, pp. 409-22.

Mcallister M. et al. (2012), Patient Empowerment: The Need to Consider it as a Measurable Patient-Reported Outcome for Chronic Conditions, in "Bmc Health Services Research", 12, 157, https://doi.org/10.1186/1472-6963-12-157.

Mcbride J. L. (2006), Effective Work Relationships: A Vital Ingredient in Your Practice, in "Family Practice Management", 13, 10, pp. 45-6.

Mccreight B. S. (2007), Narratives of Pregnancy Loss: The Role of Self-Help Groups in Supporting Parents, in "Medical Sociology Online", 2, 1, pp. 3-16.

Mclellan B. (1999), The Prostitution of Psychotherapy: A Feminist Critique, in "British Journal of Guidance and Counselling", 27, 3, pp. 325-37.

Mcmillan D. W., Chavis d. m. (1984), A Theory of Sense of Community, in "Journal of Community Psychology", 10, pp. 127-39.

Mcwhlrter E. H. (1991), Empowerment in Counseling, in "Journal of Counseling & Development", 69, 3, pp. 222-7.

Mcwhirter P. T. (2011), Differential Therapeutic Outcomes of Community-Based Group Interventions for Women and Children Exposed to Intimate Partner Violence, in "Journal of Interpersonal Violence", 26, 12, pp. 2457-82.

Mebane M. E., Benedetti M. (2022a), Community Profiling Focus Group: An Empowering Tool for Immigrant Community Groups, in "Journal of Prevention & Intervention in the Community", 50, 3, pp. 240-56.

id. (2022b), Hopes and Fears for the Future of Different Local Communities, in "Global Journal of Community Psychology Practice", 13, 2, https://www.gjcpp.org/ pdfs/MebaneBenedetti_Final.pdf.

Mebane M. E., Francescato D. (2018), Globalizzazione e innovazione tecnologiche: nuove sfide per gli psicologi di comunità del XXI secolo, in "Psicologia di comunità", 1, pp. 9-19.

Mebane M. E. (2019), Polarizzazione politica e attivismo delle donne militanti, in "La Camera blu", 20, pp. 29-39, http://www.serena.unina.it/index.php/camerablu/article/view/6159/7322.

Mebane W. (2020), Global Warming and Coronavirus Are not Distant Cousins, in "Wall Street International Journal", 25 March.

Menegatto M., Zamperini A. (2018), Coercizione e disagio psichico. La contenzione tra dignità e sicurezza, Il Pensiero Scientifico, Roma.

Meringolo P. (ed.) (2020), Preventing Violent Radicalisation in Europe: Multidisciplinary Perspectives, Springer, Chaim.

Meringolo P., Cecchini C., Donati C. (2022), Migrants as Suspects? A Participatory Consensus Conference to Promote Well-Being and Inclusion, in "Journal of Prevention & Intervention in the Community", 50, 3, pp. 224-39.

Meringolo P, Volpi C., Chiodini M. (2019), Community Impact Evaluation: Telling a Stronger Story, in "Community Psychology in Global Perspective", 5, 1, pp. 85-106.

Mertens D. M. (2009), Transformative Research and Evaluation, Guilford Press, New York.

Messina p. (2011), Policies for Strategic Territorial Development: Inter-Municipality Association as a Form of Network Governance. The Italian Experience, in "Eastern Journal of European Studies", 2, pp. 111-28.

Migliorini L., Tartaglia S. (2021), Reti e sostegno sociale, in Arcidiacono et al. (2021), vol. 1, pp. 160-76.

Mignolo W. D., Walsh C. E. (2018), On Decoloniality: Concepts, Analytics, Praxis, Duke University Press, Durham (nc).

Mikulincer M., Shaver P. R. (2009), An Attachment and Behavioral Systems Perspective on Social Support, in "Journal of Social and Personal Relationships", 26, 1, pp. 7-19.

Mills R. c., Kelly j. g. (1972), Cultural Adaptation and Ecological Analogies: Analysis of Three Mexican Villages, in S. E. Golann, C. Eisdorfer (eds.), Handbook of Community Mental Health, Appleton-Century-Crofts, New York, pp. 157-205.
Mintzberg h. (2010), Il lavoro manageriale, FrancoAngeli, Milano.

Misuraca l. (2020), I medici di Bergamo ai colleghi stranieri: "Evitate gli errori fatti in Lombardia", in "Il salvagente", 25 marzo, https://ilsalvagente.it/2020/03/25/imedici-di-bergamo-ai-colleghi-stranieri-evitate-gli-errori-fatti-in-lombardia/.
Moavero Milanesi e. (2022), L'Europa ritrovi lo slancio e la concretezza di Shuman, in "Corriere della Sera", 8 maggio.

Molina O. et al. (2009), Divorcing Abused Latina Immigrant Women's Experiences with Domestic Violence Support Groups, in "Journal of Divorce & Remarriage", 50, 7, pp. 459-71.

Montero M. (2007), The Political Psychology of Liberation: From Politics to Ethics and Back, in "Political Psychology", 28, 5, pp. 517-33.

Montero M., Sonn . C. C., Burton M. (2017), Community Psychology and Liberation Psychology: A Creative Synergy for an Ethical and Transformative Praxis, in M. A. Bond, I. Serrano-García, C. B. Keys (eds.), apa Handbook of Community Psychology, vol. 1: Theoretical Foundations, Core Concepts, and Emerging Challenges, American Psychological Association, Washington dc, pp. 149-67.

Moos R. H., Mitchell R. E. (1982), Social Networks Resources and Adaptation: A Conceptual Framework, in T. A. Wills (ed.), Basic Processes in Helping Relationships, Academic Press, New York, pp. 213-32.

Moos R. H., Moos B. S. (2006), Participation in Treatment and Alcoholics Anonymous: A 16-Year Follow-Up of Initially Untreated Individuals, in "Journal of Clinical Psychology", 62, 6, pp. 735-50.

Mora M. A. et al. (2022), Definitions, Instruments and Correlates of Patient Empowerment: A Descriptive Review, in "Patient Education and Counseling", 105, 2, pp. 346-55.

Morales-campos D., Casillas M., McCurdy S. A. (2009), From Isolation to Connection: Understanding a Support Group for Hispanic Women Living with Gender-Based Violence in Houston, Texas, in "Journal of Immigrant Minority Health", 11, pp. 57-65.

Moretti M. et al. (2012), Interpersonal Support Evaluation List (isel). Un contributo alla validazione e all'applicazione nel contesto italiano, in "Psicologia sociale", 7, 3, pp. 447-69.

Morgan G. (1986), Images of Organization, sage, Thousand Oaks (ca).

id. (1994), Images. Le metafore dell'organizzazione, FrancoAngeli, Milano.

id. (1996), Immaginizzazione. Un modo nuovo per agire nelle organizzazioni, Franco Angeli, Milano.

Morganti M (1998), Non profit: produttività e benessere. Come coniugare efficienza e solidarietà nelle organizzazioni del terzo settore, FrancoAngeli, Milano.

Morrison E. W. (2002), Newcomers' Relationships: The Role of Social Network Ties during Socialization, in "Academy of Management Journal", 45, 6, pp. 1149-60.

Mucchielli R. (1970), La dinamica di gruppo, Elledici, Torino.

id. (1980), Le travail en équipe, esf, Paris.

id. (1986), Come condurre le riunioni. Teoria e pratica, ElleDiCi, Torino.

Murray C. E. et al. (2015), Survivors of Intimate Partner Violence as Advocates for Social Change, in "Journal for Social Action in Counseling & Psychology", 7, 1, pp. 84-100.

Murrell S. (1973), Community Psychology and Social Systems: A Conceptual Framework and Intervention Guide, Behavioral Publications, New York.

Muti P. (1986), Il lavoro di gruppo. Aspetti teorici e pratici per una diagnosi dei problemi di gruppo nelle aziende, FrancoAngeli, Milano.

Meyers S. A., Goodboy a. k. (2005), A Study of Grouphate in a Course on Small Group Communication, in "Psychological Reports", 97, 2, pp. 381-6.

Nagayama Hall G. C. (2005), Introduction to the Special Section on Multicultural and Community Psychology: Clinical Psychology in Context, in "Journal of Consulting and Clinical Psychology", 73, 5, pp. 787-9.

National mental health consumers' self-help clearinghouse (2011), Starting a Self-Help/Advocacy Group, in https://www.mhselfhelp.org/technicalassistance-guide/2011/4/1/starting-a-self-helpadvocacy-group-pdf.html. national research council (ed.) (2001), Envisioning the National Health Care Quality Report, The National Academies Press, Washington dc.

Nayar K. R., Kyobutungi C., Razum O. (2004), Self-Help: What Future Role in Health Care for Low and Middle-Income Countries?, in "International Journal for Equity in Health", 3, 1, pp. 1-10.

Ndlovu-Gatsheni S. J. (2020), Decolonization, Development and Knowledge in Africa: Turning over a New Leaf, Routledge, London.

Neal Z. P., Neal J. W. (2017), Network Analysis in Community Psychology: Looking back, Looking forward, in "American Journal of Community Psychology", 60, 1-2, pp. 279-95.

Nelson G., Prilleltensky I. (2005), Community Psychology: In Pursuit of Liberation and Well-Being, Palgrave Macmillan, New York.

Neri C. (2021), Il gruppo come cura, Raffaello Cortina Editore, Milano.

Nichols C. (2021), Self-Help Groups as Platforms for Development: The Role of Social Capital, in "World Development", 146, https://doi.org/10.1016/j.worlddev.2021.105575.

Nickel S., Trojan A., Kofahl C. (2017), Involving Self-Help Groups in HealthCare Institutions: The Patients' Contribution to and Their View of "Self-Help Friendliness" as an Approach to Implement Quality Criteria of Sustainable Co-Operation, in "Health Expectations: An International Journal of Public Participation in Health Care and Health Policy", 20, 2, pp. 274-87.

Nielsen K. S. et al. (2020), How Psychology Can Help Limit Climate Change, in "American Psychologist", 76, 1, pp. 130-44.

Nikkhah H., Redzuan M. (2009), Participation as a Medium of Empowerment in Community Development, in "European Journal of Social Sciences", 11, 1, pp. 170-6.

Nisbet R. A. (1960), Moral Values and Community, in "International Review of Community Development", 5, pp. 77-87.

Norwood R. (1985), Donne che amano troppo, Feltrinelli, Milano.

Novaco R. W., Monahan j. (1980), Research in Community Psychology: An Analysis of Work Published in the First Six Years of the American Journal of Community Psychology, in "American Journal of Community Psychology", 8, pp. 131-45.

Novaga M., Borsatti G. (1979), Il lavoro di gruppo, Patron, Bologna.

Noventa a. (1996), Gruppi di auto-mutuo-aiuto, dall'approccio familiare all'approccio di comunità, in "Il seme e l'albero", 8, pp. 39-41.

Noventa A., Nava R., Oliva F. (1990), Self-help. Promozione della salute e gruppi di auto-aiuto, Edizioni Gruppo Abele, Torino.

Nowell B., Boyd N. (2010), Viewing Community as Responsibility as Well as Resource: Deconstructing the Theoretical Roots of Psychological Sense of Community, in "Journal of Community Psychology", 38, 7, pp. 828-41.

id. (2014), Sense of Community Responsibility in Community Collaboratives: Advancing a Theory of Community as Resource and Responsibility, in "American Journal of Community Psychology", 54, 3-4, pp. 229-42.

Obst P., Smith S. G., Zinkiewicz L. (2002), An Exploration of Sense of Community, part 3: Dimensions and Predictors of Psychological Sense of Community in Geographical Communities, in "Journal of Community Psychology", 30, 1, pp. 119-33.

Ockene J. K. et al. (2007), Integrating Evidence-Based Clinical and Community Strategies to Improve Health, in "American Journal of Preventive Medicine", 32, 3, pp. 244-52.

Ogbe E. et al. (2021), The Potential Role of Network-Oriented Interventions for Survivors of Sexual and Gender-Based Violence among Asylum Seekers in Belgium, in "bmc Public Health", 21, https://doi.org/10.1186/s12889-020-10049-0.

Oliva F., Croce M., Merlo R. (1995), Appunti di metodo per un intervento di rete con approccio egocentrato, in aa.vv., L'intervento di rete, Edizioni Gruppo Abele, Torino, pp. 68-78.

OMS (2021), Quali sono le evidenze sul ruolo delle arti nel miglioramento della salute e del benessere? Una scoping review. Rapporto completo, https://www.dors.itdocumentazione/testo/202108/report2019OMSartisalute_20210727.pdf.

O'neill D. W. et al. (2018), A Good Life for All within Planetary Boundaries, in "Nature Sustainability", 1, 2, pp. 88-95.

Orford J. (1992), Psicologia di comunità. Aspetti teorici e professionali, Franco Angeli, Milano.

id. (1996), Verso una teoria generale per la psicologia di comunità. Requisiti per una teoria generale, in Arcidiacono C., Gelli B., Putton A. (1996).

id. (1998), Towards a General Theory for Community Psychology, Paper presented at the ii European Conference of Community Psychology, Lisbon.

id. (2008), Community Psychology: Challenges, Controversies and Emerging Consensus, Wiley, Chichester.

id. (2018), Power and Addiction, in H. Pickard, S. H. Ahmed (eds.), The Routledge Handbook of Philosophy and Science of Addiction, Routledge, London, pp. 209-20.

Ornelas J. (1997), Psicologia comunitária. Origens, fundamentos e áreas de intervenção, in "Análise psicológica", 15, 2, pp. 375-88.

id. (2000), Diversidade e desenvolvimento comunitário, in J. Ornelas, S. Maria (eds.), Diversidade e multiculturalidade, Instituto Superior de Psicologia Aplicada, Lisboa, p. 384.

id. (2008), Psicologia Comunitária, Fim de Século, Lisboa.

id. (2022), Desinstitucionalização. Origem do modelo ecológico e colaborativo em saúde mental, Lição de Jubilação, ispa-Instituto Universitário, Lisboa.

Ornelas J., Esposito F., Sacchetto B. (2014), Contributions of a Community-Based Organization for the Transformation of the Mental Health System in Portugal, in "Rivista di psicologia clinica", 1, pp. 220-39

Ornelas J. et al. (2014), Housing First: An Ecological Approach to Promoting Community Integration, in "European Journal of Homelessness, 8, 1, pp. 29-56.

O'shaughnessy B. et al. (2021), Home as a Base for a Well-Lived Life: Comparing the Capabilities of Homeless Service Users in Housing First and the Staircase of Transition in Europe, in "Housing, Theory and Society", 38, 3, pp. 343-64.

Ozaralli N. (2003), Effects of Transformational Leadership on Empowerment and Team Effectiveness, in "Leadership & Organization Development Journal", 24, 6, pp. 335-44.

Pagano M. E., Post S. G., Johnson s. m. (2010), Alcoholics Anonymous-Related Helping and the Helper Therapy Principle, in "Alcoholism Treatment Quarterly", 29, 1, pp. 23-34.

Page L. (1998), The Crisis in Mental Health Theory, in "International Journal of Mental Health", 27, 1, pp. 33-61.

Palmonari A., Zani B. (1980), Psicologia sociale e di comunità, Il Mulino, Bologna.

Paris j. (2014), Modernity and Narcissistic Personality Disorder, in "Personality disorders: Theory, Research, and Treatment", 5, 2, pp. 220-6.

Park E. et al. (2014), Cognition and Needs on the Patients for the Activation of Oral Health Education Using Smart Phone Applications, in "Journal of Korean Academy of Dental Administration", 2, 1, pp. 45-59.

Park R. E. (1952), Human Communities: The City and Human Ecology, The Free Press, Glencoe (il).

Parker i. (1997), Psychoanalytic Culture: Psychoanalytic Discourse in Western Society, sage, London.

Parkman T. J., Loyd C., Splisbury K. (2015), Self-Help Groups for Alcohol Dependency: A Scoping Review, in "Journal of Groups in Addiction & Recovery", 10, 2, pp. 102-24.

Perkins d. d. (1995), Speaking Truth to Power: Empowerment Ideology as Social Intervention and Policy, in "Community Psychology", 23, 5, pp. 765-94.

id. (2010), Empowerment, in R. A. Couto (ed.), Political and Civic Leadership: A Reference Handbook, sage, Thousand Oaks (ca), pp. 207-18.

Perkins D. D., Zimmermann m. (1995), Empowerment Theory, Research and Application, in "American Journal of Community Psychology", 23, 5, pp. 569-79.

Perry J. (2000), Bringing Society In: Toward a Theory of Public-Service Motivation, in "Journal of Public Administration Research and Theory", 10, 2, pp. 471-88.

Peterson N. A., Speer P. W., Mcmillan D. W. (2008), Validation of a Brief Sense of Community Scale: Confirmation of the Principal Theory of Sense of Community, in "Journal of Community Psychology", 36, 1, pp. 61-73.

Peterson n. a., Zimmerman m. a. (2004), Beyond the Individual: Toward a Nomological Network of Organizational Empowerment, in "American Journal of Community Psychology", 34, 1-2, pp. 129-45.

Peterson N. A. et al. (2008), Community Organizations and Sense of Community: Further Development in Theory and Measurement, in "Journal of Community Psychology", 36, 6, pp. 798-813.

Pettit J. (2012), Empowerment and Participation: Bridging the Gap between Understanding and Practice, in "United Nations Headquaters", 10, 6, p. 39.

Piaget J. (1967), Lo sviluppo mentale del bambino, Einaudi, Torino.

Piccardo C. (1992), Empowerment, una nuova parola d'ordine per lo sviluppo organizzativo degli anni '90, in "Sviluppo e organizzazione", 134, pp. 21-31.

id. (1995), Empowerment. Strategie di sviluppo organizzativo centrate sulla persona, Raffaello Cortina Editore, Milano.

Pietrantoni L., Prati G. (2008), Empowerment psicologico: contributo alla validazione italiana della scala di Spreitzer, in "Risorsa Uomo", 14, 3, pp. 325-38.

Pinderhuhes E. B. (1983), Empowerment of Our Clients and for Ourselves, in "Social Casework: Journal of Contemporary Social Work", 64, pp. 331-8.

Pistrang N., Barker C., Humphreys K. (2008), Mutual Help Groups for Mental Health Problems: A Review of Effectiveness Studies, in "American Journal of Community Psychology", 42, 1-2, pp. 110-21.

Plante T. G. (1999), Contemporary Clinical Psychology, Wiley, New York.

Pluchino A. et al. (2011), Accidental Politicians: How randomly Selected Legislators Can Improve Parliament Efficiency, in "Physica A: Statistical Mechanics and Its Applications", 390, 21-22, pp. 3944-54.

Poli S. (1998), La promozione della salute nel processo di riforma della scuola, in R. Piccione, A. Grispini (a cura di), Prevenzione e salute mentale. Fondamenti, pratiche, prospettive, Carocci, Roma.

Pollina C. P., Magatti p. (2013), Gruppo di lavoro, gruppo operativo. Guida al coordinamento dei gruppi, Guerini e Associati, Milano.

Prati G., Albanesi C., Pietrantoni L. (2016), The Reciprocal Relationship between Sense of Community and Social Well-Being: A Cross-Lagged Panel Analysis, in "Social Indicators Research", 127, 3, pp. 1321-2.

Prati G., Cicognani E ., Albanesi c. (2017), Psychometric Properties of a Multidimensional Scale of Sense of Community in the School, in "Frontiers in Psychology",   8,   https://doi.org/10.3389/fpsyg.2017.01466.

id. (2018), The Influence of School Sense of Community on Students' Well-Being: A Multilevel Analysis, in "Journal of Community Psychology", 46, 7, pp. 917-24.

Pretty G. H., Mccarthy M., Catano V. (1992)., Psychological Environments and Burnout: Gender Considerations in the Corporation, in "Journal of Organizational Behavior", 13, pp. 701-11

Prezza M., Costantini S. (1998), Sense of Community and Life Satisfaction: Investigation in Three Different Contexts, in "Journal of Community and Applied Social Psychology", 8, 3, pp. 181-94.

Prezza M., Drahorad c., Tomai m. (1993), I gruppi di autoaiuto e il sistema formale di cura: quale collaborazione possibile?, in Francescato, Leone, Traversi (1993), pp. 273-95.

Prezza m., Pacilli M. G. (2002), Perceived Social Support from Significant Others, Family and Friends and Several Socio-Demographic Characteristics, in "Journal of Community & Applied Social Psychology", 12, 6, pp. 422-9

Prezza M., et al. (1999), La scala italiana del senso di comunità, in "Psicologia della salute", 3-4, pp. 135-59.

Prezza M., et al. (2001a), Sense of Community Referred to the Whole Town: Its Relations with Neighboring, Loneliness, Life Satisfaction and Area of Residence, in "Journal of Community Psychology", 29, 1, pp. 29-52.

Prezza M., et al. (2001b), The Influence of Psychosocial and Environmental Factors on Children's Independent Mobility and Relationship to Peer Frequentation, in "Journal of Community & Applied Social Psychology", 11, 6, pp. 435-50.

Prezza M., et al. (2009), The mtsocs: A Multidimensional Sense of Community Scale for Local Communities, in "Journal of Community Psychology", 37, 3, pp. 305-26.

Prilleltensky I., Gonick L. (1996), Politics Change, Oppression Remains, in "Political Psychology", 17, pp. 127-48.

Prilleltensky I., Nelson G. (2000), Promoting Child and Family Wellness: Priorities for Psychological and Social Interventions, in "Community & Applied Social Psychology", 10, 2, pp. 85-105.

id. (2009), Community Psychology: Advancing Social Justice, in D. Fox, I. Prilleltensky, S. Austin (eds.), Critical Psychology: An Introduction, sage, Los Angeles (ca)London, pp. 126-43.

Prilleltensky I., Prilleltensky O. (2006), Promoting Well-Being: Linking Personal, Organizational, and Community Change, Wiley, Hoboken (nj).

Prilleltensky I. et al. (2015), Assessing Multidimensional Well-Being: Development and Validation of the i coppe Scale, in "Journal of Community Psychology", 43, 2, pp. 199-226.

Prince P. N., Gerber G. J. (2005), Subjective Well-Being and Community Integration among Clients of Assertive Community Treatment, in "Quality of Life Research", 14, 1, pp. 161-9.

Privitera G. (1987), Il trattamento delle informazioni. Costruzione di griglie per l'analisi sociale, Aies, Firenze.

Procentese F., De Carlo F., Gatti F. (2019), Civic Engagement within the community and Sense of Responsible Togetherness, in "tpm: Testing, Psychometrics, Methodology in Applied Psychology", 26, 4, pp. 513-25.

Procentese F., Gatti F. (2021), Sense of Responsible Togetherness, Sense of Community, and Civic Engagement Behaviours: Disentangling an Active and Engaged Citizenship, in "Journal of Community & Applied Social Psychology", 32, 2, pp. 187-96.

Procentese F., Marta E. (2021), Ricerca azione partecipata, in Arcidiacono et al. (2021), vol. ii, pp. 98-191.

Procentese f. et al. (2021), Gruppo di lavoro su setting di intervento online in psicologia di comunità, in "Psicologia di comunità", 17, 1, pp. 113-22.

Procentese F. et al. (2023), Individual and Community-Related Paths to Civic Engagement: A Multiple Mediation Model Deepening the Role of Sense of Responsible Togetherness, Community Trust, and Hope, in "Community Psychology in Global Perspective", 9, 1, pp. 64-83.

Puddifoot J. E. (1996), Some Initial Considerations in the Measurement of Community Identity, in "Journal of Community Psychology", 24, 4, pp. 327-6.

Putman A. O. (1991), Empowerment: In Search of Viable Paradigm, in "Performance Improvement Quarterly", 4, 4, pp. 4-11.

Putnam R. D. (2023), Comunità contro individualismo. Una parabola americana, Il Mulino, Bologna.

Putton A. (1999), Empowerment e scuola. Metodologie di formazione nell'organizzazione educativa, Carocci, Roma.

Qaddumi B. et al. (2021), The Factors Affecting Team Effectiveness in Hospitals: The Mediating Role of Using Electronic Collaborative Tools, in "Journal of Interprofessional Education & Practice", 24, https://doi.org/10.1016/j.xjep.2021.100449.

Quaglino G. P. (1985), Fare formazione, Il Mulino, Bologna.

id. (1996), Psicodinamica della vita organizzativa. Competizione, difese, ambivalenza nelle relazioni di lavoro, Raffaello Cortina Editore, Milano.

Quaglino G. P., Casagrande S., Castellano A. (1992), Gruppo di lavoro, lavoro di gruppo. Un modello di lettura della dinamica di gruppo. Una proposta di intervento nelle organizzazioni, Raffaello Cortina Editore, Milano.

Quaglino G. P., Cortese C. G. (2003), Gioco di squadra. Come un gruppo di lavoro può diventare una squadra eccellente, Raffaello Cortina Editore, Milano.

Racamier P. C. (1997), Una comunità di cura psicoterapica, in "Psichiatrie française", 1, pp. 20-32.

Rapkin b. d., Mulvey e. p. (1990), Towards Excellence in Quantitative Community Research, in P. P. Tolan et al. (eds.), Researching Community Psychology: Issue of Theory and Methods, American Psychological Association, Washington dc.

Rappaport J. (1977), Community Psychology: Values, Research, and Action, Holt, Rinehart, & Winston, New York.

id. (1980), The President's Column Informing Social Ecology, in "Division of Community Psychology Newsletter", 13, 3, pp. 1-2

id. (1981), In Praise of Paradox: A Social Policy of Empowerment over Prevention, in "American Journal of Community Psychology", 9, 1, pp. 1-26.

id. (1984), Seeking Justice in the Real World: A Further Explication of Value Context, in "Journal of Community Psychology", 12, 3, pp. 208-16.

id. (1985), The Power of Empowerment Language, in "Social Policy", 15, pp. 15-21.
id. (1990), Research Methods and the Empowerment Social Agenda, in P. Tolan et al. (eds.), Researching Community Psychology: Issues of Theory and Methods, American Psychological Association, Washington dc, pp. 51-63.

id. (1995), Empowerment Meets Narrative: Listening to Stories and Creating Settings, in "American Journal of Community Psychology", 23, 5, pp. 795-807.

id. (2000), Community Narratives: Tales of Terror and Joy, in "American Journal of Community Psychology", 28, 1, pp. 1-24.

id. (2004), On Becoming a Community Psychologist: The Intersection of Autobiography and History, in "Journal of Prevention & Intervention in the Community", 28, 1-2, pp. 15-39.

Rappaport J., Seidman E. (eds.) (2000), Handbook of Community Psychology, Springer, New York.

Raworth k. (2017), L'economia della ciambella. Sette mosse per pensare come un economista del xxi secolo, Ambiente, Milano.

Reale E. (2000), Vita quotidiana delle donne: rischi di violenza e disagio psichico, in P. Romito (a cura di), Violenza alle donne e risposte delle istituzioni. Prospettive internazionali, FrancoAngeli, Milano, pp. 49-64.

Reblin M., Uchino B. N. (2008), Social and Emotional Support and Its Implication for Health, in "Current Opinion in Psychiatry", 21, 2, pp. 201-5.

Redman W. K., Cullari S., Farris H. E. (1985), An Analysis of Some Important Tasks and Phases in Consultation, in "Journal of Community Psychology", 13, pp. 375-86.

Reich S. et al. (eds.) (2007), International Community Psychology: History and Theories, Springer, New York.

Repucci N. (1981), Entry Restriction by Licensure to Community Psychology: Problems with the Strategy, in "Division of Community Psychology Newsletter", 15, 1, pp. 6-8.

Reynolds J. S., Perrin n. a. (2004), Mismatches in Social Support and Psychosocial Adjustment to Breast Cancer, in "Health Psychology", 23, 4, pp. 425-30.

Reynolds P., Kaplan G. A. (1990), Social Connections and Risk for Cancer: Prospective Evidence from the Alameda County Study, in "Behavioral Medicine", 16, 3, pp. 101-10.

Ridolfi L. (2013a), La Community care come modello di integrazione sociosanitaria a livello territoriale, in "Professioni infermieristiche", 66, 4, https://www.profinf.net/ pro3/index.php/IN/article/view/44

id. (2013b), La Community care come modello di rete sociosanitaria, in C. Clemente, P. Guzzo (a cura di), Sistemi sociosanitari regionali tra innovazioni e spendibilità, Cacucci, Bari, pp. 213-22.

Riemer M., Harré N. (2017), Environmental Degradation and Sustainability: A Community Psychology Perspective, in M. A. Bond et al. (eds.), apa Handbook of Community Psychology, vol. 2: Methods for Community Research and Action for Diverse Groups and Issues, American Psychological Association, Washington dc, pp. 441-55.

Riessman F. (1965), The "Helper" Therapy Principle, in "Social Work", 10, 2, pp. 27-32.

Rimé B. (1993), Le partage social des émotions, in B. Rimé, K. Schere (éds.), Les émotions, Delachaux et Niestlé, Paris, pp. 271-300.

Rizalar s. et al. (2014), Effect of Perceived Social Support on Psychosocial Adjustment of Turkish Patients with Breast Cancer, in "Asian Pacific Journal of Cancer Prevention", 15, 8, pp. 3429-34.

Roberti F. (2017), La funzione psicologico clinica in casi di assistenza domiciliare rivolta a minori. Funzione integrativa scuola-famiglia, in "Quaderni della Rivista di psicologia clinica", 1, pp. 69-79.

Roberts B. W., Helson r. (1997), Changes in Culture, Changes in Personality: The Influence of Individualism in a Longitudinal Study of Women, in "Journal of Personality and Social Psychology", 72, 3, pp. 641-51.

Robinson W. L. et al. (2017), Advancing Prevention Intervention From Theory to Application: Challenges and Contributions of Community Psychology, in M. A. Bond, I. Serrano-García, C. B. Keys (eds.), apa Handbook of Community Psychology: Methods for community research and action for diverse groups and Issues, apa, Washington dc, vol. 2, pp. 193-214.

Robles T. F. et al. (2014), Marital Quality and Health: A Meta-Analytic Review, in "Psychological Bulletin", 140, 1, pp. 140-87.

Rogers C. R. (1971), I gruppi d'incontro, Astrolabio, Roma.

Rosen J., Painter G. (2019), From Citizen Control to Co-Production: Moving beyond a Linear Conception of Citizen Participation, in "Journal of the American Planning Association", 85, 3, pp. 335-47.

Rosenthal T. (2018), Immigration and Acculturation: Impact on Health and WellBeing of Immigrants, in "Current Hypertension Reports", 20, 8, pp. 1-8.

Ross A., Searle M. (2019), Age Related Differences in Neighborhood Sense of Community: Impacts of the Neighborhood Environment and Leisure Time Physical Activity, in "International Journal of Community Well-Being", 2, 1, pp. 41-59.

Ross A., Talmage C. A., Searle M. (2019), Toward a Flourishing Neighborhood: The Association of Happiness and Sense of Community, in "Applied Research Quality Life", 14, 5, pp. 1333-52

Rossi P.H. Freeman H.E, (1982), Evaluation, sage, Beverly Hills (ca). rothman j. (1974), Three Models of Community Organization Practice, in F. Cox et al. (eds.), Strategies of Community Organization: A Book of Readings, Peacock, Ithaca (ny).

Roulston K. (2019), Preparing Researchers to Conduct Interdisciplinary, MultiMethod Qualitative Research, in "The Qualitative Report", 24, 9, pp. 2259-92.

Rowthorn A., Rowthorn J. (2018), God's Good Earth: Praise and Prayer for Creation, Liturgical Press, Collegeville (mn).

Rözer J. J., Poortman A. R., Mollenhorst G. (2017), The Timing of Parenthood and Its Effect on Social Contact and Support, in "Demographic Research", 36, 1, pp. 1889-916.

Ruiu M. L. (2015), The Effects of Cohousing on the Social Housing System: The Case of the Threshold Centre, in "Journal of Housing and the Built Environment", 30, 4, pp. 631-44.

Rutten L. J. F. et al. (2005), Information Needs and Sources of Information among Cancer Patients: A Systematic Review of Research (1980-2003), in "Patient Education and Counseling", 57, 3, pp. 250-61.

Ryan W. (1971), Blaming the Victim, Vintage Books, New York.

Salas E., Rosen M. A., King H. (2007), Managing Teams Managing Crises: Principles of Teamwork to Improve Patient Safety in the Emergency Room and beyond, in "Theoretical Issues in Ergonomics Science", 8, 5, pp. 381-94.

Salas E., Bagl K. C., Burke C. S. (2004), 25 Years of Team Effectiveness in Organizations: Research Themes and Emerging Needs, in "International Review of Industrial and Organizational Psychology", 19, pp. 47-92.

Salzer M. S., Rappaport J., Segre I. (2001), Mental Health Professionals' Support of Self-Help Groups, in "Journal of Community & Applied Social Psychology", 11, 1, pp. 1-10.

Sánchez Vidal A. (2007), Manual de psicología comunitaria. Un enfoque integrado, Ediciones Piramide, Madrid.

Sánchez Vidal A., Musitu G. (1996), Intervención comunitaria. Aspectos científicos, técnicos y valorativos, eub, Barcelona.

Santinello M., Crespi I., Vieno A. (a cura di) (2000), La prevenzione nella scuola e nella comunità. Esperienze a confronto, Atti del convegno, Edizioni Unipress, Padova.

Santinello M., Vieno A., Martini M. C. (2006), La diffusione e i predittori dell'uso di alcol e tabacco in preadolescenza, in "Giunti Organizzazioni Speciali", 249, pp. 3-15.

Santinello M. et al. (2018), Photovoice nei luoghi di lavoro: quali caratteristiche organizzative influenzano il lavoro con le persone senza dimora?, in "Psicologia di comunità", 2, pp. 37-49.

Sanyal P. (2009), From Credit to Collective Action: The Role of Microfinance in Promoting Women's Social Capital and Normative Influence, in "American Sociological Review", 74, 4, pp. 529-50.

Sarason I. G. et al. (1983), Assessing Social Support: The Social Support Questionnaire, in "Journal of Personality and Social Psychology", 44, 1, pp. 127-39.

Sarason S. B. (1974), The Psychological Sense of Community: Prospects for a Community Psychology, Jossey-Bass, San Francisco (ca).

Sarbin T. R., Kitsuse I. i. (1994), Constructing the Social, sage, Thousand Oaks (ca).

Sbarra D. A., Law R. W., Portley R. M. (2011), Divorce and Death: A Meta-Analysis and Research Agenda for Clinical, Social, and Health Psychology, in "Perspectives on Psychological Science", 6, 5, pp. 454-74.

Scabini E., Marta E. (2007), Famiglia e comunità: promuovere e rigenerare legami, reti, generatività sociale, in "Psicologia di comunità", 1, pp. 9-27.

Schetter C. D. (2017), Moving Research on Health and Close Relationships Forward – A Challenge and an Obligation: Introduction to the Special Issue, in "American Psychologist", 72, 6, pp. 511-6.

Schmutz J. D., Meier L. L., Manser t. (2019), How Effective Is Teamwork Really? The Relationship between Teamwork and Performance in Healthcare Teams: A Systematic Review and Meta-Analysis, in "bmj Open", 9, 9, doi: 10.1136/bmj open-2018-028280.

Schoenborn C. A. (1986), Health Habits of us Adults, 1985: The "Alameda 7" Revisited, in "Public Health Reports", 101, 6, pp. 571-80.

Schulz-Nieswandt F. (2011), Gesundheitsselbsthilfegruppen und Selbsthilfeorganisationen in Deutschland, Nomos, Baden-Baden.

Schwartz J. (1997), Meaning vs. Medical Necessity: Can Psychoanalytic Treatments Exist in a Managed Care World?, in "Psychotherapy: Theory, Research, Practice, Training", 34, 2, pp. 115-23.

Schwartz M. B., Brownell k. d. (2007), Actions Necessary to Prevent Childhood Obesity: Creating the Climate for Change, in "Journal of Law, Medicine & Ethics", 35, 1, pp. 78-89.

Segal S. P., Silverman C. J., Temkin T. L. (2010), Self-Help and Community Mental Health Agency Outcomes: A Recovery-Focused Randomized Controlled Trial, in "Psychiatric Services", 61, 9, pp. 905-10.

Seligman M. (1990), Imparare l'ottimismo. Come cambiare la vita cambiando il pensiero, Giunti, Firenze.

Seligman M., Csíkszentmihályi m. (2000), Positive Psychology: An Introduction, in "American Psychologist", 55, 1, pp. 5-14.

Senge P. (2016), The Learning Organization, in "Creative and Knowledge Society", 6, 1, pp. 1-13.

Serrano-garcía I. (1982a), The Future of Education in Community Psychology, in "Division of Community Psychology Newsletter", 15, 3, pp. 6-7.

id. (1982b), Community Psychology Is Alive and Well in the Caribbean, in "Division of Community Psychology Newsletter", 15, 4, p. 25.

Setia M., Singh Tandon M., Brijpal (2017), Impact Study of Women Empowerment through Self-Help Groups: A Study of Haryana, in "Global Journal of Enterprise Information System", 9, 50, doi: 10.18311/gjeis/2017/16010.

Sgarro M. (1988), Il sostegno sociale, Kappa, Roma.

Shah S. K., Corley k. g. (2006), Building better Theory by Bridging the Quantitative-Qualitative Divide, in "Journal of Management Studies", 43, 8, pp. 1821-35.

Shelton R. C. et al. (2019), Use of Social Network Analysis in the Development, Dissemination, Implementation, and Sustainability of Health Behavior Interventions for Adults: A Systematic Review, in "Social Science & Medicine", 220, pp. 81-101.

Shepherd M. D. et al. (1999), Continuum of Professional Involvement in Self-Help Groups, in "Journal of Community Psychology", 27, 1, pp. 39-53.

Sherbourne C. D., Stewart a. l. (1991), The Moos Social Support Survey, in "Social Science & Medicine", 32, 6, pp. 705-14.

Shin S., Park H. (2017), Effect of Empowerment on the Quality of Life of the Survivors of Breast Cancer: The Moderating Effect of Self-Help Group Participation, in "Japan Journal of Nursing Science", 14, 4, pp. 311-9.

Shipper F., Manz C. C. (1992), Employee Self-Management without Formally Designated Teams: An Alternative Road to Empowerment, in "Organizational Dynamics", 20, 3, pp. 48-61.

Shor E., Roelfs D. J. (2015), Social Contact Frequency and All-Cause Mortality: A Meta-Analysis and Meta-Regression, in "Social Science & Medicine", 128, pp. 76-86.

Siegel C. et al. (2006), Tenant Outcomes in Supported Housing and Community Residences in New York City, in "Psychiatric Services", 57, 7, pp. 982-91.

Silberfeld M. (1978), Psychological Symptoms and Social Supports, in "Social Psychiatry", 13, pp. 11-7.

Silverman P. R. (1989), I gruppi di mutuo aiuto. Come l'operatore sociale li può organizzare e sostenere, Erickson, Trento.

Sinek S. (2014), Leaders Eat Last: Why Some Teams Pull Together and Others Don't, Penguin, London.

Skelcher C. (1993), Involvement and Empowerment in Local Public Service, in "Public Money and Management", July-September, pp. 13-20.

Skovholt T. M. (1974), The Client as Helper: A Means to Promote Psychological Growth, in "Counseling Psychologist", 4, 3, pp. 58-64.

Sloan T. S. (2016), Damaged Life The Crisis of the Modern Psyche, Routledge, London-New York.

Smith-Merry j. et al. (2019), Social Connection and Online Engagement: Insights from Interviews with Users of a Mental Health Online Forum, in "jmir Mental Health", 6, 3, doi: 10.2196/11084.

Sohi K. K., Singh p., Bopanna k. (2018), Ritual Participation, Sense of Community, and Social Well-Being: A Study of Seva in the Sikh Community, in "Journal of Religion and Health", 57, 6, pp. 2066-78.

Sokolovsky J. et al. (1978), Personal Networks of Ex-Mental Patients in a Manhattan sro Hotel, in "Human Organization", 37, 1, pp. 5-15.

Solimeno A. et al. (2008), The Influence of Students and Teachers Characteristics on the Efficacy of Face-to-Face and Computer Supported Collaborative Learning, in "Computers & Education", 51, 1, pp. 109-28.

Somhlaba N. Z., Wait J. W. (2008), Psychological Adjustment to Conjugal Bereavement: Do Social Networks Aid Coping Following Spousal Death?, in "Omega", 57, 4, pp. 341-66.

Sonn C. C. et al. (2022), Fostering and Sustaining Transnational Solidarities for Transformative Social Change: Advancing Community Psychology Research and Action, in "American Journal of Community Psychology", 69, 3-4, pp. 269-82.

Sorensen S. M. (1981), Group-Hate: A Negative Reaction to Group Work, Paper presented at the Annual Meeting of the International Communication Association (Minneapolis, May 21-25), https://eric.ed.gov/?id=ED204821.

Sowa M. et al. (2018), Assessment of Quality of Life in Women Five Years after Breast Cancer Surgery, Members of Breast Cancer Self-Help Groups – Non-Randomized, Cross-Sectional Study, in "Contemporary Oncology", 22, 1, pp. 20-6.

Spaltro E. (1969), Gruppi e cambiamento, Etas Kompass, Milano

id. (1977), Il check-up organizzativo. Diagnosi dei comportamenti e dei climi organizzativi, isedi, Milano.

id. (1982a), I gruppi e l'organizzazione, in M. Bruscaglioni, E. Spaltro (a cura di), La psicologia organizzativa, FrancoAngeli, Milano, pp. 424-48.

id. (1982b), Il lavoro di gruppo e le sue modalità, in M. Bruscaglioni, E. Spaltro (a cura di), La psicologia organizzativa, FrancoAngeli, Milano, pp. 450-65.

id. (1985), Pluralità. Manuale di psicologia di gruppo, Patron, Bologna.

id. (1990), Complessità. Introduzione alla psicologia delle organizzazioni complesse, Patron, Bologna.

id. (1999), Il gruppo. Sintesi e schemi di psichica plurale, Pendragon, Bologna. speck r. v., attneave c. l. (1976), La terapia di rete. Un nuovo approccio alla terapia nel contesto sociale, Astrolabio, Roma.

Speer P. et al. (1992), In Search of Community: An Analysis of Community Psychology Research from 1984-1988, in "American Journal of Community Psychology", 20, 2, pp. 195-209.

Speer P. et al. (2013), The Influence of Participation, Gender and Organizational Sense of Community on Psychological Empowerment: The Moderating Effects of Income, in "American Journal of Community Psychology", 51, 1-2, pp. 103-13.

Spielberger C., Iscoe I. (1972), Graduate Education in Community Psychology, in S. Golann, E. Eisdorfer (eds.), Handbook of Community Mental Health, AppletonCentury-Crofts, New York.

Spreitzer G. M. (1995), Psychological Empowerment in the Workplace: Dimensions, Measurement, and Validation, in "Academy of Management Journal", 38, 5, pp. 1442-65.

Stame N. (2001), Tre approcci principali al tema della valutazione: distinguere e combinare, in M. Palumbo (a cura di), Il processo di valutazione, FrancoAngeli, Milano, pp. 21-46.

Stansfeld S. A. et al. (2011), Repeated Exposure to Socioeconomic Disadvantage and Health Selection as Life Course Pathways to Mid-Life Depressive and Anxiety Disorders, in "Social Psychiatry and Psychiatric Epidemiology", 46, 7, pp. 549-58.

Stark W. (1995a), Gruppi di auto-aiuto. Migliorare la qualità dei servizi psicosociali, in aa.vv., Psicologia di comunità oggi. Progetti, ricerche, esperienze, Atti del i congresso europeo di psicologia di comunità (cnr, Roma, 25-27 maggio), Fondazione laboratorio Mediterraneo, Napoli.

id. (1995b), Il profilo dell'esperienza tedesca, in aa.vv., Psicologia di comunità oggi. Progetti, ricerche, esperienze, Atti del i congresso europeo di psicologia di comunità (CNR, Roma, 25-27 maggio), Fondazione laboratorio Mediterraneo, Napoli.

id. (2011), Community Psychology as a Linking Science: Potentials and Challenges for Transdisciplinary Competencies, in E. Almeida (ed.), International Community Psychology: Community Approaches to Contemporary Social Problems, Universidad Interamericana, Puebla, pp. 123-36.

Stein C. H., Mankowski e. s. (2004), Asking, Witnessing, Interpreting, Knowing: Conducting Qualitative Research in Community Psychology, in "American Journal of Community Psychology", 33, 1, pp. 21-35.

Steiner c. (1971), Radical Psychiatry: Principles, in J. Agel (ed.), The Radical Therapist, Ballatine, New York, pp. 18-26.

Stern T. (2019), Participatory Action Research and the Challenges of Knowledge Democracy, in "Educational Action Research", 27, 3, pp. 435-51.

Stevens G. (2018), Raced Repetition: Perpetual Paralysis or Paradoxical Promise, in "International Journal of Critical Diversity Studies", 1, 2, pp. 42-57.

Stevens G., Sonn C. C. (eds.) (2021), Decoloniality and Epistemic Justice in Contemporary Community Psychology, Springer, Berlin.

Stewart M. et al. (2001), Promoting Positive Affect and Diminishing Loneliness of Widowed Seniors through a Support Intervention, in "Public Health Nursing", 18, 1, pp. 54-63.

Stokes J. P. (1983), Predicting Satisfaction with Social Support from Social Network Structure, in "American Journal of Community Psychology", 11, 2, pp. 141-52.

Strom J. L., Egede L. (2012), The Impact of Social Support on Outcomes in Adult Patients with Type 2 Diabetes: A Systematic Review, in "Current Diabetes Reports", 12, 6, pp. 769-81.

Sukwatjanee A. et al. (2009), Enhancing Self-Care Ability and Quality of Life among Rural-Dwelling Thai Elders with Type 2 Diabetes through a Self-Help Group: A Participatory Action Research Approach, in "The Journal of Behavioral Science", 4, 1, pp. 70-6.

Sullivan C. M. (2012). Support Groups for Women with Abusive Partners: A Review of the Empirical Evidence, National Resource Center on Domestic Violence, Harrisburg (pa).

Swan J. A. (1970), Response to Air Pollution: A Study of Attitudes and Coping Strategies of High School Youths, in "Environment and Behavior", 2, 2, pp. 127-52.

Swartz L. P., Gibson k., Gelman t. (eds.) (2002), Reflective Practice: Psychodynamic Ideas in the Community, hsrc, Cape Town.

Swift C., Levin G. (1987), Empowerment: An Emerging Mental Health Technology, in "Journal of Primary Prevention", 8, 1-2, pp. 71-94.

Taleb N. N. (2008), The Black Swan: The Impact of the Highly Improbable, Penguin Books, Harlow-Chicago (il).

id. (2014), Antifragile: Things That Gain from Disorder, Random House, New York.

Talò C., Mannarini T., Rochira A. (2014), Sense of Community and Community Participation: A Meta-Analytic Review, in "Social Indicators Research", 117, 1, pp. 1-28.

Tancredi M. (1981), Un modello dinamico del processo aziendale, in "Direzione Aziendale", 2, pp. 73-8.

Tancredi M., Francescato D. (1989), L'analisi organizzativa come strumento di formazione, in "Rivista AIF, 3, pp. 14-8.

Tani F., Castagna V. (2017), Supporto sociale materno, qualità dell'esperienza di nascita e depressione post-partum nelle donne primipare, in "The Journal of Maternal Fetal & Neonatal Medicine", 30, 6, pp. 689-92.

Tappis H. et al. (2016), Effectiveness of Interventions, Programs and Strategies for Gender-Based Violence Prevention in Refugee Populations: an Integrative Review, in "PLoS Currents", 8, doi: 10.1371/currents. dis.3a465b66f9327676d61eb8120eaa5499.

Taverna R. (2014), Il sociale che lavora in rete, in "Igiene e cultura medico-sanitaria", 2, pp. 1-

Taylor G. S., Spencer b. a. (1989), An Empirical Technique for Multilevel Analysis in Organizational Research, in "Behavioral Science", 34, 1, pp. 61-9.

Tebes J. K. (2017), Foundations for a Philosophy of Science of Community Psychology: Perspectivism, Pragmatism, Feminism, and Critical Theory, in M. A. Bond, I. Serrano-García, C. B. Keys (eds.), apa Handbook of Community Psychology: Methods for Community Research and Action for Diverse Groups and Issues, American Psychological Association, Washington dc, pp. 21-40.

Teo T. (2005), The Critique of Psychology: From Kant to Postcolonial Theory, Springer, New York.

id. (2012), Critical Psychology, in R. Rieber (ed.), Encyclopedia of the History of Psychological Theories, Springer, New York, pp. 236-48.

Terry R., Townley G. (2019), Exploring the Role of Social Support in Promoting Community Integration: An Integrated Literature Review, in "American Journal of Community Psychology", 64, 3-4, pp. 509-27.

Testoni I. et al. (2019), Forgiveness and Blame among Suicide Survivors: A Qualitative Analysis on Reports of 4-Year Self-Help-Group Meetings, in "Community Mental Health Journal", 55, 2, pp. 360-8.

Thoits p. a. (2011), Mechanisms Linking Social Ties and Support to Physical and Mental Health, in "Journal of Health and Social Behavior", 52, 2, pp. 145-61.

Thomas N. et al. (2016), Promoting Personal Recovery in People with Persisting Psychotic Disorders: Development and Pilot Study of a Novel Digital Intervention, in "Frontiers in Psychiatry", 7, 196, doi: 10.3389/fpsyt.2016.00196.

Thunberg G. (2019), No One Is too Small to Make a Difference, Penguin, London. id. (2022), The Climate Book, Mondadori, Milano.

Timimi S. (2010), The McDonaldization of Childhood: Children's Mental Health in Neo-Liberal Market Cultures, in "Transcultural Psychiatry", 47, 5, pp. 686-706.

Tolsdorf C. C. (1976), Social Networks, Support and Coping: An Exploratory Study, in "Family Process", 15, 4, pp. 407-17.

Tomai M., Lauriola M. (2022), Separate but Related: Dimensions of Healthcare Provider Social Support in Day-Treatment Oncology Units, in "Frontiers in Psychology", 13, doi: 10.3389/fpsyg.2022.773447.

Tomai M., Lauriola M., Caputo A. (2019), Are Social Support and Coping Styles Differently Associated with Adjustment to Cancer in Early and Advanced Stages?, in "Mediterranean Journal of Clinical Psychology", 7, 1, https://doi.org/10.6092/22821619/2019.7.1983.

Tomai M. et al. (2017), Promoting the Development of Children with Disabilities through School Inclusion: Clinical Psychology in Supporting Teachers in Mozambique, in "Mediterranean Journal of Clinical Psychology", 5, 3, https://doi.org/10.6092/2282-1619/2017.5.1671

Tönnies F. (1963), Comunità e società, Edizioni di Comunità, Milano.

Tosini D., Fraccaro D. (2022), "Like Climbing a Glass Wall": Suicide Survivors in an Italian Province, in "Death Studies", 46, 4, pp. 987-95.

Townley G., Kloos B., Wright P. (2009), Understanding the Experience of Place: Expanding Methods to Conceptualize and Measure Community Integration of Persons with Serious Mental Illness, in "Health Place", 15, 2, pp. 520-31.

Townsend P. B., Whitehead M., Davidson N. (eds.) (1992), Inequalities in Health: The Black Report & the Health Divide, Penguin, London (3 rd ed.).

Triandis H. C. (1995), Individualism and Collectivism, Routledge, London.

Trickett E. J. (2009a), Community Psychology: Individuals and Interventions in Community Context, in "Annual Review of Psychology", 60, 1, pp. 395-419.

id. (2009b), Multilevel Community-Based Culturally Situated Interventions and Community Impact: An Ecological Perspective, in "American Journal of Community Psychology", 43, 3-4, pp. 257-66.

Trickett E. J. et al. (2011), Advancing the Science of Community-Level Interventions, in "American Journal of Public Health", 101, 8, pp. 1410-9.

Tojan A., Nickel S. (2011), Selbsthilfefreundlichkeit – ein Qualitätsziel auch für den ögd, in "Blickpunkt öffentliche Gesundheit", 3, pp. 4-5.

Trojan A., Nickel S., Kofahl C. (2014), Implementing "Self-Help Friendliness" in German Hospitals: A Longitudinal Study, in "Health Promotion International", 31, 2, pp. 303-13.

Tsai J., Mares A., Rosenheck R. (2012), Does Housing Chronically Homeless Adults Lead to Social Integration?, in "Psychiatric Services", 63, 5, pp. 427-34.

Tsai R., Rosenheck r. (2012), Conceptualizing Social Integration among Formerly Homeless Adults with Severe Mental Illness, in "Journal of Community Psychology", 40, 4, pp. 456-67.

Tsemberis S. (2010), Housing First: The Pathways Model to End Homelessness for People with Mental Illness and Addiction Manual, Hazelden, Center City (mn).

Tuominen T. (2018), Multi-Method Research, in E. Di Giovanni, Y. Gambier (eds.), Reception Studies and Audiovisual Translation, Benjamins, Amsterdam-Philadelphia (pa), pp. 69-90.

Tutty L. M., Babins-Wagner R., Rothery M. A. (2016), You're not Alone: Mental Health Outcomes in Therapy Groups for Abused Women, in "Journal of Family Violence", 31, 4, pp. 489-97.

Twenge J. M., Campbell W. K. (2009), The Narcissism Epidemic: Living in the Age of Entitlement, Free Press, New York.

Tyler A. (1962), Freedom's Ferment: Phases of American Social History from the Colonial Period to the Outbreak of the Civil War, Harper and Row, New York.

united nations general assembly (2012), Report of the Special Rapporteur on Violence against Women, Its Causes and Consequences on Her Mission to Italy

(15-26 January), in https://documents-dds-ny.un.org/doc/UNDOC/GEN/ G12/136/00/ PDF/G1213600.pdf?OpenElement.

Van Dam H. A. et al. (2005), Social Support in Diabetes: A Systematic Review of Controlled Intervention Studies, in "Patient Education and Counseling", 59, pp. 1-12.

Vaughn C. E., Ieff J. P. (1981), Patterns of Emotional Response in Relatives of Schizophrenic Patiens, in "Schizophrenia Bulletin", 7, 1, pp. 43-4.

Venkatesh S., Weatherspoon L. (2013), Social and Health Care Provider Support in Diabetes Self-Management, in "American Journal of Health Behavior", 37, 1, pp. 112-21.

Vieno A. et al. (2008), Antisocial Behavior and Depressive Symptoms: Longitudinal and Concurrent Relations, in "Adolescence", 43, 171, pp. 649-60.

Vieno A. et al. (2013), Sense of Community, Unfairness, and Psychosomatic Symptoms: A Multilevel Analysis of Italian Schools, in "Journal of Adolescent Health", 53, 1, pp. 142-5.

Vila J. (2021), Social Support and Longevity: Meta-Analysis-Based Evidence and Psychobiological Mechanisms, in "Frontiers in Psychology", 12, https://doi.org/10.3389/ fpsyg.2021.717164.

Vissenberg C. et al. (2016), Impact of a Social Network-Based Intervention Promoting Diabetes Self-Management in Socioeconomically Deprived Patients: A Qualitative Evaluation of the Intervention Strategies, in "bmj Open", 6, 4, doi: 10.1136/bmj open-2015-010254.

Walker C., Zlotowitz S., Zoli A. (2022), The Palgrave Handbook of Innovative Community and Clinical Psychologies, Palgrave Macmillan (e-book).

Walker K. N., Macbride A., Vachon M. L. (1977), Social Support Networks and the Crisis of Bereavement, in "Social Science & Medicine", 11, 1, pp. 35-41.

Wallerstein N. (1999), Power between Evaluator and Community: Research Relationships with New Mexico's Healthier Communities, in "Social Science and Medicine", 49, 1, pp. 39-53.

id. (2006), What Is the Evidence on Effectiveness of Empowerment to Improve Health?, in https://apps.who.int/iris/handle/10665/364209.

Wandersman A., Florin P. (2003), Community Interventions and Effective Prevention, in "American Psychologist", 58, 6-7, pp. 441-8.

Wang X. et al. (2014), Social Support Moderates Stress Effects on Depression, in "International Journal of Mental Health Systems", 8, 1, pp. 1-5.

Wardian J., Sun F. (2014), Factors Associated with Diabetes-Related Distress: Implications for Diabetes Self-Management, in "Social Work in Health Care", 53, 4, pp. 364-81.

Warner E., Sutton E., Andrews F. (2020), Cohousing as a Model for Social Health: A Scoping Review, in "Cities & Health", https://doi.org/10.1080/2374883 4.2020.1838225.

Warren D. I. (1981), Helping Networks: How People Cope with Problems in the Urban Community, University of Notre Dame Press, Notre Dame (in).

Watzlawick P., Beavin J. H., Jackson D. D. (1971), Pragmatica della comunicazione umana. Studio dei modelli interattivi, delle patologie e dei paradossi, Astrolabio, Roma.

Weick K. (1997), Senso e significato nell'organizzazione. Alla ricerca delle ambiguità e delle contraddizioni nei processi organizzativi, Raffaello Cortina Editore, Milano.

Weine S. et al. (2008), Evaluating a Multiple-Family Group Access Intervention for Refugees with ptsd, in "Journal of Marital and Family Therapy", 34, 2, pp. 149-64.

WHO– World Health Organization (1986), Ottawa Charter for Health Promotion, in https://www.who.int/publications/i/item/WH-1987#.

id. (1997), World Health Report: 1997. Conquering Suffering, Enriching Humanity, in https://apps.who.int/iris/handle/10665/41900.

id. (2001), The World Health Report 2001: Mental Health. New Understanding, New Hope, in https://apps.who.int/iris/handle/10665/42390.

id. (2006), What Is the Evidence on Effectiveness of Empowerment to Improve Health?, in https://apps.who.int/iris/handle/10665/364209.

id. (2012), Good Health Adds Life to Years: Global Brief for World Health Day 2012, in https://apps.who.int/iris/handle/10665/7085

Wiley J., Camacho T. (1980), Lifestyle and Future Health: Evidence from the Alameda County Study, in "Preventive Medicine", 9, 1, pp. 1-21.

Wilkinson R. G., Pickett k. e. (2009), Income Inequality and Social Dysfunction, in "Annual Review of Sociology", 35, pp. 493-511.

Wilson M., Baglioni A., Downing D. (1989), Analyzing Factors Influencing Readmission to a Battered Women's Shelter, in "Journal of Family Violence", 4, 3, pp. 275-84.

Wohlwill j. f., Carson D. H. (1972), Environment and the Social Sciences: Perspectives and Applications, American Psychological Association, Washington dc.

Wollert R. (1987), Self-Help Clearing-Houses in North America: A Survey of Their Structural Characteristics and Community Health Implications, in "Health Promotion International", 2, 4, pp. 377-86.

Woodall J. et al. (2010), Empowerment & Health and Well-Being: Evidence Review, Centre for Health Promotion Research, Leeds Metropolitan University, Leeds.

Wright K. B. (2016), Communication in Health-Related Online Social Support Groups/Communities: A Review of Research on Predictors of Participation, Applications of Social Support Theory, and Health Outcomes, in "Review of Communication Research", 4, pp. 65-87.

Wright K. B., Bell S. B. (2003), Health-Related Support Groups on the Internet: Linking Empirical Findings to Social Support and Computer-Mediated Communication Theory, in "Journal of Health Psychology", 8, 1, pp. 39-54.

Wrzus C. et al. (2013), Social Network Changes and Life Events across the Life Span: A Meta-Analysis, in "Psychological Bulletin", 139, 1, pp. 53-80.

Wu X. et al. (2020), Air Pollution and covid-19 Mortality in the United States: Strengths and Limitations of an Ecological Regression Analysis, in "Science Advances", 6, 45, doi: 10.1126/sciadv.abd4049.

Yağmur Y., Duman M. (2016), The Relationship between the Social Support Level Perceived by Patients with Gynecologic Cancer and Mental Adjustment to Cancer, in "International Journal of Gynecology and Obstetrics", 134, 2, pp. 208-11.

Yalom I. D. (1978), Alcoholics in Interactional Group Therapy, in "Archives of General Psychiatry", 35, pp. 419-25.

Yancey W. L. (1971), Architecture, Interaction, and Social Control: The Case of a Large-Scale Public Housing Project, in "Environment and Behavior", 3, 1, pp. 3-21.

Yang Z., Xin Z. (2016), Community Identity Increases Urban Residents' In-Group Emergency Helping Intention, in "Journal of Community and Applied Social Psychology", 26, 6, pp. 467-80.

Yanos P., Stefanic A., Tsemberis S. (2012), Psychological Community Integration among People with Psychiatric Disabilities, in "Journal of Community Psychology", 39, 4, pp. 390-401.

Yen I. H., Kaplan G. A. (1999), Neighborhood Social Environment and Risk of Death: Multilevel Evidence from the Alameda County Study, in "American Journal of Epidemiology", 149, 10, pp. 898-907.

Yetim N., Yetim U. (2014), Sense of Community and Individual Well-Being: A Research on Fulfillment of Needs and Social Capital in the Turkish Community, in "Social Indicators Research", 115, 1, pp. 93-115.

Yildirim H. et al. (2020), Determining the Correlation between Social Support and Hopelessness of Syrian Refugees Living in Turkey, in "Journal of Psychosocial Nursing and Mental Health Services", 58, 7, pp. 27-33.

Yukl G. A., Becker W. S. (2006), Effective Empowerment in Organizations, in "Organization Management Journal", 3, 3, pp. 210-31.

Yun M., Song M. (2013), A Qualitative Study on Breast Cancer Survivors' Experiences, in "Perspectives in Nursing Science", 10, 1, pp. 41-51.

Zagrebelsky G. (1995), Libertà e democrazia, in "La Stampa", 4 marzo.

id. (2002), Diritto per: valori, principi o regole?, in "Quaderni fiorentini per la storia del pensiero giuridico moderno", 31, 2, pp. 865-97.

Zani B. (2012), Psicologia di comunità. Prospettive, idee, metodi, Carocci, Roma.

Zani B., Cicognani e. (2012), Sense of Community in the Work Context: A Study on Members of a Co-Operative Enterprise, in "Global Journal of Community Psychology Practice", 3, 4, https://www.gjcpp.org/pdfs/2012-Lisboa-079.pdf.

Zani B., Cicognani e., Albanesi c. (2001), Adolescent's Sense of Community and Feeling of Unsafety in the Urban Environment, in "Journal of Community and Applied Social Psychology", 11, pp. 475-89.

Zani B., Palmonari a. (a cura di) (1996), Manuale di psicologia di comunità, Il Mulino, Bologna.

Zarkin G. A. et al. (2005), Cost Methodology of combine, in "Journal of Studies on Alcohol", Supplement, 15, pp. 50-5.

Zemore S. E., Kaskutas l. a., amon l. n. (2004), In 12-Step Groups, Helping Helps the Helper, in "Addiction", 99, 8, pp. 1015-23.

Zettel L. A., Rook K. S. (2004), Substitution and Compensation in the Social Networks of Older Widowed Women, in "Psychology and Aging", 19, 3, pp. 433-43.

Zimet G. D. et al. (1988), The Multidimensional Scale of Perceived Social Support, in "Journal of Personality Assessment", 52, 1, pp. 30-41.

Zimmerman M. A. (1990a), Taking Aim on Empowerment Research: On the Distinctions between Individuals and Psychological Conceptions, in "American Journal of Community Psychology", 18, 1, pp. 169-77.

id. (1990b), Toward a Theory of Learned Hopefulness: A Structural Model Analysis of Participation and Empowerment, in "Journal of Research in Personality", 24, 1, pp. 71-86.

id. (1995), Psychological Empowerment: Issues and Illustrations, in "American Journal of Community Psychology", 23, pp. 581-99.

id. (1999), Empowerment e partecipazione della comunità. Un'analisi per il prossimo millennio, in "Animazione sociale", 29, 2, 1999, pp. 10-24.

id. (2000), Empowerment Theory, in Rappaport, Seidman (2000), pp. 43-63.

Zimmerman M. A., Rappaport j. (1988), Citizen Participation, Perceived Control, and Psychological Empowerment, in "American Journal of Community Psychology", 16, 5, pp. 725-50.

Zimmerman M. A., Warschausky s. (1998), Empowerment Theory for Rehabilitation Research: Conceptual Methodological Issue, in "Rehabilitation Psychology", 43, 1, pp. 3-16.

Zimmerman M. A. et al. (1992), Further Explorations in Empowerment Theory: An Empirical Analysis of Psychological Empowerment, in "American Journal of Community Psychology", 19, pp. 707-27.

Zoli M. et al. (2018), Neuronal and Extraneuronal Nicotinic Acetylcholine Receptors, in "Current Neuropharmacology", 16, 4, pp. 338-49.

Acknowledgments

Donata Francescato

The extensive list of references reflects the many individuals who have contributed to the content of this book, and I extend my gratitude to all the community psychologists—a small but precious minority whose values I deeply share. I am particularly thankful to a few people not directly mentioned in the book. First, my English teacher at the European College in Stresa, John Cerhan, a Hungarian refugee in the 1950s, who encouraged me to apply for an American Field Service scholarship. Thanks to him, in 1962, I spent a year at a high school in Vinton, Iowa, traveled with hundreds of foreign students, and shook hands with the inspiring President Kennedy, who urged us to strive for a better world. Tragically, he was assassinated the following November while I was in Germany, where I had just met a young Texan, William Mebane, who has since become my loving and creative partner, standing by me in both sickness and health.

I am also grateful to several professors in the United States who guided my academic journey. Dale Johnson at the University of Houston, and Professors Wann and Topazio at Rice University, supported me in earning a Master's degree in 1970 in clinical psychology and French literature, allowing me the time to carefully consider my next steps toward a PhD program.

I owe special thanks to Philip Slater, Morris Schwartz, and Jacky Doyle, who created a year-long program for activists dedicated to changing the world. Through this program, I learned to integrate political and feminist goals with personal psychological growth, while forming lifelong friendships with Jenny Mansbridge and Ginger Goldner. My deepest gratitude, however, goes to Ira Goldberg at Harvard and to Stan Sacon, Cynthia Ganung, and John Pappajohn, who mentored me during my internship at South Shore Mental Health Center in Quincy, Massachusetts, in 1971–1972, teaching me what it means to be a community psychologist.

Special thanks also go to my European colleagues, including José Ornelas, Wolfgang Stark, Jim Orford, Piero Amerio, Maria Vargas Moniz, David Fryer, Caterina Arcidiacono, Bruna Zani, Patrizia Meringolo, and Laura Migliorini. I dedicate this book to the next generation of community psychologists, who will build a better future.

Manuela Tomai

I am deeply grateful to Donata Francescato, who introduced me to the field of community psychology and has been an invaluable guide throughout my exploration of its rich knowledge. I also thank Renzo Carli for developing theories and methods that have been essential in analyzing social contexts, and Viviana Langher, my collaborator in pioneering research methods that integrate community psychology with psychodynamics.

I am particularly thankful to Lauriola for guiding me in the challenging task of creating tools to measure the constructs I hold dear. Finally, my heartfelt thanks go to my family for their unwavering support.

9 781940 387222